CALCULUS
for a
NEW CENTURY
A PUMP, NOT A FILTER

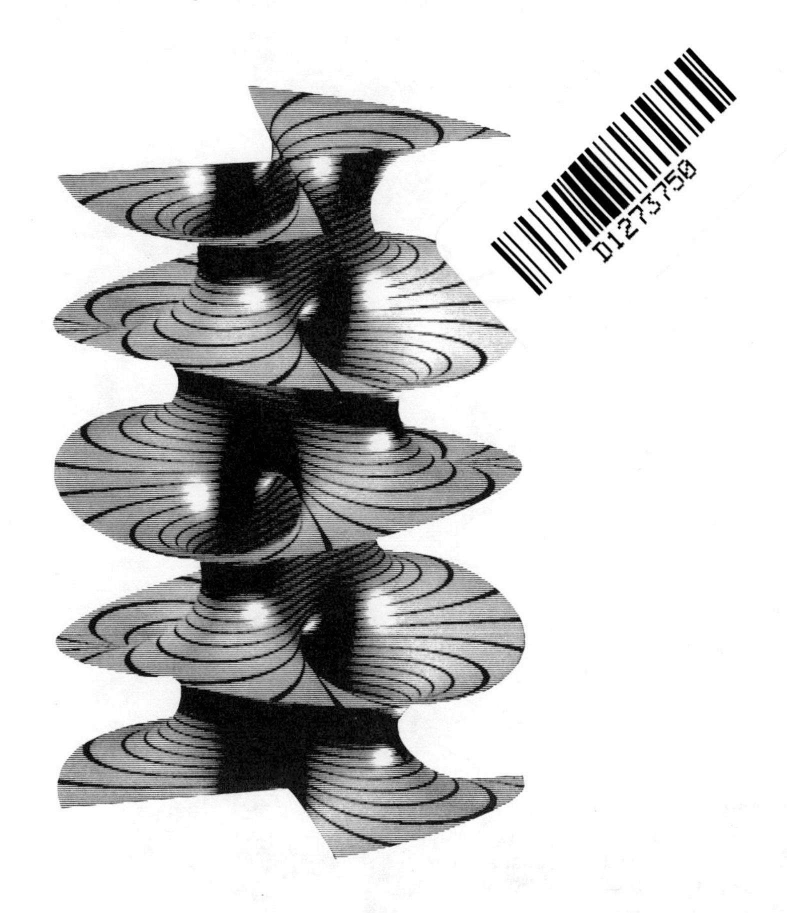

A National Colloquium

October 28-29, 1987

Edited by Lynn Arthur Steen for
The Board on Mathematical Sciences and
The Mathematical Sciences Education Board of
The National Research Council

National Academy of Sciences
National Academy of Engineering
Institute of Medicine
Mathematical Association of America

The image used on the cover of this book and on the opening pages of each of its seven parts shows a periodic minimal surface discovered in 1987 by Michael Callahan, David Hoffman, and Bill Meeks III at the University of Massachusetts. This computer generated image was created by James T. Hoffman, ©1987.

The questions from the AP calculus test in the Examination section are reproduced with the permission of The College Entrance Examination Board and of Educational Testing Service, the copyright holder.

Library of Congress Card Number 87-063378
ISBN 0-88385-058-3

Printed in the United States of America

Current printing (last digit) 6 5 4 3 2 1

Calculus for a New Century

A Pump, Not a Filter

Reports

Issues

Innovation

Science, Engineering, and Business

Teaching and Learning

Institutional Concerns

Mathematical Sciences

Examinations

Readings

Participants

National Research Council
Task Force on Calculus

Ronald G. Douglas (Chairman), Dean, College of Physical Sciences and Mathematics, State University of New York, Stony Brook.

Lida K. Barrett, Dean, College of Arts and Sciences, Mississippi State University.

John A. Dossey, Professor of Mathematics, Illinois State University.

Andrew M. Gleason, Hollis Professor of Mathematicks and Natural Philosophy, Harvard University.

Jerome A. Goldstein, Professor of Mathematics, Tulane University.

Peter D. Lax, Professor of Mathematics, Courant Institute of Mathematical Sciences, New York University.

Arthur P. Mattuck, Professor of Mathematics, Massachusetts Institute of Technology.

Seymour V. Parter, Professor of Mathematics, University of Wisconsin.

Henry O. Pollak, Bell Communications Research, (Retired).

Stephen B. Rodi, Chair, Division of Mathematics and Physical Sciences, Austin Community College.

Supporting Staff

Lawrence H. Cox, Staff Director, Board on Mathematical Sciences, National Research Council.

Frank L. Gilfeather, Consultant to the Board on Mathematical Sciences, University of Nebraska.

Therese A. Hart, Research Associate, Mathematical Sciences in the Year 2000, National Research Council.

Bernard L. Madison, Director, Mathematical Sciences in the Year 2000, National Research Council.

Peter L. Renz, Associate Director, Mathematical Association of America.

Lynn Arthur Steen, Consultant to the Task Force on Calculus, St. Olaf College.

Marcia P. Sward, Executive Director, Mathematical Sciences Education Board, National Research Council.

Alfred B. Willcox, Executive Director, Mathematical Association of America.

Board on Mathematical Sciences

Phillip Griffiths (Chairman), Provost, Duke University.

Peter Bickel, Professor of Statistics, University of California, Berkeley.

Herman Chernoff, Professor of Statistics, Harvard University.

Ronald Douglas, Dean, College of Physical Sciences and Mathematics, State University of New York, Stony Brook.

E.F. Infante, Dean, Institute of Technology, University of Minnesota.

William Jaco, Professor of Mathematics, Oklahoma State University, Stillwater.

Joseph J. Kohn, Professor of Mathematics, Princeton University.

Cathleen S. Morawetz, Director, Courant Institute of Mathematical Sciences, New York University.

Alan Newell, Chairman, Department of Mathematics, University of Arizona.

Ronald Pyke, Professor of Statistics, University of Washington.

Guido Weiss, Professor of Mathematics, Washington University.

Shmuel Winograd, Mathematical Sciences Department, IBM T.J. Watson Research Center.

Mathematical Sciences Education Board

Preface

On October 28-29, 1987, over six hundred mathematicians, scientists, and educators gathered in Washington to participate in a Colloquium, Calculus for a New Century, sponsored by the National Academy of Sciences and the National Academy of Engineering. Centering on that Colloquium and containing 75 separate background papers, presentations, responses, and other selected readings, *Calculus for a New Century: A Pump, Not a Filter* conveys to all who are interested the immense complexity of issues in calculus reform.

Conducted by the National Research Council in collaboration with the Mathematical Association of America, the Colloquium is a part of Mathematical Sciences in the Year 2000 (MS 2000), a joint project of the Board on Mathematical Sciences and the Mathematical Sciences Education Board. The National Research Council is the principal operating agency of the National Academy of Sciences and the National Academy of Engineering and is jointly administered by both academies and the Institute of Medicine.

The National Academy of Sciences is a private, non-profit, self-perpetuating society of distinguished scholars engaged in scientific and engineering research, dedicated to the furtherance of science and technology and to their use for the general welfare. Upon the authority of the charter granted to it by the Congress in 1863, the Academy has a mandate that requires it to advise the federal government on scientific and technical matters.

The National Academy of Engineering was established in 1964, under the charter of the National Academy of Sciences, as a parallel organization of outstanding engineers. It is autonomous in its administration and in the selection of its members, sharing with the National Academy of Sciences the responsibility for advising the federal government. The National Academy of Engineering also sponsors engineering programs aimed at meeting national needs, encourages education and research, and recognizes the superior achievement of engineers.

The Institute of Medicine was established in 1970 by the National Academy of Sciences to secure the services of eminent members of appropriate professions and the examination of policy matters pertaining to the health of the public. The Institute acts under the responsibility given to the National Academy of Sciences by its congressional charter to be an advisor to the federal government and, upon its own initiative, to identify issues of medical care, research, and education.

The National Research Council was organized by the National Academy of Sciences in 1916 to associate the broad community of science and technology with the Academy's purposes of furthering knowledge and advising the federal government. Functioning in accordance with general policies determined by the Academy, the Council has become the principal operating arm of both the National Academy of Sciences and the National Academy of Engineering in providing services to the government, the public, and the scientific and engineering communities. Frank Press and Robert M. White are chairman and vice chairman, respectively, of the National Research Council.

The Mathematical Association of America is an organization of about 26,000 members dedicated to the improvement of mathematics, principally at the collegiate level. It has played a long-term role in improving mathematics education through the work of its committees and its publications, of which this volume is one example.

The Board on Mathematical Sciences of the Commission on Physical Sciences, Mathematics, and Resources was established in 1984 to maintain awareness and active concern for the health of the mathematical sciences and to serve as the focal point in the National Research Council for issues connected with the mathematical sciences. The Mathematical Sciences Education Board was established in 1985 to provide a continuing national overview and assessment capability for mathematics education. These boards, joint sponsors of MS 2000, selected a special Task Force on Calculus, chaired by Ronald G. Douglas, to direct the planning of the colloquium Calculus for a New Century.

Calculus for a New Century was made possible by a grant from the Alfred P. Sloan Foundation to the National Research Council. Support for the umbrella project, MS 2000, was provided by the National Science Foundation. This volume was prepared at St. Olaf College under the direction of editor Lynn Arthur Steen, and the publishing was directed for the Mathematical Association of America by Peter L. Renz.

Bernard L. Madison
MS 2000 Project Director

Introduction

Nearly one million students study calculus each year in the United States, yet fewer than 25% of these students survive to enter the science and engineering pipeline. Calculus is the critical filter in this pipeline, blocking access to professional careers for the vast majority of those who enroll. The elite who survive are too poorly motivated to fill our graduate schools; too few in number to sustain the needs of American business, academe, and industry; too uniformly white, male, and middle class; and too ill-suited to meet the mathematical challenges of the next century.

These facts led Robert White, President of the National Academy of Engineering, to suggest that calculus must become a pump rather than a filter in the nation's scientific pipeline. Others used a different metaphor—to become a door, not a barrier. To make calculus a pump is a challenge to educators and scientists; to walk through the door that calculus opens is a challenge to students. Regardless of the metaphor, calculus must change so that students will succeed.

All One System

In a narrow sense, calculus can be viewed simply as a sequence of courses in the mathematics curriculum, of concern primarily to those who teach high school and college mathematics. But in fact, calculus is a dominating presence in a number of vitally important educational and social systems. Calculus is:

- A capstone for school mathematics, the culmination of study in the only subject (apart from reading) taught systematically all through K-12 education.
- A pre-requisite to the majority of programs of study in colleges and graduate schools.
- The dominant college-level teaching responsibility of university departments of mathematics, intimately linked to the financial support of graduate education in mathematics.
- A course whose techniques are rapidly being subsumed by common computer packages and pocket calculators.
- An important component of liberal education, part of the core learning that is the hallmark of an educated person.

These interlocking systems make calculus extraordinarily resistant to change. Calculus has immense inertia that is rooted in tradition, reinforced by client disciplines, and magnified by masses of students. Yet calculus is failing our students: no one is well served by the present course. Those who apply calculus want students to have more mathematical power rather than mere mimicry skills; those concerned for education fear that for far too many able students, calculus is the end of ambitions rather than the key to success; still others foresee that computers will make much of what we now teach irrelevant:

- Computers and advanced calculators can now do most of the manipulations that students learn in a typical calculus course.
- College administrators report that calculus is a lightning rod for students' complaints.
- Computer scientists, one of the largest of mathematics' many clients, advocate discrete mathematics rather than calculus as a student's first college mathematics course.
- In many large universities, fewer than half of the students who begin calculus finish the term with a passing grade.

- Little in the typical calculus course contributes much to the aims of general education, although for most students, calculus is the last mathematics course they ever take.

For all these reasons, a broad coalition of organizations has now undertaken an effort to revitalize calculus. Calculus for a New Century is a first visible public step in this long but crucially important process.

Issues and Controversies

Calculus for a New Century: A Pump, Not a Filter is intended to be a resource for calculus reform, rooted in the October 28-29 Colloquium, but including much additional material. Over 80 authors from mathematics, science and engineering convey in 75 separate contributions (totalling over 165,000 words) a very diverse set of opinions about the shape of calculus for a new century. The authors agree on the forces that are reshaping calculus, but disagree on how to respond to these forces; they agree that the current course is not satisfactory, yet disagree about new content emphases; they agree that neglect of teaching must be repaired, but do not agree on the most promising avenues for improvement. Readers must judge for themselves how they will respond to this diverse yet realistic sample of informed views on calculus:

Colloquium: A record of presentations prepared for the first of the two-day Colloquium. These include four plenary presentations and prepared remarks from two panels.

Responses: A collage of reactions to the Colloquium by a variety of individuals representing diverse calculus constituencies. These commentaries, each a mini-editorial, were written in response to what was said and unsaid during the formal proceedings.

Reports: Summaries of sixteen discussion groups that elaborate on particular themes of importance to reform efforts. Each report was prepared jointly by the discussion leader and reporter.

Issues: A series of background papers providing context for the calculus Colloquium. The first two, on Innovation, are journalistic analyses based on extensive interviews with many mathematicians, educators, and scientists. The remaining sixteen, in four sections, provide background on issues in Science, Engineering, and Business; Teaching and Learning; Institutional Perspectives; and Mathematical Sciences.

Examinations: A selection of final examinations from Calculus I, II, and III from universities, colleges, and two-year colleges around the country. These exams are intended to illustrate calculus as it is today, to document exactly what students are currently expected to achieve.

Readings: A collection of reprints of documents related to calculus. Many of these papers are referred to either in the Colloquium talks or in the background papers; they are reproduced here as a convenience to the reader.

Participants: A complete list of names, addresses, institutions, and telephone numbers from the registration list of the October 28-29 Colloquium, provided here to facilitate future correspondence among those with similar interests.

Process of Renewal

Calculus for a New Century is itself a middle chapter in a long process of calculus reform. It builds on a much smaller workshop organized by Ronald G. Douglas at Tulane University in January, 1986 which led to the much-cited publication *Toward a Lean and Lively Calculus* (MAA Notes No. 6, 1986). Shortly afterwards, the Mathematical Association of America appointed a Committee on Calculus Reform chaired by Douglas to make plans for an appropriate follow-up to the Tulane meeting. Then in January 1987, the National Science Foundation proposed to Congress a major curriculum initiative in reform of calculus. To help frame a national agenda for calculus reform, and to insure broad participation of the scientific and engineering communities, the Douglas Committee recommended that the National Academies of Science and Engineering sponsor a national colloquium on calculus reform.

At the same time, the two mathematics boards of the National Research Council—the Board on Mathematical Sciences and the Mathematical Sciences Education Board—jointly launched Mathematical Sciences in the Year 2000 (MS 2000), a project led by Bernard L. Madison to provide a comprehensive assessment of collegiate and graduate education in the mathematical sciences, analyzing curricular issues, resources, personnel needs, and links to science, engineering, and industry. Since calculus is a central ingredient in the agenda of MS 2000, the Calculus Colloquium became the first undertaking of that project.

Calculus for a New Century

Three hundred years ago, precisely, the first edition of Newton's *Principia Mathematica* was published. Two hundred years ago, more or less, calculus was first offered as a regular subject in the university curriculum. One hundred years ago the mathematical revolution launched by Newton gave birth in the New World to what is now the American Mathematical Society.

"Calculus for a New Century" celebrates these three centenaries, not by looking back but by looking forward to the 21st Century, when today's students will be our scientific and mathematical leaders. Calculus determines the flow of personnel in the nation's scientific pipeline. To fill this pipeline, we must educate our youth for a mathematics of the future that will function in symbiosis with symbolic, graphical, and scientific computation. We must interest our students in the fascination and power of mathematics—in its beauty and in its applications, in its history and in its future. *Calculus for a New Century: A Pump, Not a Filter* offers a vision of the future of calculus, a future in which students and faculty are together involved in learning, in which calculus is once again a subject at the cutting edge—challenging, stimulating, and immensely attractive to inquisitive minds.

ACKNOWLEDGEMENTS: Details of Calculus for a New Century were handled efficiently by Bernard Madison and Therese Hart, staff members of Project MS 2000. *Calculus for a New Century: A Pump, Not a Filter* was prepared in an extraordinarily short time—three weeks for preparation of the camera-ready copy—using a TEX system at St. Olaf College. Mary Kay Peterson deserves special thanks for typing and correcting the entire volume—all 1,083,000 characters—on such a very tight timetable.

Lynn Arthur Steen
St. Olaf College
Northfield, Minnesota
November 20, 1987

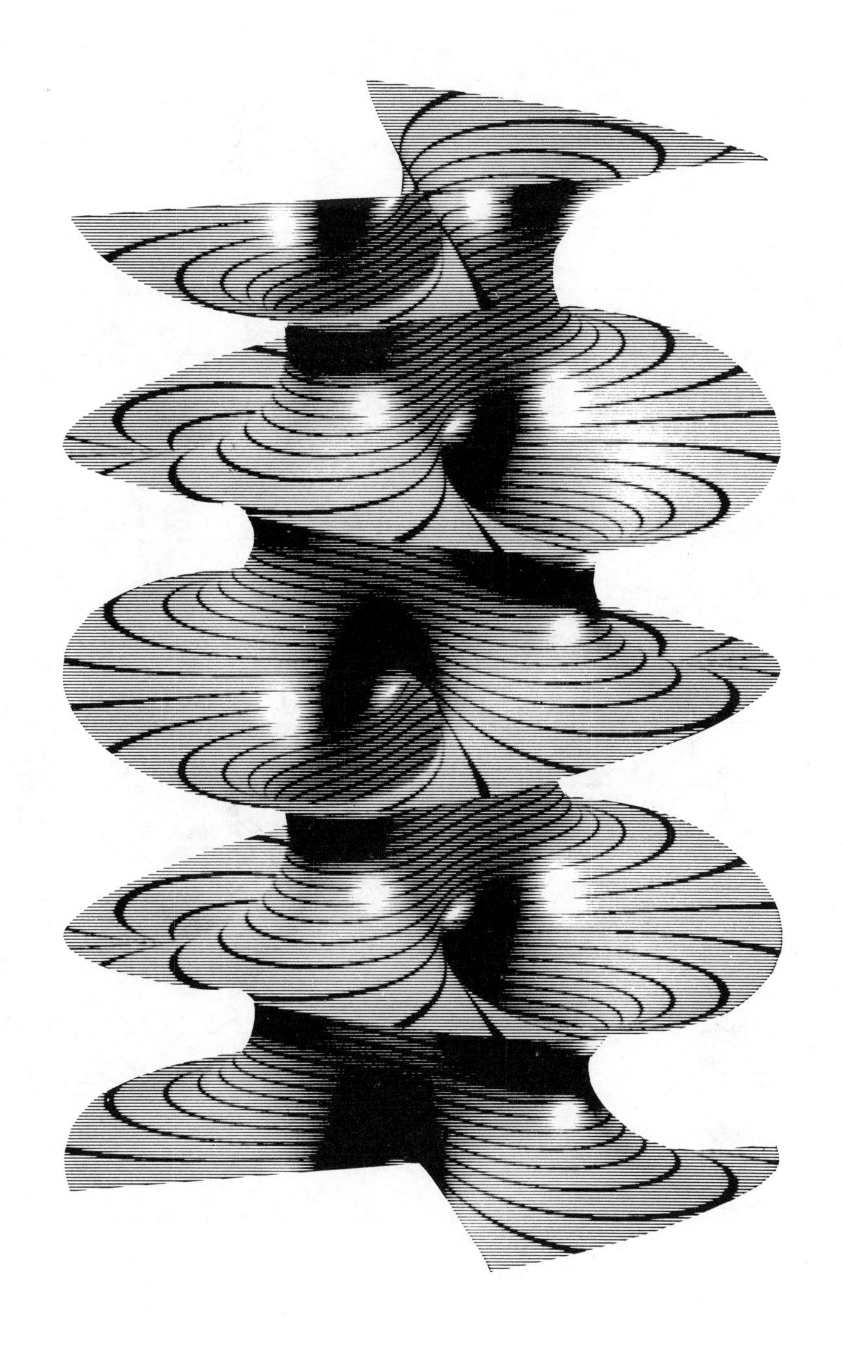

COLLOQUIUM

GENERAL COMMENTS FROM PARTICIPANTS

- Even average students want to see the excitement in a subject.
- In this age of Velcro, digital watches, and the HP28C, must one still learn to tie shoes, read two-handed clocks, and solve quadratic equations?
- Calculus is <u>not</u> in need of major changes.
- The new calculus is <u>more difficult</u> than the present course. "Reform" in this direction cannot succeed unless students are better prepared.
- Innovation requires creativity. Creativity threatens the status quo.
- I would have liked to have heard from someone who was in favor of the status quo.

Calculus for a New Century

Frank Press

NATIONAL ACADEMY OF SCIENCES

Good morning. I would like to welcome all of you to this National Colloquium on Calculus For A New Century.

Last night I welcomed to this auditorium another audience for a preview of a film called "The Infinite Voyage," which will be shown all over the country in prime-time television, both commercial and public broadcasting. It's a family-oriented program that will continue for three years. Last night's program featured many things, but computational mathematics was an essential feature of all of the science components in the film.

I would like to say a few words about the strong interest of our own organization here in mathematics education. I guess that everyone recognizes that mathematics education is centrally important as a foundation for science, for engineering, and also for the social sciences. Here at the National Research Council where we work in all of these fields, we certainly recognize that, as you must too in your own schools and organizations.

Mathematics is increasing in importance, not only in the fields that I just mentioned, but also in the service sector and in banking and finance. (In fact, some mathematical geniuses may be responsible for programming those computers that caused the automatic selling that contributed to the stock market crash. They should have had academic careers ....)

Why are we losing so many bright people—losing them from mathematics and science in favor of Wall Street and business schools and law schools? Mathematics is a critical filter in that it can knock you out of the pipeline, permanently, particularly for women and minorities who might otherwise have careers in the professions until their mathematics courses destroy their hopes and aspirations.

I, myself, had a traumatic experience. I was knocked out in the sense that I took the most advanced undergraduate course in mathematics and received the only A in the course. Yet the professor, after giving me my grade, said, "I don't want you to major in mathematics." So I was knocked out of mathematics into physics—and it had a traumatic effect on me.

We are very proud and pleased that our Mathematical Sciences Education Board is providing a new type of national leadership in mathematical education. It is appropriate for us to have this emphasis in our own organization. It is appropriate for the country to take initiatives as represented by this Colloquium because of the importance of mathematics in so many different ways. We are pleased also that the Mathematical Sciences Education Board is interested in the whole range of mathematics, from the youngest children through college.

This particular conference is under the initiative of our Board on Mathematical Sciences which works in partnership with our Mathematical Sciences Education Board. Together these two Boards launched the Project MS2000, a comprehensive review of mathematics education through college and university systems. This Colloquium on the calculus is the lead activity of this three-year project. We are putting together our committee that will be responsible for MS2000 and have just received the acquiescence of a major industrialist to serve as chairman of the committee that will supervise MS2000.

As I have said, calculus is an important foundation for the scientific, engineering, and research communities. It is also an important foundation for anyone who wants a good solid education in the modern era. Later this morning you will hear from Robert White, the President of our sister organization, the National Academy of Engineering, who will certainly emphasize the importance of calculus to our technological society.

I know that you have a lot of work to do in these two days, with a program and workshops that deal with many different groups covering essentially everybody who will need calculus for their training. I look forward to hearing the results of your Colloquium.

FRANK PRESS is President of the National Academy of Sciences. A former Science Advisor to President Carter, Dr. Press served for many years as head of the Department of Earth and Planetary Sciences at the Massachusetts Institute of Technology. He received a Ph.D. in geophysics from Columbia University.

Castles in the Sand

Ronald G. Douglas

STATE UNIVERSITY OF NEW YORK AT STONY BROOK

I want to welcome you to this Colloquium on behalf of the Calculus Task Force. I am delighted that you have come and that so many people are interested in improving the teaching of calculus. There are over six hundred people here today, about five hundred of whom are mathematicians. About one hundred are from research universities, four hundred from four-year colleges, fifty from two-year colleges, and fifty from high schools. There are over a hundred people from other ("client") disciplines; some are publishers and others journalists.

I am delighted that ... so many people are interested in improving the teaching of calculus.

Why are we all here? Why is this Colloquium being held? Let me try to explain this with a little personal history.

Five years ago, I became Chair of the Mathematics Department at my university for the second time. I found a large difference between this time and the last, which had been about ten years earlier. Teaching was different! Although everyone still did their teaching, morale was low and there was an overwhelming sense of futility, felt by all. Since calculus involved the largest number of students, much of this feeling was centered in the calculus.

About the same time I was confronted with the issue of the continuing relevance of the subject of calculus. Debate on the rising importance of discrete mathematics was sweeping across university campuses, and I often was called on to defend the role of calculus in these discussions.

My interest in calculus was treated as a curiosity. No one ever talked about teaching. Teaching was something we had to do and get over with.

For both of these reasons I was forced to think about calculus and calculus teaching. Moreover I began to ask questions about calculus when I travelled to other universities.

My interest in calculus was treated as a curiosity. *No one* ever talked about teaching. Teaching was something we had to do and get over with. No one said that mathematics teaching was good; everyone believed implicitly that since nothing could be done about it, then why talk about it.

Most curricular and teaching activities had stopped around 1970 in the "post post-Sputnick" era. But the need for change had not. Now there were more and different students; faculty cutbacks had led to larger classes, and to more teaching assistants and adjunct instructors. And finally there were more powerful hand-held calculators; computers had gone from a building somewhere across campus to a room down the hall to your desk top.

A few "oddballs" persevered. ... But they were not able to get others to join the effort and ultimately the sea overwhelmed the castles in the sand which they had so painstakingly built.

I found in my travels, however, that not all faculty had given up on curriculum and teaching reform. A few "oddballs" had persevered. Through nearly superhuman efforts they had mounted experimental programs with different methods of teaching and new curricula. Perhaps surprisingly, they often achieved considerable success.

But they were not able to get others to join the effort and ultimately the sea overwhelmed the castles in the sand which they had so painstakingly built. Therefore, I concluded that isolated innovations are not the answer to the problems in mathematics teaching. This posed a real dilemma since change certainly cannot be dictated from the top down.

What was needed was a strategy to place the improvement of mathematics teaching and curriculum on the national agenda. Support had to be provided for local efforts, and this didn't mean just money. Further, a mechanism needed to be provided to coordinate and network the results of innovation, both the good and the bad.

Change certainly cannot be dictated from the top down.

To start to accomplish all of this I asked the Sloan Foundation to fund a Calculus Workshop, which was held at Tulane University two years ago. There a diverse group agreed that both the teaching and content of calculus *should* change, and that it *could* change. We reached substantial agreement on some changes to make. All this is reported in *Toward a Lean and Lively Calculus* [1].

However, changing the calculus is an enormous and complex undertaking. Calculus is taught to over three quarters of a million students a term; about a half billion dollars a year is spent on tuition for teaching calculus; and calculus is a prerequisite for more than half of the majors at colleges and universities. Almost everyone has a stake in calculus.

This Colloquium has been organized to discuss how to proceed, not to ratify some preordained plan. We hope to plant many seeds and to help provide the sun and the rain necessary for these seeds to grow.

Changing the calculus is an enormous and complex undertaking. ...Almost everyone has a stake in calculus.

A number of excellent position papers have already been prepared to provide background and to set the stage. Robert White will provide an engineer's perspective on the role and importance of calculus. Lynn Steen will discuss calculus as it is today and Thomas Tucker will discuss some of the possibilities for the calculus of the future. The morning session concludes with a description of the new NSF calculus initiative by Judith Sunley and Robert Watson.

After lunch there will be two panel discussions. The first, moderated by Cathleen Morawetz, will allow representatives from the "client disciplines" to discuss their disciplines. The second, moderated by Andrew Gleason, will allow campus administrators to provide a view of calculus from the campus. Finally, all participants will get a chance to contribute their ideas in the discussion groups tomorrow.

Isolated innovations are not the answer to the problems in mathematics teaching.

Any change in the way calculus is taught, or in the way mathematics is taught, will depend on all of us, on those here today and on the thousands of others who teach calculus and all the rest of mathematics in the colleges (both two-year and four-year), in the high schools, and in the universities.

My title, based on my early experiences, refers to castles in the sand. I had intended to close by stating that we wanted to avoid building more castles in the sand. However, another metaphor, also based on the sea, might be better. The number of people at this Colloquium, and the interest that has been manifested in this topic suggest that a bigger phenomenon is involved. So I conclude by urging all of us to "catch the wave" and try to direct it to the Calculus for a New Century that we will choose.

Reference

[1] Douglas, Ronald G. (ed.) *Toward a Lean and Lively Calculus.* MAA Notes Number 6. Washington, D.C.: Mathematical Association of America, 1986.

RONALD G. DOUGLAS is Dean of the College of Physical Sciences and Mathematics at the State University of New York at Stony Brook and Chairman of the National Research Council's Task Force on the Reformation of the Teaching of Calculus. A specialist in operator theory, Douglas is a past holder of both Sloan and Guggenheim fellowships. He is a member of the Board on Mathematical Sciences of the National Research Council and Chairman of the American Mathematical Society's Science Policy Committee. He received a Ph.D. degree in mathematics from Louisiana State University.

Calculus of Reality

Robert M. White

NATIONAL ACADEMY OF ENGINEERING

Let me extend to all of you a welcome on behalf of the National Academy of Engineering. I have rarely seen this auditorium as filled as it is today. It is a good sign that the topic we will be discussing today has generated great concern throughout the mathematics and the non-mathematics community also.

I am probably one of the few non-mathematicians in this auditorium today. I am one of the "clients," as you call them. I would like to talk to you from the viewpoint of a client. I have views that stem from my personal experiences in science and engineering, and from my experience in trying to make scientific and technological institutions function.

We were asked to have faith in what we were being taught ... and faith that sometime in the future what we were learning would have application.

My introduction to the calculus was an introduction to the mysteries. It was assumed, when I was going through school, that we would learn even without understanding. We were asked to have faith in what we were being taught, faith in our instructors, teachers, and professors, and faith that sometime in the future what we were learning would have application.

I graduated many years ago from the Boston Public Latin School. It's the oldest public school in the country, founded in 1635. Harvard University was founded in 1636. Those of us who went to the Boston Public Latin School always claimed that Harvard was founded as a place where Latin School graduates could go after graduation.

Today, however, students will not take education in any field on faith.

My introduction to the calculus was as a freshman at Harvard. They didn't teach calculus at Boston Public Latin School in my days. My instructor, although he may not have been famous then, is well recognized these days as a famous mathematician. He was Oscar Zariski. He might have been a great mathematician, but he was not the greatest instructor of freshman calculus. I survived.

Later, at MIT, when I became involved in the atmospheric and oceanographic sciences, it became clear in hindsight why calculus was important and why it was so necessary for me as a person interested in geophysical fluid dynamics.

If ever a field needed to be brought out of mystery to reality, it is the calculus.

Today, however, students will not take education in any field on faith. They want to know why what they're studying is important and how it's going to help them. If ever a field needed to be brought out of mystery to reality, it is the calculus.

Calculus now is more important than ever. Calculus, as the mathematics of change, is the skeleton on which the flesh of our modern industrial society grows. The public does not understand this fundamental role of the calculus. But as people who are responsible for imparting the calculus, you must understand it, and you must understand the consequences of failure to impart an understanding of this field of mathematics.

Restoring Economic Growth

There are many realities that we need to face in this country, not the least being the recent events in the stock market. But the most serious reality we face today is the need to harness science and technology for economic growth. And harnessing science and technology for economic growth means harnessing the calculus.

We are "a nation at risk," as that famous report indicated. However, we are at risk in many ways—not only in our educational system for which the phrase was first introduced. Science and technology are the touchstones of economic growth which is fundamental to the standard of living, job creation, health care, provision of a good environment, and much more. During the past century this country has done very well in harnessing science and technology. Investment in science and technology has created new industries of all kinds, e.g., in semiconductors and biotechnology. Many of the most progressive and successful companies depend on new products that are based on science and technology for their economic growth. For example, I recently visited the 3M Company and learned that 25% of its revenues are based on products that did not exist just five years ago and were the result of their research and development activities.

We have done well in this country because over the past century we have had far-sighted policies that led to an educational system that has produced the talent we need to run our industrial, academic, and governmental enterprises. We have made major investments in research and development enterprises, both those supported by the government and by private industry, to produce scientific and engineering knowledge. We have an economic system that has provided the incentives and rewards for innovation and application.

Harnessing science and technology for economic growth means harnessing the calculus.

Nevertheless, in recent years that picture has been changing, as anyone who reads the headlines or watches nightly television news will know. Industry has moved many production systems abroad. Many industries have been severely hurt in terms of jobs. At the end of World War II, the United States accounted for 40% of the world's GNP; today we account for 20%. We now run a $200 billion trade deficit.

We are in a continuing competitive battle with industries in other countries:

- In aviation, the European Airbus now has over 30% of the world market share in large transport aircraft.
- On U.S. highways, 30% of the automobiles are foreign imports.
- In consumer electronics, it is difficult to find products produced in the United States.
- Even in heavy machinery, the last manufacturer of large steam turbines has just given up.

We are in a competitive battle, across the board. In industry it is a battle for market share, but it is a new kind of a battle. It is a battle that is fought with tariffs, wage rates, and economic policies. But above all else, it is a battle that is fought with trained people. Lack of an adequate pool of trained people will, in the long run, lose that battle for us. And losing that battle means a loss of jobs and a lowering of standards of living. It becomes a central responsibility for all of us in education, especially those in science and engineering education, to make sure that this country has an adequate pool of trained talent.

We are in a continuing competitive battle with industries in other countries. ...Above all else, it is a battle that is fought with trained people.

There are many disturbing signs. Some have been highlighted by the work of the Mathematical Sciences Education Board. There appears to be a lessening of interest among citizens of the United States in careers in science and technology. We know, on the basis of surveys, that the mathematical attrainment of our students are inferior, at least in K-12 grades, to those of most of our industrial allies.

Schoolhouse to the World

At the college level, the United States has become schoolhouse to the world. In 1986 we trained 340,000 foreign students in our universities. A very large number of these, about 130,000, go into engineering.

That's most welcome. I think we should train foreign students. The problem is that at the very highest levels, at the Ph.D. levels in schools of engineering, 50% of the students are not United States citizens. Now that's not necessarily bad, because about 60% of the foreign-born students eventually remain in the United States. They become productive participants in the industrial and governmental apparatus of this country.

We are not getting adequate response out of our own pool of talent in this country.

This country has been built on immigrants. But a high percentage of foreign students in science and technology is an indication of the fact that we are not getting adequate response out of our own pool of talent in this country. We are coming to depend more and more on the in-flow of talent from other countries.

Intellectual Capital

The availability of intellectual capital is now worldwide. Just as the economy has become international, with the economy of one nation dependent on the economy of others, so the scientific and engineering enterprise has also become internationalized. Since the end of World War II, we have seen the growth of centers of excellence in many countries of the world with very substantial capabilities in science and technology. The issue needs to be framed in terms more familiar in industry where investment in capital equipment and physical plant is accepted. We now need to think in terms of investment in intellectual capital.

Development of intellectual capital requires training, so training must therefore be looked upon as an investment. We must begin to think of our investments in education as being just that—investments in building the intellectual capital of this country on which our industry, our government, and our universities can draw.

Intellectual capital needs nurturing, it needs protection, and it needs renewal. Yet our ability to form intellectual capital is decreasing in many ways. The technological workforce in America is graying. We are all familiar with the effects of decreasing birth rates, with the changing demographics and evolving ethnic composition of the work force. We need to be deeply concerned about how we are going to build this intellectual capital in the face of a changing and diminishing pool of individuals on whom this intellectual capital is based.

Intellectual capital needs nurturing, it needs protection, and it needs renewal. ...As teachers of mathematics, you are in the front lines of building this intellectual capital.

As teachers of mathematics, you are in the front lines of building this intellectual capital. The question is: How can we repair our failings, how can we buttress our strengths? As you know better than I, one of our failings is in teaching mathematics. Buttressing our strengths is one of the purposes of MSEB.

Making Calculus Exciting

Among the difficult parts of teaching mathematics is teaching calculus. As it is now taught, and as you appreciate better than I, it tends to be a barrier to students. Many of them drop out; many of them fail; many lose interest. What we need to do is to find ways to encourage and not discourage students, to keep them in the pipeline. I don't think it is necessary for us to reduce standards to maintain students in the pipeline. I think instead there need to be new approaches to teaching. You have heard about some of these already, and you will hear about many more ideas in the days ahead.

Calculus is really exciting stuff, yet we are not presenting it as an exciting subject.

Calculus is really exciting stuff, yet we are not presenting it as an exciting subject. At MIT, when I was taking my graduate degree, I went through the calculus course for engineers. By that time I understood why I was taking calculus and what it would be used for. It is clear to all of you, but not entirely clear to all the people who need to make decisions, that engineering and scientific applications just cannot exist without the calculus. Whether it is the design of a bridge, an electronic circuit, aircraft, or calculations of chemical processes, the calculus is at the core. Without calculus we would revert back to the engineering empiricism of a century ago. And that we don't want.

Weather Forecasting

My own experience is, I think, illustrative. I am a sometime weatherman. I used to be chief of the United States Weather Bureau. The field of weather forecasting is a good illustration of the fundamental importance of the calculus. It dominates work in this field.

Back in 1904 the idea that one could forecast the weather on the basis of physical law was first broached by Vilhelm Bjerknes in Norway. But of course at that time there was no way to apply the laws of motion. It wasn't until 1922 that an Englishman named Lewis Fry Richardson set those partial differential equations up in finite difference form, and sought to calculate by hand the time changes that would occur given a knowledge of the initial state of the atmosphere at a grid of points. It was the first application in my field of finite difference methods. He did it by hand, if you can imagine that, and the results were just all wrong.

In the 1940's and 1950's, when I earned my spurs as a weather forecaster, the best we could do was to use perturbation theory in trying to understand the growth of disturbances in a fluid system. But that only gave you initial tendencies of the growth of these disturbances.

It wasn't until 1955 that the use of digital machines to calculate the weather was attempted. First applications were run on ENIAC, one of the first digital computers developed by the Signal Corps. Under the overall supervision of John von Neumann of the Institute for Advanced Study at Princeton, the first experiments were conducted. That was the beginning of the transformation of weather forecasting from an art to a science.

The ability to apply ... equations transformed an entire science. The value to the world of modern weather calculations is enormous. The central role of calculus ... to produce important practical results is evident.

The ability to apply those partial differential equations in finite difference form transformed an entire science. The value to the world of modern weather calculations is enormous. The central role of calculus in dealing with these problems to produce important practical results is evident.

Teaching for Flexibility

The issues that this Conference must address are well laid out in your background documents. As I look at it, what we need to do now is to teach calculus in a way that provides a body of understanding which contributes to the flexibility and adaptability required of scientists and engineers, of social scientists and managers.

What we need to build into our students (and eventually into the people in our work force) is an ability to move from field to field. ... To do that you need ... an appreciation of the calculus.

The reason why flexibility is important is that in an era of very rapid technological change, with newly emerging fields of all kinds, what we need to build into our students (and eventually into the people in our work force) is an ability to move from field to field. To do that you need the kind of understanding that comes from an appreciation of the calculus.

We need to teach the calculus in a way that facilitates complex and sophisticated numerical computation in an age of computers. Somehow or other you have to make calculus exciting to students. The question as to the role of calculus in an age of computational mathematics is one that clearly this Colloquium needs to address.

In the National Spotlight

We confront a real challenge. It is clear that there is growing appreciation of the role of mathematics. Enrollments in mathematics departments have increased. We do have some problems with teaching assistants—over half come from outside the United States. But we can't meet these challenges just by arm waving or by generalized statements about what needs to be done—that high schools should prepare students better, or that students should work harder. A Colloquium like this has a real opportunity to come up with suggestions for how to attack this problem.

Calculus is a critical way-station for the technical manpower that this country needs. It must become a pump instead of a filter in the pipeline.

The national spotlight is turning on mathematics as we appreciate its central role in the economic growth of this country. The linkage between mathematics and economic growth needs to be made, and needs to be made stronger than it has been to date. Calculus is a critical way-station for the technical manpower that this country needs. It must become a pump instead of a filter in the pipeline. It is up to you to decide how to do that.

ROBERT M. WHITE in President of the National Academy of Engineering. He has served under five U. S. Presidents in leadership positions concerning science and technology policy. He continues to be active in an advisory capacity to the United States Government, and has also retained an active role in academic affairs. He received his Sc.D. degree in meteorology from the Massachusetts Institute of Technology.

Calculus Today

Lynn Arthur Steen

ST. OLAF COLLEGE

It is clear from the size of this colloquium that the calculus enterprise we are embarking on is of immense interest. One of the things I would like to do today is to document with some figures that it is also of immense proportions.

Calculus is our most important course .... The future of our subject depends on improving it.

I begin with a quotation from Gail Young, from the background paper he prepared for this colloquium. It is a summary assessment by a long-time leader of the mathematics community, speaking to mathematicians:

> Calculus is our most important course .... The future of our subject depends on improving it.

In support, I offer this evidence:

- *Three-quarters of collegiate-level mathematics is calculus.* (By "collegiate level" I refer to those parts of higher mathematics that are calculus-level and above, since all other courses taught in colleges and universities are really school-level mathematics.)

Calculus is the last mathematics course taken by our national leaders.

- *Calculus is the last mathematics course taken by our national leaders.* This is a very important issue. If you think about the career patterns of our national leaders, and of what students study in universities, the best students who enter universities by and large *do* take calculus, either in high school or in college, whether or not they are going to be scientists or engineers. Future lawyers, doctors, clergy, public school leaders—all professional leaders seem to take a little bit of calculus. But for virtually everybody, it is the *last* mathematics course they take. The entire public image of leaders of the United States concerning the nature of mathematics, and of the mathematical enterprise, is set by the last course that they take, which is calculus.
- *Calculus is among the top five collegiate courses in annual enrollment.* Calculus not only dominates the mathematics curriculum, but it dominates the entire university curriculum. Also in that top five is precalculus. When you put the two of them together, those two enrollments make up a substantial fraction of enrollment in higher education.
- *Most of what students learn in calculus is irrelevant to the workplace.* This observation came out of many of the background papers: an awful lot of what students actually learn in the current calculus course is no longer relevant to the way mathematics is used in science or industry.

Calculus Enrollments

Here is a crude portrait of calculus enrollment, rounded to the nearest hundred thousand. At the high school level, there are about 300,000 students enrolled in calculus courses of some kind. Only about 15-20% of these students are in AP calculus, so there is a large number of students—well over 200,000—who go through calculus in a once-over in high school.

There are 100,000 calculus enrollments in two-year colleges, and another 600,000 in the four-year colleges and universities. According to recent data from a special calculus survey conducted by Richard Anderson and Donald Loftsgaarden [2], about half of these 600,000 are in mainstream "engineering" calculus, with the remainder in non-mainstream ("soft") calculus or various summer and extension courses.

Only one-fourth of calculus-level enrollments (or one-eighth of mathematics enrollments in higher education) are in courses at or above calculus.

Looking at mathematics vertically, a little more than half of total college mathematics enrollments are below the calculus level. Of the remaining half, 75% are in calculus. So only one-fourth of calculus-level enrollments (or one-eighth of mathematics enrollments in higher education) are in courses at or above calculus.

Characteristics of Calculus

The 1985-86 CBMS survey [1] shows that the average section size in calculus is about 34. That's no surprise. What *is* surprising is that only about 7% of calculus courses use computers. The more recent survey [2] reveals that 3% of the calculus courses *require* the use of computers.

In a sample of final examinations that appears in the proceedings of this colloquium, we asked about whether calculators are permitted on final examinations. It seems to split about 50-50, in a way that is not correlated with the type of institution. So there is a real division in the community on that issue.

Only about 7% of calculus courses use computers.

Since these sample examinations also contain information on grade distribution, I looked at the data to determine the percentage of students who withdrew or failed—those who were enrolled at the two or three week mark, but who did not finish the course with a passing grade. Looking at it institution by institution—with an unscientific but very diverse sample—it looks like a uniform distribution for withdrawal and failure of 5% to 60%. For comparison, in the recent Anderson-Loftsgaarden study, of the 300,000 students who took mainstream calculus, only 140,000 finished the year with a grade of D or higher.

Ron Douglas and others have conjectured that success in calculus is correlated with feedback on homework. So we looked at what percentage of courses correct homework regularly. The recent survey shows that 55% rarely or never pick up homework and grade it.

Exams

When I looked at the questions on the sample of final examinations representing colleges of every type, from community colleges to ivy-league institutions, easily 90% of the questions were asking students to

Solve	Evaluate
Sketch	Determine
Find	Calculate
Graph	What is?

Most questions asked for straightforward calculations or posed template problems that are taught over and over again in the course and that are in the textbook in nice boxed examples. Anybody who is wide awake and pays attention ought to be able to figure out how to do these kinds of problems.

About 10% of the questions posed higher-order challenges; most of those were template word problems. Those of you who teach calculus know what that means—problems that fit a standard pattern. Some institutions and some courses have dramatically different patterns, but the mainstream examinations are like this: 90% calculation, 10% thought.

You find very rarely—only one problem in 1 out of 20 examinations—the kind of question that used to be very common 20 and 30 years ago: "State-and-prove ...." Problems dealing with the theory of calculus or with rigorous calculus have simply vanished from American calculus examinations.

Problems dealing with the theory of calculus ... have simply vanished from American calculus examinations.

Look at what students are asked to do for 90% of their examination problems; look at the verbs *solve, sketch, find, evaluate, determine, calculate, graph, integrate, differentiate*. What these commands correspond to, more or less, are the buttons on an HP *28C*. What we are actually examining students on, what we really expect them to learn, and what they know we really expect despite whatever the general goals of calculus are claimed to be, is the ability to do precisely the kinds of things that calculators and computers are now doing.

Pushing buttons, whether mechanical or mental, is one of the things we have to look at very carefully, to figure out how we are going to adapt calculus to better meet the needs of students.

What we are actually examining students on, what we really expect them to learn, and what they know we really expect ... is the ability to do precisely the kinds of things that calculators and computers are now doing.

Ron Douglas talked about riding the wave as another calculus metaphor from the sea. If you pay attention to the general concerns that are coming out of discussions of higher education, there is a great deal of concern about making sure that freshmen courses and other courses that are taken as part of a student's general education make a significant contribution to the broad aims of undergraduate education—that they help students learn to think clearly, to communicate, to wrestle with complex problems ..., etc.

There is very, very little in the calculus course of today that does any of these things. Word problems are a small step in that direction, but they are very rare and most of us who teach calculus know that if you put too many of those problems on your test, you are at great risk in the student evaluations.

Innovation: Technology

One other item that will appear in the proceedings of this colloquium is a paper by Barry Cipra based on interviews with many people who are now doing experimental things with calculus. These are innovations already under way, out of the mainstream, at least a little bit. The areas in which people are currently working fall into two broad categories: technology and teaching.

- *Disks.* A lot of textbooks, as well as individual authors, offer supplementary PC or Macintosh disks to help illustrate what is going on in calculus. Because of the power of this equipment, these packages tend to be exclusively numerical and graphing, since they do not have enough power to handle symbolic manipulation.
- *Symbolic Algebra.* These systems perform the manipulative routines of algebra and calculus in purely symbolic form–as we teach them in class. Packages like MACSYMA, SMP, and Maple run on large systems or workstations (e.g., VAX or Sun). Some of these are being compressed to fit in to the small desktop machines.
- *Programming.* In some places, instructors embed the teaching of programming into calculus, usually in Basic, in many cases now in TrueBasic; rarely is it in Pascal. It still is pretty uncommon anywhere to expect calculus students to also do programming.
- *SuperCalculators.* Current top-line calculators from HP, Sharp, and Casio are in the $80-200 price range, but we all know that soon they will be one-fourth that price. All can do graphs; some can do symbolic manipulation; most can do a great deal of what students normally accomplish in their freshman course.
- *Electronic Blackboards.* Some experiments use technology to make lecturing and presentation more dynamic. With a good classroom setup with a computer and a screen, you can do more examples, more realistic examples, and more dynamic examples. Calculus is the study of change; with an electronic blackboard you can actually demonstrate that change in real time, so students can see what these concepts are all about.
- *Electronic Tutors.* At the further out research level, there are people in artificial intelligence who are trying very hard to adapt techniques from symbolic algebra and the electronic blackboard and put it all together into a sophisticated program that would amount to an electronic tutor for calculus.

Now a lot of people have been working on this for school mathematics—for algebra and geometry. I'm frankly skeptical that this work will ever come to much, since it seems to me that the subtleties that are involved in learning calculus are probably a few generations beyond the ability of the artificial intelligence community to catch up with it. But I know there are people in artificial intelligence who believe that I am wrong, people who believe that in five years they will have these tutors really working. Some of them are probably in the audience right now ....

Innovation: Teaching

If you move away from the technology arena, there is not too much else going on. Technology is certainly what has captured the most interest. Here are three important areas related to teaching where serious work is taking place.

Teaching Assistants. Many universities are trying to devise means of incorporating TA's into the teaching of the calculus in a way that makes the experience for the students more satisfactory. As you all know, the budget structure of the major universities essentially requires heavy dependence on teaching assistants in calculus in some form or other, so there is a lot of experimenting going on to figure out what forms are better than others.

Even in a calculus course that is very well done ... students can go through ... getting a grade of B, maybe even a grade of A, and never write a complete sentence.

Writing. Some people have taken up the task of integrating into calculus the objectives of teaching students to write and communicate by making writing an important component of calculus. It certainly is the case now that even in a calculus course that is very well done—with good lecturing and small classes, where students ask questions and the instructor answers them, where instructors or assistants help students, where students turn in homework and take regular examinations—it is probably the case that students can go through such a course getting a grade of B, maybe even a grade of A, and never write a complete sentence in the entire semester, and probably never even talk at length about calculus with anyone.

There are people who are trying to correct that. If you look in the Tulane Proceedings, *Toward a Lean and Lively Calculus,* there are a set of objectives for calculus that emphasize expanding its goals beyond core mathematics goals to include a lot more general education objectives. Some far-sighted instructors are working on broad issues like that.

Constructivism. There is a movement in educational psychology that is actually led by physicists, that points out (with good supporting data) that students do not approach the study of subjects like mathematics and physics—or probably anything else—with a blank slate on which we teachers can fill in details by writing them on the blackboard and expecting students to xerox them into their brain.

Students do not approach the study of mathematics with a blank slate. ... When we instruct them, it is like pushing on a gyroscope. The student moves in a different direction than we push.

Students come with their own preconceptions about what mathematics should be, with their own repertoire of means of coping with mathematics as they encounter it. In many cases it may be evasive behavior, but it is part of a variety of prior experiences that students have.

When we instruct them, it is like pushing on a gyroscope. The student moves in a different direction than we push. So in order to teach students what we want them to learn, we have to understand the interaction that goes on when students construct their own images of mathematics which are quite likely different than the ones we have in our minds or that we are trying to convey to them.

Options for the Future

Let me close with an outline of issues for Tom Tucker, who will be telling you his view of what calculus may be like in the future.

When I talked with Tom about these presentations, I suggested that he could view this in a manner that fits his own expertise as Chairman of the AP Examining Committee which sets the most widely-used multiple choice examination for calculus. So I gave Tom a multiple choice question which he will answer shortly. I would like to conclude by telling you what the question is, so you can think about how *you* would answer it.

Where is calculus headed? Here are five choices for calculus tomorrow:

A. It will disappear completely as client disciplines discover that they can teach students to run computers better than the mathematics department can.

B. It will become the first modern classic—a scholarly refuge, like Latin, in which arcane insights of a past age are rehashed for those who wish to understand the history of our present culture.

C. It will remain totally unchanged due to the inability of forces acting from different directions to move an object with such large mass.

D. It will grow to double its present gargantuan mass, under pressure from the many client disciplines who want students who enter college knowing nothing to learn everything before they are sophomores.

E. It will explode into a supernova, with every discipline teaching calculus its own way.

We will shape the answer to this question—not today, but in the next few years—and in so doing respond to Gail Young's challenge. Reforming calculus *is* our most important task.

References

[1.] Albers, Donald J. *et al. Undergraduate Programs in the Mathematical and Computer Sciences: The 1985-1986 Survey.* MAA Notes No. 7. Mathematical Association of America, 1987.

[2.] Anderson, Richard D. and Loftsgaarden, Donald O. "A Special Calculus Survey: Preliminary Result." *Calculus for a New Century,* Mathematical Association of America, 1987.

LYNN ARTHUR STEEN is Professor of Mathematics at St. Olaf College. He is Past President of the Mathematical Association of America, a member of the Executive Committee of the Mathematical Sciences Education Board, and Chairman-Elect of the Conference Board on Mathematical Sciences. He also holds offices in the Council of Scientific Society Presidents and the American Association for the Advancement of Science. He received a Ph.D. degree in mathematics from the Massachusetts Institute of Technology.

Calculus Tomorrow

Thomas W. Tucker

COLGATE UNIVERSITY

Is there a "crisis" in calculus? If I were to ask that of students in one of my calculus classes, they would all say "yes," but their interpretation of "crisis" would be at a more personal level, especially with that hour exam coming up next week.

Leonard Gillman, in the most recent issue of *Focus* [1], writes that the phrase "crisis in calculus" reminds him of a 1946 murder trial in Indiana. The case should have been open and shut, but by continually referring to the murder as "an unfortunate accident" ("Where were you the afternoon of that unfortunate accident?"), the seasoned defense attorney managed to get his client off with only 2 1/2 years for manslaughter.

Maybe Gillman is right. Maybe there is no crisis in calculus. Maybe it's just ... an unfortunate accident. My students like that phrase too. They use it a lot themselves: "Professor Tucker, about that unfortunate accident of mine on problem 3."

No, there *is* a crisis today in mathematics, and in science. The crisis lies in the infrastructure of science, which is fed by the undergraduate mathematics curriculum. Like it or not, calculus is the entry for the entire undergraduate program in mathematics as well as the foundation for the sciences.

We have in our calculus classes a captive audience. ... If we cannot produce from this audience the mathematicians and scientists the country needs, we must ask "Why not?"

We spend enormous amounts of time and effort teaching calculus to masses of students. We have in our calculus classes a captive audience, at least for the time being. If we cannot produce from this audience the mathematicians and scientists the country needs, we must ask "Why not?" Are we doing the right things in this course? Can we change what we teach and how we teach it? What will calculus be like tomorrow?

Business as Usual

I propose three pictures of the future for calculus. The first is the obvious one: business as usual. Textbooks will continue to get bigger; using the logistic equation with data 640 grams in 1934 (Granville, Smith and Longley), 1587 grams in 1960 (Thomas, Third Edition), and 2617 grams in 1986 (Grossman, Third Edition), I get a limiting mass of 3421 grams. The content, however, will be unchanged.

Nearly half of all calculus students will be enrolled in classes of size 80 or more. Many smaller classes and recitation sections will be taught by graduate students whose native language is not English. Calculators and computers will be banned from most examinations. Pencil-and-paper algebraic manipulation will be the order of the day. Students will fail or withdraw in large numbers. And no one will complain because, after all, calculus is calculus. It's too familiar, too respected, too comfortable, and too big to change.

Calculus in High School

The second picture is that the mainstream college calculus course, like a river in the Great Basin, will gradually disappear into a number of sinkholes. It will seep away into the secondary school curriculum. Already 60,000 students a year are taking the Advanced Placement calculus exams and that number has been growing steadily at 10% a year ever since 1960. (In fact, it jumped 20% last year.)

It would not be surprising to see 200,000 AP Calculus exams in year 2000. And that is only the tip of the iceberg. From surveys, it appears that fewer than half of the students enrolled in an AP course actually take the exam, and even more students are taking non-AP calculus. Within a decade, there could be more students taking calculus in secondary school than presently take it in college.

High school calculus can be very, very good. ... The hundreds of AP teachers I have met ... embarrass me with their dedication, enthusiasm, and expertise.

Before we wring our hands over this state of affairs, let me say that high school calculus can be very, very good. It should be. They have 150 meetings to cover what we in college do in 50. Teaching calculus in secondary school is viewed as a pleasure and a reward; in college, it is a burden. The hundreds of AP teachers I have met through my involvement in the AP program embarrass me with their dedication, enthusiasm, and expertise. And of course they teach to the select few in secondary school, while we teach to the masses in college. It is no wonder that in my fall Calculus II class, the incoming freshman always outperform the sophomores who took our own Calculus I the previous semester.

But can secondary schools also teach calculus to the masses? In Russia, all 16 and 17-year olds are supposed to learn calculus. In Europe, general education takes place in secondary school, not in the universities. Many, students, perhaps a majority, who are enrolled in our mainstream calculus courses are there to get a good general education: a calculus course on the transcript is the sign of an educated person.

If the calculus of the future takes place in secondary school, what will college mathematics departments have to teach?

There is but one question: if the calculus of the future takes place in secondary school, what will college mathematics departments have to teach? Already I am seeing more and more students arrive on campus with AP Calculus credit and never set foot in a mathematics classroom again. This could be an omen.

Calculus Across the Curriculum

Mainstream calculus may also disappear into the client disciplines. Mathematicians are not the only people smart enough to teach calculus, and it shouldn't be surprising if other users wish to tailor a calculus course to their own needs. Some of the most innovative textbooks to appear in recent years have been for alternative tracks—for business, for life sciences, even for computer science. When we were making syllabi recommendations at the Tulane Calculus Conference in 1986, we found that many of our suggestions had already been adopted in a nonstandard text published independently by the Institute for Electrical and Electronic Engineering (IEEE).

The mathematics community could learn something from alternative courses. If it doesn't, mainstream calculus may find its flow of students diverted more and more. Already more than one in three college calculus students is enrolled in an alternative course, and, as with secondary school calculus, this number will grow.

The truth of the matter is that our clients have been remarkably tolerant of mainstream calculus. At my institution, we don't even teach exponential growth in first semester calculus, and yet the economics department, which urges students to take at least one semester of calculus, doesn't seem to notice. Out of sight, out of mind, perhaps; but we shouldn't count on *laissez-faire* forever.

The truth of the matter is that our clients have been remarkably tolerant of mainstream calculus.

Finally, mainstream calculus may disappear into computers and calculators. Long division, root extraction, use of log and trig tables are all fading from the precollege curriculum (not as fast, however, as one would expect: even though every student has a $10 scientific calculator close at hand, most textbooks still include tables of values of sines, cosines, natural logs, and exponentials).

Many traditional calculus topics such as curve sketching, relative maxima and minima, even formal differentiation and integration may become just as obsolete in the face of symbolic manipulation and curve plotting on computers and calculators. How long will mathematics faculty be able to maintain discipline in the ranks of students "digging and filling intellectual ditches," as Lynn Steen so aptly put it in a recent article [2] in the *Chronicle of Higher Education*?

What happens when our calculus clients find we are still teaching the moral equivalent of long division while they simply want their students to know how to push buttons intelligently? Of course we would find it barbaric if students could only recognize $\int_0^1 1/(1+x^2)\,dx$ as .785398, but what do we do when their calculators actually answer "$\pi/4$"? That will happen, you know, and as usual, before we're ready for it.

What happens when our calculus clients find we are still teaching the moral equivalent of long division while they simply want their students to know how to push buttons intelligently?

In the future, we may only need a few people who know the inner workings of the calculus, a cadre of the same size as the numerical analysts who keep our calculators going today, and I guarantee you that number is a lot less than 600,000 students a year.

Cleaning House

In contrast to this second picture of calculus tomorrow, which we might call the twilight of mainstream calculus, let us consider a third picture, a vista of a more conceptual, intuitive, numerical, pictorial calculus. The first step in that direction is honesty, with our students and with ourselves. If we teach techniques of integration because it builds character, let's admit it to ourselves and to our students. If we like $\pi/4$ as an answer rather than .785398 because it is beautiful rather than useful, let's tell our students that. If complex numbers are both too useful and too beautiful to ignore, then let's include them.

Do we need l'Hôpital's rule to know that e^x grows faster than x^{100}? Do we ever need to mention the cotangent and cosecant? Do we really want to know the volume of that solid of revolution? Are all functions encountered in real life given by closed algebraic formulas? Are any?

Are all functions encountered in real life given by closed algebraic formulas? Are any?

We should be asking these questions. In fact, we should have asked these questions long ago. And when we have figured out really why we teach what we do, let's tell it to our students.

After we clean house, what will the new calculus be like? I hope that it uses calculators and computers, not for demonstrations but as tools, tools that raise as many questions as they answer. The first time one plots the graph of a random polynomial on a computer, one learns that polynomials lead very dull lives most of the time, just going straight up or straight down, and that singularities are just that, singular. That is an important lesson, and one which is lost on most calculus students.

I remember when a colleague arranged a classroom computer demonstration to show the limit of quotients $\Delta y/\Delta x$ is the derivative dy/dx and was shocked and embarrassed to find after a few iterations that the quotients diverged. (I warned him about roundoff error but he didn't believe me.) Finding good numerical answers can be just as difficult, just as instructive, and just as rewarding as slick algebraic manipulation.

Computers and calculators can also be used as a tool to infuse new mathematics into a staid course. We complain that the lay view of mathematics is that there is nothing to study—it was all finished off a long time ago. That shouldn't be surprising if the calculus we teach today we could have taught one hundred years ago, maybe even two hundred.

What have mathematicians been doing for the last century? Our calculus classes say the answer is "Nothing."

For example, leafing through those boxed-in biographies of mathematicians in a well-known text, I could not find a single mathematician active after Riemann's death in 1866. What have mathematicians been doing for the last century? Our calculus classes say the answer is "Nothing."

Computers could change that. We could play with contemporary mathematics: the dynamics of functional iteration, fractals, stability, three-dimensional graphics, optimization, maybe even minimal surfaces. I know, I know: just because it's new, doesn't mean it's good. But a little "live" mathematics in a lean and lively calculus wouldn't hurt, even if it's only a commercial.

I hope the use of computers and calculators will also teach students to think about the reasonableness of their answers. As it is, they work so hard to get solutions that they never even give their solutions a second thought. I remember a problem we graded on an AP exam, which asked for the largest possible volume for a water tank meeting certain restrictions. We kept track of the largest and smallest answers. The largest was somewhat bigger than the universe, and the smallest was much less than an atom (not counting, of course, all the negative answers).

We may even end up in the future not only with "machines who think" but also with "students who think."

Students who just push buttons and write down the answer will find out quickly that that is not enough. We may even end up in the future not only with "machines who think" but also with "students who think."

The Day After Tomorrow

Suppose this third picture of calculus tomorrow comes true. It's a nice picture, an exciting picture, but there is still the day after tomorrow. Reforms have ways of becoming undone. In the late 1960's the NSF-supported Committee on the Undergraduate Program in Mathematics (CUPM) recommended that multivariable calculus be taught in the full generality of n dimensions with a whole semester of linear algebra as a prerequisite. By the early 1970's, textbooks for such a course had been published and many, if not the majority, of colleges and universities taught their multivariable calculus that way.

But something happened in the next decade without any urgings or direction from on high. A recent survey for the AP program revealed that 90% of the respondents now teach multivariable calculus in two- and three-dimensions only, out of a standard thick calculus textbook, with no linear algebra prerequisite. Mathematicians tried to twist the calculus sequence and came back later to find it, just like those metals with memory, back in the same old shape again.

There are strong forces that molded today's calculus, and those forces will still exist tomorrow. Students will always follow the principle of least action. Textbook publishers will still be guided by the laws of the marketplace. Faculty will still have limited time to devote to calculus teaching in a system which rewards more glamorous professional activity. The reform movement must take these forces into account.

Students will always follow the principle of least action. Textbook publishers will still be guided by the laws of the marketplace.

Unfortunately, some forces are societal and beyond our control. The conference held here last January on international comparisons of mathematics achievement was particularly depressing in this regard. As much as one might like to blame American shortcomings on spiral curricula which circle instead, or on middle school mathematics which is mostly remedial, it seems clear that the real problems are much more deeply rooted in our society.

Japanese students think that mathematics is hard, but that anyone can learn it by working enough. American students think mathematics is a knack only few are born with, and if you don't have it, extra work won't help. Japanese parents are intensely involved with their children's education and are generally critical of the academic program in their local schools. American parents think their children are doing fine, even when they are not, and are generally happy with the academic program in their local schools, even when they should not be.

If American children spend their after-school hours working for spending money at fast-food franchises, it cannot be surprising that their mathematics achievement might suffer. On my campus, every student wants now to become an investment banker—at least they did until last week; the old favorite, pre-med, is dwindling because it's too much science, too much "academics," too much hard work.

The most visible rewards in our society go to entertainers, athletes, and corporate raiders, but I have yet to hear of our nation being at risk because of a shortage of, say, TV personalities. If our calculus students do not learn, if their attention wanders, or if they do not even show up in the first place, we should not burden ourselves with *all* the blame.

I am not arguing that we should not try to make tomorrow's calculus different. I am sure we can do better and doing better could have a dramatic effect on the infrastructure of mathematics and science. We cannot do it alone, however. Like any other educational enterprise today, calculus reform needs broad support, from government, from private industry, from colleges and schools, from professional societies, from the media, from teachers, from students, from parents. Changing calculus may be more of a battle than we would ever imagine, but it is a battle worth fighting.

References

[1] Gillman, Leonard. "Two Proposals for Calculus." *Focus,* 7:4 (September 1987), p. 3, 6.

[2] Steen, Lynn Arthur. "Who Still Does Math with Paper and Pencil?" *The Chronicle of Higher Education,* October 14, 1987, p. A48.

THOMAS W. TUCKER is Professor of Mathematics at Colgate University and author of the Content Workshop Report at the Tulane Conference on Calculus. He chaired the College Board's Advanced Placement Calculus Committee from 1983-1987 and currently serves on a number of committees for the Mathematical Association of America, the Committee on the Undergraduate Program in Mathematics, the College Board, and the Educational Testing Service. He received a Ph.D. degree in mathematics from Dartmouth College.

Views from Client Disciplines

Cathleen S. Morawetz

Courant Institute of Mathematical Sciences

Welcome. We are here to hear a panel representing the client disciplines. They are the users of the students we mathematicians turn out. I am here to represent applied mathematics in the channel connecting mathematics to other disciplines.

As a professor at the Courant Institute, every once in a while I have taught the calculus. But I have to admit, not very often. I am a user, and as a user I feel that I belong with the client disciplines.

Someone mentioned this morning that Leonard Gillman wrote in *Focus* that things were all right with calculus. Why fix what isn't broke?

I would like to read to you a little bit of what Peter Lax wrote to Gillman in answer to that. Peter is sorry that he isn't here to give the message himself.

> Dear Lenny:
>
> At one time the historian-essayist-moralist Carlyle was irritated by a friend who didn't believe in the existence of the devil. So Carlyle took him to the gallery of the House of Commons, and after listening to the goings-on, turned to his friend and remarked, "D'ye believe in the devil noo?"
>
> When I encounter a skeptical colleague, I feel like showing him today's most widely used calculus text and ask him "D'ye believe in the crisis in the calculus noo?"

I think this story has another moral too, which is that we should do a lot of listening as well as talking about the calculus. Many of you have a first love—perhaps topology, or algebra, or some other very abstract field. You do not use the calculus today, not as I do. So I am a client and I represent the link that binds core mathematics to the client disciplines.

Someone once said that I am a card-carrying applied mathematician. If someone else wants to dispute that, and some people will, then I'll see them afterwards.

As a card carrying member, I form a link to physics and to engineering. I am interested in the links to biology, although I have to confess that it was only rather late in life as a mother of an economics student that I learned of the importance of calculus to economics. I also come from an institution which in its research specializes in this linkage. So I am here today to help pose questions for my panelists who come from the client disciplines.

Everyone who can learn calculus should learn calculus.

I have a profound interest in what gets taught in the calculus. I think I understand the needs of industry and the other sciences in continuing the scientific education of students beyond the calculus. I would like to suggest a slogan for the future: Everyone who *can* learn calculus *should* learn calculus.

Cathleen S. Morawetz is Director of the Courant Institute of Mathematical Sciences, New York University. She is a member of the Board on Mathematical Sciences of the National Academy of Sciences.

Calculus for Engineering Practice

W. Dale Compton

National Academy of Engineering

In considering the Calculus for a New Century, with its obvious emphasis on the next century, it is important that we be sensitive to the context in which the student will be studying this important subject. The future environment will be determined, in a significant way, by the competitive position of the U.S. in the world marketplace. We must consider, therefore, the role that calculus can have in helping this nation achieve an improved competitiveness.

While many measures could be given of our current competitiveness, it is sufficient to remind ourselves of the impact of the large trade deficit and of the changes in employment levels that have resulted from the movement of manufacturing off-shore. With roughly *seventy* percent of our current manufacturing output facing direct foreign competition, we can expect the competitive pressures on the manufacturing sector to continue. Even the service sector is not immune to these pressures. One example of this is the increasing fraction of engineering services being contracted to off-shore companies.

Calculus should encourage students to proceed to an engineering career—not by being easy, but by being exciting.

Many factors determine our competitiveness and many actions will be required to improve it. Most people agree, however, that one of the principal tools for recovering our competitive position will be a more effective use of technology. This requires that the practice of engineering be more effective. To affect the practice of engineering one must focus on industry, the employer of about 75% of all engineers. Therefore, in considering the future capability of industry to effectively use technology, an important issue becomes the availability and qualifications of future engineers.

If an adequate supply of qualified engineers is critical for industrial competitiveness, it is important to examine the trends in engineering enrollment. First, student demographics predict a sharp drop in the college-age population, Second, a decreasing percentage of new college students are indicating an interest in science and engineering. Finally, a larger proportion of entering college students will be minorities and females, a group that have not been strongly attracted to engineering as a profession. An important conclusion is that we must find a way to reduce the fraction of students who withdraw from the study of engineering, even though they may have expressed an early interest in the field and are qualified to study it.

We must find a way to reduce the fraction of students who withdraw from the study of engineering.

It is here that calculus becomes so important. Mathematical skills are a prerequisite to the successful practice of engineering, and calculus is the first major step in acquiring the skills needed for an engineering career. Calculus should encourage students to proceed to an engineering career—not by being easy, but by being exciting. It must not be an artificial barrier that is used to discourage students from proceeding. It should encourage students to explore the possibilities further. It should help convey to the student the sense of excitement that the practitioner of engineering experiences.

Engineering, in the words of the Accreditation Board for Engineering and Technology (ABET), is "the profession in which knowledge of the mathematical and natural sciences gained by study, experience, and practice is applied with judgment to develop ways to utilize, economically, the materials and forces of nature for the benefit of mankind."

The operative words in this statement are knowledge and judgment. Whereas mathematics has most often been considered as a requirement for *knowledge,* it is time for us to begin to consider its role in *judgment.* The calculus course can be a place to start creating this sense of judgment and a sense of the excitement of the field through the examples that are used.

Whereas mathematics has most often been considered as a requirement for knowledge, it is time for us to begin to consider its role in judgment.

Consider the following possibilities. Engineering deals with systems. Many systems are large and thus complex. Most systems are non-linear. Hence, approximate solutions are required to many system problems. It follows that the practitioner must have a good sense of the reasonableness of a solution.

It is my guess that students would react positively to a calculus that includes examples that require the exercise of good judgment. What better time to introduce the student to a sense of engineering than through examples of this type. What better way to introduce some excitement into calculus.

W. DALE COMPTON is currently a Senior Fellow at the National Academy of Engineering. Previously, he served a total of 16 years with Ford Motor Company, first as Director of the Chemical and Physical Sciences Laboratory and from 1973-1986 as Vice President of Research. From 1961-1970 Dr. Compton was a Professor of Physics at the University of Illinois at Urbana, serving as Director of the Coordinated Sciences Laboratory from 1965-1970. He received a Ph.D. degree in physics from the University of Illinois.

Calculus in the Biological Sciences

Henry S. Horn

PRINCETON UNIVERSITY

After describing my own deplorable formal mathematics background, I would like to make an idiosyncratic list of the techniques that my own research and teaching now require. Then I shall condense that list into a recommendation for the kind of calculus course that I wish I had had as an undergraduate.

My perspective comes from doing empirical research in ecology, studying the behavior of birds, butterflies, and trees, with a strong conceptual bias, and with interests in population genetics, evolution, development, and biomechanics.

My formal courses started with developing geometric intuition and the notion of rigorous proofs in high school geometry, and it ended in college with baby-calculus through an introduction to linear differential equations. My fanciest mathematics, namely complex phase plane analysis, was learned in an engineering course called "Electrical Engineering for Engineers who Aren't Electrical."

The help that I currently give my children on middle school and high school homework should also count as formal coursework in higher math. This is partly because of their appalling textbooks and the last vestiges of the New Math, but it is also partly because of the depth and breadth of substance in the current secondary school mathematics curriculum. My research and that of close colleagues has required far more mathematics than I have learned formally.

Meeting even the elementary mathematical needs of my area of biology requires more than the usual attention to the disciplines neighboring calculus: geometry, difference equations, nonlinear qualitative analysis, linear algebra, and statistics.

In my own research I use infinite series to calculate the photosynthesis of layers of leaves in a forest. I look up standard derivatives and integrals in the *Handbook of Chemistry and Physics* to model butterfly movements as a diffusion process. I use phase-plane analysis of systems of differential equations to discover population consequences of dispersal behavior, and, in the privacy of my own bedroom, I calculate the dispersion of eigenvalues of a transition matrix to explore the speed and repeatability of field-to-forest succession. Multivariate calculus and linear algebra are useful in statistics.

Reality, breadth, and substance are crucial if examples are to hold the motivation for their own solutions.

Colleagues studying populations, genetics, and neural networks cite crucial differences between differential and difference equations, with respect to stability and chaotic behavior. They also use combinatorics in the construction and analysis of genetic sequences, evolutionary trees, and the like. Propagation of noise spectra through differential and difference equations is widely practiced in biology.

A topic that is far more different than is usually recognized is the study of heterogeneous nonlinear systems in which it is necessary to carry the full complexity through the analysis and plot the distribution of the result. An appropriate choice among these techniques is needed to study effects of varying environments on population dynamics, of variation among individuals on parameters of population or behavior, of sensory filtering in physiology, and of error in estimation of parameters in general.

Meeting even the elementary mathematical needs of my area of biology requires more than the usual attention to the disciplines neighboring calculus: geometry, difference equations, nonlinear qualitative analysis, linear algebra, and statistics. In addition it requires a perspective more like applied mathematics or engineering than pure mathematics.

An ideal calculus course from my perspective would have the following properties:

1. Start with its historical origin as solutions to problems in physical dynamics, but move on at least to the qualitative behavior of solutions of difference and nonlinear differential equations, propagation of variance, and propagation of qualitative heterogeneity through complex systems. Here I echo the enthusi-

asm of many symposium speakers for discrete mathematics.

Even qualitative contrasts with the traditional calculus are of use to me, to say nothing of approximate solutions. These are traditionally considered to be advanced topics, but they are the necessary rudiments to treat the real world beyond idealized Newtonian physics. (Incidentally, I have recently enjoyed just the kind of playing with advanced concepts that Thomas Tucker extolled this morning, using semitutorial graphical microcomputer programs that are being developed by folks in my field.)

2. Develop and emphasize proofs, not from first principles, but from agreed-on intuitive principles. Unfortunately this depends on common sense, which as Voltaire observed is not so common.
3. Use plug-in problems and template word problems to drill on substantive questions from a variety of academic disciplines. A partial list of disciplines whose examples would have direct relevance to my own is: ecology, molecular biology, physics, structural engineering, fluid dynamics, fractal geometry, economics, and politics. Reality, breadth, and substance are crucial if examples are to hold the motivation for their own solutions.

HENRY S. HORN, Professor of Biology at Princeton University, specializes in ecology and the behavior of birds, butterflies, and trees. The adaptive significance in biology of geometrical components is of particular interest to him. He is a member of several professional organizations including the Ecological Society of America and the Animal Behavior Society. Dr. Horn received a Ph.D. degree in ecology from the University of Washington.

Calculus for Management: A Case Study

Herbert Moskowitz

PURDUE UNIVERSITY

My views on calculus for management will be based on my experience and observations of our undergraduate and graduate professional programs in management at the Krannert School of Management at Purdue University. As background, I will overview relevant aspects of the curricula in each of these programs and relate these to the need for calculus. Then I will state several issues regarding the nature of the calculus courses taken by our students. From this, inferences and conclusions will be drawn regarding whether and how the instruction in calculus should change to meet the immediate and future needs of students in schools of management, or perhaps, whether it is really needed at all!

Undergraduate Management Programs

There are three undergraduate programs in management at the Krannert School: Industrial Management (IM), Management (M), and Accounting (A). The mathematics requirement in each of these programs are high and demanding compared to other comparable universities, far exceeding AACSB guidelines.

Pre-Management students are required to take 2-3 semesters of calculus; the 3 semesters applying specifically to our IM students, who must minor in a physical or engineering science discipline. In addition, all pre-Management students must take a mathematical statistics course. Once in the management program, students additionally take a managerial statistics course (calculus-based) as well as a management science course whose primary emphasis is on optimization.

The Calculus Requirement. As the Krannert School's undergraduate programs are currently structured, calculus is essential:

1. It is a pre-requisite for satisfying courses constituting the minor in the IM program. It is also a prerequisite or co-requisite for such required courses as Introduction to Probability, Quantitative Methods (statistics and optimization), Micro and Macro Economics, and such functional area courses as Operations Management and Marketing Management.
2. It serves as a "mild" filter, in the sense that grades in the course, along with other courses, are used to determine whether a student can transit successfully from the Pre-Management Division into the Management Division.
3. It is a virtual necessary condition for acceptance (viz, "license") into a quality graduate professional or Ph.D. program in Management.

The Calculus Sequence. There is an option for one of the following two sequences, which is contingent upon a student's prior mathematical preparation and ability, determined through testing and review of a student's records:

1. The standard 3-course sequence oriented towards the physical sciences consisting of differential, integral, and multivariate calculus, respectively;
2. A nonstandard sequence for the less prepared student composed of a fundamental course in algebra and trigonometry, a less rigorous differential calculus course, and a less rigorous integral calculus course.

More than 80% of all management students begin with the algebra and trigonometry course. Seventy-five percent then take the nonstandard calculus sequence (all pre-IM's, however, must take the calculus sequence).

Fifty percent of our students obtain a grade of A or B as compared to 40% in the university in the standard calculus sequence. Moreover, almost 60% of our students obtain an A or B in the nonstandard calculus sequence as compared to about 40% in the university. This high success rate is attributed in part to an initial evaluation of the student's background and ability, remedial training, and placement in the appropriate calculus sequence (standard vs. nonstandard) for our students.

Nature of Calculus Courses. All calculus courses are taught by the mathematics department. The courses are highly structured and automated, in part to cope with delivering mass-produced training efficiently. The instructional approach is algorithmically oriented, i.e., "how to differentiate and integrate." Little or no emphasis is focused on conceptualization, modeling, or relevant and meaningful management applications. Class sizes are very large (large lecture halls) for calculus courses in the standard sequence and approximately 40 for courses in the nonstandard sequence. Typically, instructors are either not the best mathematics faculty or are foreign Ph.D. students majoring in mathematics. Neither, presumably, are familiar with calculus applications in management.

Observations and Implications. Relatively speaking, although students are exposed rather extensively to the calculus and must use it in their economics, engineering and management courses, they have considerable difficulty, particularly in problem and model conceptualization and formulation. Moreover, even the well-drilled procedures of differentiating and integrating are forgotten much too quickly and must be reviewed. Hence, the educational impact should and must be improved.

Graduate Professional Programs

There are two predominant masters in management programs at the Krannert School: the Master of Science in Industrial Administration (MSIA), and the Master of Science in Management (MS) programs. The MSIA is an 11-month general management program designed for students with technical degrees. The MS is a two (or 1-1/2) year general management program, also requiring a specialization in a functional area of management.

Historically, virtually 100% of the students in the MSIA program had technical backgrounds, while about 80% of such students entered the MS program. Today, the composition of the students in both programs has changed considerably, the trend being towards more students with degrees in business, economics, and the liberal arts. Concommitantly, so have their mathematical backgrounds changed.

The average GMAT of entering students is in the 90th percentile. Both programs place strong emphasis on the use of quantitative (and computing) skills for problem solving and decision making, hence students are screened carefully for the quantitative aptitudes in the admissions process (background in calculus appears to be independent of the GMAT quantitative score).

Calculus is rarely used in any course ...due to the personal computer, which has been delegated the task of performing computation and analysis.

Due, in part, to demographics, only about 70% of currently enrolled students have at least one semester of college calculus on entering the program. In the not too distant past, *all* students were well trained in calculus, and this was reflected in the Masters program coursework.

For example, our statistics course used to be a calculus-based course in mathematical statistics. However, today, calculus is rarely used in any course in our Masters in Management Programs, including our statistics course. This is, in part, due to the lack of mathematical background of students in our program, which is more quantitatively rigorous than most MBA programs in the nation. But it is due even more so to the personal computer, which has been delegated the task of performing computation and analysis.

Now, with computers, increasing effort can and is focused on solving large-scale, real-world problems with emphasis on a problem's "front end" (problem definition, formulation, modeling) and "rear end" (performing "what if" and "what's best" analyses, and interpreting the results from a managerial viewpoint). To illustrate the trend, the simplex method of linear programming is no longer taught in our management science course, but is solved by an appropriate software package integrated to a spreadsheet. Calculus in the MBA program, for all intents and purposes, has essentially vanished—much to the delight of students.

Calculus in the MBA program, for all intents and purposes, has essentially vanished—much to the delight of students.

Is Calculus Needed? The concepts of calculus are clearly relevant to management in obviously pervasive and important ways. Hence, they should be taught to management students; probably to all college students. However, the nature and focus of what is being taught in a calculus course must be improved. This can be accomplished in a variety of ways including the following:

1. Incentive systems must be established to reward innovative and solid calculus instruction.
2. Calculus classes perhaps ought to be partitioned broadly into the physical, engineering, and social sciences to at least allow the possibility for focusing more on applications that are relevant to students in these respective disciplines.
3. Increased coordination and collaboration are necessary between mathematics departments and, for example, schools of management, to maximize topical relevance and develop meaningful management applications of concepts.
4. Following the innovations in management and engineering schools in particular, the computer can assume a significant instructional role, via development and implementation of appropriate interactive software and graphics. This will allow instructors to focus more heavily on calculus concepts, modeling, and interpretation, relegating computational work to the computer.
5. Under conditions of mass-produced training, in particular, intelligent tutoring systems could be developed to "teach" novices to become experts in a given topical domain. Dissertations could and should be encouraged to develop such software systems, perhaps in joint collaboration with faculty in computer science, engineering, and management. Computer laboratories for experimentation in interactive instruction should also be established to try out novel and imaginative instructional technologies. Such efforts would simultaneously make both research as well as teaching contributions.

What are the alternatives? Perhaps business as usual with its well-known, predictable result; perhaps, schools of management teaching calculus to their own students; perhaps, no calculus at all!

HERBERT MOSKOWITZ is the James Brooke Henderson Professor of Management and is a past director of Graduate Professional Programs in Management at the Krannert Graduate School of Management at Purdue University. His area of specialization is management science and quantitative methods with interests in judgment, decision-making, and quality control. He received a Ph.D. in management from the University of California at Los Angeles.

Calculus and Computer Science

Anthony Ralston

STATE UNIVERSITY OF NEW YORK AT BUFFALO

I stand second to no one in a belief that the teaching of calculus needs to be changed considerably if American university students are going to be well-served by departments of mathematics. But the title of this conference epitomizes one of the things that is wrong with calculus teaching today. It implies that the place of calculus in the mathematics firmament is still just what it has been for the last century or so, namely the root of the tree from which all advanced mathematics—at least all advanced applied mathematics—must be approached.

You don't have to agree with me that calculus and discrete mathematics should be coequal in the first two years of college mathematics to recognize that *some-*

thing is happening out there which implies a need for the teaching of calculus to adapt to changes in the way mathematics is applied and in the clientele for college mathematics. This need includes, at least, a requirement that new subject matter be considered for the first two years of college mathematics and that there be some integration of the discrete and continuous points of view in those first two years.

How ironic that there has been a steady, perhaps accelerating trend in recent years for college calculus to be dominated by the teaching of just those symbol manipulations which humans do poorly and computers do well.

Let me make a few remarks about the impact of computer and calculator technology on calculus and how it should be taught. Hand-held calculators can now perform all but a very few of the manipulations of K-14 mathematics. Within a very few years the "all" will not have to be qualified and, moreover, the devices will cost no more than about the price of a book. How ironic, therefore, that there has been a steady, perhaps accelerating trend in recent years for college calculus to be dominated by the teaching of just those symbol manipulations which humans do poorly and computers do well and whose mastery, I believe, does not aid the ability to apply calculus or to proceed to advanced subject matter.

Too many mathematicians act like mechanical engineers when teaching calculus—they focus on crank-turning.

Too many mathematicians act like mechanical engineers when teaching calculus—they focus on crank-turning. This must be changed. Calculus must become a mathematics course again, one which focuses on mathematical understanding and on intellectual mastery of the subject matter and not on producing symbol manipulators. The technology available must be integrated (pun intended) with the teaching of calculus in order to get rote mastery of out-dated skills out of the syllabus and to get mathematics back in.

Except perhaps for engineering students, computer science students are now the largest single potential client population for departments of mathematics. Nevertheless, too often today departments of mathematics have, in effect, encouraged computer science departments to teach their own mathematics because they have not been willing to teach discrete mathematics themselves. While I would argue that all undergraduates should be introduced to the calculus since it is one of the great artifacts created by humankind, there is little disciplinary reason for computer science undergraduates to study calculus.

There is littlc disciplinary reason for computer science undergraduates to study calculus.

With the almost sole exception of the course in analysis of algorithms, there is no standard course in the computer science undergraduate curriculum which leans more than trivially on calculus. (Yes, I know about numerical analysis—I was, after all, once a numerical analyst—but fewer and fewer computer science undergraduates take numerical analysis any longer, and almost none are required to take it.) Even in undergraduate courses in the analysis of algorithms, the use of calculus-based material is often non-existent and, even when it is not, only very elementary aspects of calculus are used. For students headed toward graduate work in computer science, one cannot be quite so unequivocal but, even at the graduate level, few computer science students have anywhere near as much use for continuous analysis as they do for discrete analysis.

•

ANTHONY RALSTON is Professor of Computer Science and Mathematics at the State University of New York at Buffalo. Dr. Ralston currently chairs the Mathematical Sciences Education Board Task Force on the K-12 mathematics curriculum and is a past president of the Association for Computing Machinery. In recent years he has been interested in the interface between mathematics and computer science education, particularly in the first two years of the college mathematics curriculum. He received a Ph.D. degree in mathematics from the Massachusetts Institute of Technology.

Calculus for the Physical Sciences

James R. Stevenson

IONIC ATLANTA, INC.

My original entry into this discussion came several years ago when I was asked by Ronald Douglas to represent the physical sciences at a workshop in New Orleans (also supported by the Sloan Foundation). At that time I was Professor of Physics and Executive Assistant to the President at Georgia Institute of Technology. Today I am emeritus in both those positions, and chief executive officer of a small start-up high technology company. Whether I have gained or lost credibility in the interim is an open question. In the few minutes today, I will emphasize a few of the points addressed at New Orleans.

Physical science is empirical. The description and the formulation of "Laws of Nature" depend on observations rather than on logical development from axioms. Mathematics has always provided a convenient framework for the description of physical phenomena. Can it do the same thing in the future? In the past the physical science community has interacted strongly with mathematicians and mathematical descriptions of Newtonian mechanics have made a smooth transition from non-relativistic to relativistic and from non-quantum to quantum descriptions. The next challenge is one of the reasons for today's discussion.

Observations over the years have demonstrated that the important features of nature are imbedded in the description of non-linear phenomena. Both physical scientists and mathematicians have been very clever at arriving at excellent approximate descriptions. Recently many scientists and mathematicians have questioned this approach, believing instead that many secrets of nature may be hidden because of our insistence to force physical observations to be described by available mathematics.

Many secrets of nature may be hidden because of our insistence to force physical observations to be described by available mathematics.

The computational approach has uncovered the worlds of "chaos" and "fractal geometry." Will we discover sufficient cause to reformulate our "Laws of Nature" in a new descriptive format? Is the invasion of mathematics by the empiricists of the computer going to result in a completely new approach to mathematics at the undergraduate level? Quo vadis calculus?

The development of intuitive thinking is a most valuable asset to the physical scientist. Calculus instruction can play an important role. A quotation from Maxwell provides insight:

> For the sake of persons of these different types, scientific truth should be presented in different forms, and should be regarded as equally scientific, whether it appears in the robust form and the vivid coloring of a physical illustration, or in the tenuity and paleness of a symbolic expression.

The physical scientist would argue that a similar statement can be made for presenting introductory calculus. The meaning of slope and curvature and their relation to the first and second derivative are important. The location of maxima, minima, and inflection points are also important, and the physical scientist must have sufficient drill to be able to look at a graph and tell immediately the sign of the first and second derivatives as well as to estimate their magnitudes without resorting to calculators or computers.

The content of introductory calculus is probably not as important to the physical scientist as the insight to this form of mathematical reasoning.

In a similar vein the area under a curve must have an intuitive relation to integration. Infinite series are used to approximate analytical functions. Some knowledge of convergence as well as truncation errors are needed on an intuitive basis prior to releasing the power of the computer to grind away and produce nonsense. Intuition and "back-of-the-envelope" calculations are still important in guiding the physical scientist to understand the significance of observations. Mathematical intuition and physical intuition are frequently interrelated.

The content of introductory calculus is probably not as important to the physical scientist as the insight to this form of mathematical reasoning. Many times a student in physical science uses techniques of mathematical analysis the student has never encountered in a mathematics course.

Principles of mathematics taught in the context of a specific application have the danger of not being recognized for the breadth of their applicability.

The important parameter with regard to content is effective communication between the mathematics faculty and the faculty teaching courses for which calculus is prerequisite. Most faculty in the physical sciences have sufficient background in mathematics so that they can direct students to appropriate sources or use their lectures to provide background coverage. The student can become a victim if communication between faculty of different disciplines is missing.

Principles of mathematics taught in the context of a specific application have the danger of not being recognized for the breadth of their applicability. Thus the mathematics faculty does have an obligation to look at content and the order in which it is presented to minimize the amount delegated to other faculty. In deciding content, order of presentation, and pedagogical approach, an effective dialogue is most important.

In looking at Calculus for a New Century, the mathematics faculty has an obligation to look at the educational content of its courses. The educational content must contain a balance between the teaching of new skills and the development of mathematical and intuitive reasoning. In addition, scientists and mathematicians must continue to examine the question of the applicability of mathematics to the description of the "Laws of Nature."

•

JAMES R. STEVENSON, Chief Executive Officer at Ionic Atlanta, Inc., also serves as consultant to the president at Georgia Institute of Technology. He is a member of several professional organizations, including the American Physical Society, the American Association of Physics Teachers, and the American Society for Engineering Education. He received a Ph.D. degree in physics from the University of Missouri.

A Chancellor's Challenge

Daniel E. Ferritor

UNIVERSITY OF ARKANSAS, FAYETTEVILLE

Since it is nearly always more fun to reform someone else's discipline than your own, I have looked forward with anticipation to this colloquium on calculus instead of one bent on reforming sociology.

The reforms contemplated here, however, are different from most. They start, it seems to me, from established strength, not from disarray. Calculus courses already command a respect (even among students who hope never to have to take one) which all college courses should command, but which few others do.

Calculus courses already command a respect ... which all college courses should command, but which few others do.

Along with respect, to be sure, calculus courses evoke fear, awe, resignation, delight, and even resentment. Some students find what they learn in our calculus classes an indispensable tool; others view it as a meaningless hurdle. To some it is the pinnacle of mathematical achievement; to others it is only a foundation course for years of further study. I sympathize with your challenge to make the experience better for students regardless of their different views and needs.

While improving the quality of mathematics education for all Americans is becoming an agenda for action throughout the country, we in Arkansas may be slightly behind the curve in recognizing its importance. Before a recent raising of educational standards in Arkansas, there was *no* requirement in mathematics for high school graduation. In fact, until recently, our own university, the strongest in the state, had no mathematics requirement in many of its degree programs. Only last year did the arts and sciences faculty include significant college-level mathematics requirements in its B.A. degree programs.

Students bring to the University of Arkansas an extreme range of mathematics skills and experiences, including, in some cases, the apparent lack of either. It has been our mission for 116 years as a land-grant school to attempt to meet the needs of students whose diverse skills and needs are ensured by an admissions policy which opens our doors to most would-be students with a high school diploma and minimal GPA or national test scores.

Because mathematics education in many of our public schools has been limited, even able students often come to us unprepared for college mathematics. At the same time, we are the only institution in Arkansas with established programs of research and doctoral study, many of our undergraduate programs are unique in the state, and many of our students come well equipped for challenges and expect us to provide an educational experience which ranks with the best available.

To meet the needs of both groups of students is no easy task. Our 2,500 entering students each year have math subscores on the ACT test ranging from 1 to 36, although the average composite score is well above the national average at nearly 21 (comparable to an SAT of 870-900). This range of skills makes initial placement quite difficult but enormously important, and we have six levels at which students may begin the study of mathematics. These range from a one-semester remedial course in algebra to beginning calculus. Placement beyond the first course in calculus is possible, but highly unusual. About 20 percent of our entering students are placed in the remedial course, and about 20 percent are placed in the first semester of calculus, with others entering at intermediate levels.

Because mathematics education in many of our public schools has been limited, even able students often come to us unprepared for college mathematics.

Our calculus courses have many of the same problems as those at other schools across the country. However, the percentage of students who fail or withdraw is not quite as high as the 50 percent often reported nationally. We feel that our placement scheme has helped increase the success rate for students in calculus as well as in other mathematics courses.

Better placement, though, is not enough. Nowhere are the differing views of calculus more obvious than in the classrooms, where the perceptions of instructor and student can be worlds apart. For most students, calculus is the highest level of mathematical knowledge to which they aspire, while to many instructors, it is the lowest legitimate level of mathematical inquiry.

For most students, calculus is the highest level of mathematical knowledge to which they aspire, while to many instructors, it is the lowest legitimate level of mathematical inquiry.

The difficulty is further complicated by the fact that some students need calculus as a working tool in subsequent mathematics courses, while others conclude their study of mathematics with calculus. Both should gain from calculus a larger vision of the mathematical landscape. One teaching approach and one syllabus, though, may not be right for both kinds of students, and therein lies a major dilemma which must be resolved as we develop a calculus for a new century.

Colleges and universities must do a better job of raising the success rate of entering students by staffing beginning mathematics courses with instructors who are especially well qualified for the demands of such courses. This would not be an easy task even with a shared philosophy and a well-defined policy, and few institutions, I suspect, have either.

Colleges and universities must do a better job of raising the success rate of entering students by staffing beginning mathematics courses with instructors who are especially well qualified for the demands of such courses.

Within our own institution we do not even agree on which instructors—those with a broad vision of the field or those committed to providing tools for specific uses—are likely to be the best teachers. Nor have we begun to consider radical approaches such as more stringent controls, supervision, and uniformity in class organization and conduct in such courses. The current debate should lead to better definition of the teaching problem and, eventually, to the improvement we all desire.

Exacerbating the problem at an institution such as ours is the interaction of our graduate and undergraduate programs, which is generally healthy but which makes instructor choice and placement a practical as well as philosophically challenging matter. In lower-division mathematics, we depend heavily on graduate assistants as assigned teachers and as teaching assistants.

As is the case nationally, our graduate students include international students for whom English is a second language. Students who experience difficulty with a course like calculus—and many do—look widely for an explanation of their difficulty. If the instructor speaks English with an accent, he or she is a likely target. While the accent may well be a factor in the student's failure, even if it is not, too often students, parents, and legislators believe it is the primary cause.

"Why can't we have American math and science teachers?" is a tough question, but it is one of the questions most frequently asked of me by Arkansas legislators.

"Why can't we have American math and science teachers?" is a tough question to handle, but it is one of the questions most frequently asked of me by Arkansas legislators. We must avoid overreacting to such criticism and remind critics of the American tradition, from our earliest beginnings, to welcome and rely on imported talent. However, by the same token, we cannot rely solely on international students and faculty. The decline in mathematics and science majors among native-born students has reached alarming proportions and should stimulate us to devise ways to attract more of our own students into such careers.

Finally, in addition to accurate placement and focused and enlightened instruction in calculus courses, I see a need on the institutional level for an increased awareness among our students and faculty of the importance of mathematics. I hope that efforts like this one at the national level will help us there. The umbrella project "Mathematical Sciences in the Year 2000" is, I understand, designed to broaden this discussion beyond calculus to the other courses, to the flow of mathematical talent, and to the issues of resources. Since the University of Arkansas programs span a broad area from remedial algebra through graduate work and research programs in mathematics and statistics, we will be looking forward to that broader discussion and assessment.

I am sure I speak for many university chancellors and presidents, as well as for my own university, when I express support for the goals of the projects here at the

National Academy. While I can't evaluate the substantive changes in calculus which are suggested, I certainly recognize the need for a revitalization of the teaching and of the learning of college mathematics.

Returning a sense of discovery and excitement to classrooms where calculus is taught will be a vital step toward university preparation of the kinds and numbers of mathematics graduates needed by U.S. science, industry, and society. It should also ensure for countless students the pleasure of accomplishment and insight which mastery of calculus is meant to bring to the liberally-educated individual. Even if Edna St. Vincent Millay was right that "Euclid alone has looked on Beauty bare," every student of mathematics should be able to catch more than a glimpse of a similar vision.

DANIEL E. FERRITOR is Chancellor of the University of Arkansas, Fayetteville, where he has also served as Provost, Vice Chancellor for Academic Affairs, and Chairman of the Department of Sociology. The author or co-author of over 40 publications in the field of sociology, Dr. Ferritor has worked for several years in national educational programs funded by grants from the federal government. He received a Ph.D. degree in sociology from Washington University in St. Louis.

National Needs

Homer A. Neal

UNIVERSITY OF MICHIGAN

I am honored to have been asked to appear on this panel today to discuss the prospects for major revisions in the undergraduate calculus curriculum.

In 1985, I chaired a National Science Board Task Committee to study the state of undergraduate science, engineering, and mathematics education. Our committee completed its work in the spring of 1986, after drawing upon published reports and information from interviews with faculty members, university presidents, vice presidents and other university officers, representatives of foundations and industry, as well as representatives of various professional societies. Included in this group was the President of the Mathematical Association of America, a former President of the American Mathematical Society, and an executive officer of the Sloan Foundation—individuals who have played a particularly significant role in advancing the cause of the conference here today.

We found in the work of our committee that there were numerous reasons to be concerned. There was widespread evidence of serious problems in the curriculum and laboratory instrumentation used in the instruction of both majors and general students. Moreover, related motivational problems existed for students and faculty. Students found many of their key courses to be dull and uninspired. Faculty were often frustrated by the lack of student interest, and the faculty themselves often found it difficult to keep abreast of the rapid developments in their fields in the absence of special provisions for them to have the time and resources to d so.

Our recommendations for action were extensive, and many have already been implemented, either directly by the National Science Foundation, or indirectly through new initiatives at the university, regional, or state level. Examples include the immediate launching of the Research Experience for Undergraduates program at the NSF, legislative hearings on the health of undergraduate science, engineering, and mathematics by states such as New York, and the President's request for increased support for the College Instrumentation Program and other related initiatives.

No other discipline is so fundamental to ensuring a talented pool of future scientists and engineers, and a technologically literate generation.

Regarding the focus of today's symposium, our task committee was frequently reminded of the critical role played by mathematics in the training of students in all disciplines. In particular, mathematics is often the determining filter for all science and engineering disciplines, not to mention for the advanced mathematics programs themselves. No other discipline is so fundamental to ensuring a talented pool of future scientists and engineers, and a technologically literate generation.

Within mathematics, there is strong evidence that one of the most urgent challenges is to reform the calculus curriculum. Experts here today have made the case for why this is so, and the impact that such reforms could have on a wide span of disciplines. I only wish to add my encouragement for this initiative and to congratulate those who have had the insight and the persistence to proceed with the formulation of strategies for addressing head-on such a massive problem.

It would be very easy to view present calculus instruction as being an invariant of nature. The way it was taught to us could be thought to be the way it must be taught forever.

There are, to be sure, numerous obstacles that lie ahead. Faculties in mathematics departments must embrace the concept that significant changes are called for in the calculus curriculum for the reforms to be successful. Colleagues in cognate departments must cooperate in providing feedback on the success of the reforms, as viewed from the perspectives of their disciplines. Deans and provosts must get on board, to provide both moral encouragement and some significant fraction of the actual financial support required to implement and monitor the revised programs. Funding agencies must commit to providing the required external support over a sufficiently long period to insure that the revisions can be implemented, studied and refined. I am particularly pleased to see the Sloan Foundation and NSF take a lead in achieving this goal.

It would be very easy to view present calculus instruction as being an invariant of nature. The way it was taught to us could be thought to be the way it must be taught forever, regardless of the fact that the technological context has changed by leaps and bounds. It takes unusual insights and courage to challenge such a tradition. What you are doing is extraordinarily important and I wish you every success.

HOMER A. NEAL is Chair of the Department of Physics at the University of Michigan. Dr. Neal was Provost at the State University of New York at Stony Brook before assuming his current position as Chairman of the Department of Physics at the University of Michigan. The recipient of Guggenheim and Sloan fellowships and a former member of the National Science Board, Dr. Neal headed the National Science Board Task Committee on Undergraduate Science, Mathematics, and Engineering Education. His research is in the area of experimental high energy physics. He received a Ph.D. degree in physics from the University of Michigan.

Now Is Your Chance

Michael C. Reed

DUKE UNIVERSITY

I would like to talk about three things. First, let me say what I think about current calculus courses and texts.

I think they are awful—but they're awful for a lot of understandable reasons. They are awful because they are too technical; they try to teach too much material; they teach very little conceptual understanding; and they have a tremendous lack of word problems. As we all know, it is the word problems that students hate the most, and yet it is the ability to do word problems that makes mathematics applicable for a physics major, a chemistry major, or an engineering major.

If you look at the section on differentiation of polynomials in the text you are using, you will undoubtedly discover at the end of that section an extremely long list of problems. I guess that none of the problems is a word problem and that not a single one of the examples at the beginning of the section—examples which are supposed to motivate why we want to know how to differentiate polynomials—is a word problem with any kind of interesting application attached to it.

It is the ability to do word problems that makes mathematics applicable.

To try to press this point home, I have to tell you a story. I was standing around the common room last year and there I saw a calculus exam which, as it turned out, had been prepared by one of our graduate students who, unfortunately, was standing near me. The first problem on the exam was to differentiate $x^{\sin x}$ and I said, "*That* is what is wrong with calculus." I was then very embarrassed because the poor student was there and a large discussion ensued. My colleagues challenged me to explain what was wrong with making students differentiate $x^{\sin x}$.

Why should I spend all my time worrying about how to differentiate stupid looking functions like that? No function like that has every occurred in the history of physics.

Here is my answer. We live in the same building as the Physics Department. I guess that not a single member of the Physics Department could differentiate that particular function and say what the answer is. Furthermore they wouldn't be embarrassed by it. They would say, "Are you crazy? Why should I spend all my time worrying about how to differentiate stupid looking functions like that? No function like that has every occurred in the history of physics, so why should I be concerned that I can't differentiate it?"

Well, it sounds like a joke, but it's not a joke. It means that the teaching of calculus has developed into a series of technical hurdles for students to go past, one after the other, bearing very little relation to what they're suppose to get out of the course.

Now we come to the reasons why this has happened. First of all, students arrive at the university with very little motivation to think about mathematics. That's because their training in K-12 is mostly plug-and-chug and we give them what they expect because we get trouble when we don't give them what they expect. It's much easier to teach plug-and-chug than it is to teach a conceptual course.

It's much easier to teach plug-and-chug than it is to teach a conceptual course.

Secondly, a great deal of calculus teaching in this country is done by non-tenure-track faculty, by graduate students, and by part-time instructors who have been hired to fill large gaps on the teaching staff. At a large university, it is the best you can hope for that all instructors teach more or less the same thing so that when students go on to the next course they will have the same background. In a situation where you have a very large number of instructors who perhaps are neither very well trained nor motivated to teach well by continuing attachment to the institution for which they teach, the best that you can ask for is an adequate, standard job. Finally, the lack of original textbooks that try to strike out in new directions is really a great hindrance.

Two years ago my colleagues told me that I should either shut up about this or go teach calculus myself. So, I taught it myself—since I didn't want to shut up about it, and still have not. I taught out of one of the standard texts; I tried as much as possible to put word problems and applications in the course. I found it *very* difficult.

Without excellent standard textual material, innovations will surely die out. That means that at the end of the projects that many of you are considering, books or other materials that every student can buy for twenty, thirty, or forty dollars at the bookstore has to come out. If you are thinking about a project, you have to figure out how at the end of the experiment you are going to produce something that can be used at other institutions.

Without excellent standard textual material, innovations will surely die out.

These are some of the reasons why I think the projects that many of you are considering will encounter real obstacles to success. Even if the curriculum changes, much of the teaching is going to be done by non-tenure track people. Many of the students are going to arrive not wanting to take your new interesting course. Finally, there's the question of when these wonderful text books are really going to arrive so that you can use them. When are the publishers going to agree to cooperate with you—as individuals or as groups—to produce such textbooks.

There's a third aspect of this issue which I would like to address, not to my fellow administrators here on the panel, but to my colleagues, the mathematicians. That's what I like to call the G.H. Hardy syndrome.

I trace a lot of the evils in calculus instruction to G.H. Hardy. There is a common attitude very well expressed by his posture in the picture on the front of that book (*A Mathematician's Apology*), that mathematics has little to do with the rest of the world, and, in fact, should properly be contemptuous of the world.

This attitude has served us rather badly, I think, in the last thirty or forty years. It has bred an attitude among mathematicians not to talk to their colleagues; it has made narrow mathematicians who have very few interests outside of their own discipline; and this has produced mathematicians who are not very capable of enlarging their courses with appropriate and interesting applications.

So that is a problem within the mathematics community, not a problem for administrators. However, there is another issue which is related to administrators, and that's this: Mathematicians as a group are terrible entrepreneurs. We are really bad. Mathematicians are embarrassed to stand up for their subject, to say that it's important, and to fight for it in their home institutions.

Mathematicians are embarrassed to stand up for their subject.

Many mathematicians do not realize that their colleagues in physics, chemistry, and biology could not exist unless they were excellent entrepreneurs, because their laboratories depend on their skill in administration, as well as on their scientific skills. Those guys from those other departments are harassing their deans and their provosts all the time for money on behalf of their research, on behalf of their teaching programs, and for everything else.

Mathematics departments are often afraid to do that. They think they will not be well thought of; they're too timid to do it; they feel it's like P.R.; it's self-serving. In many cases, they secretly believe that mathematics isn't so important anyway.

Here is my message: This is your opportunity. Here you are at a national colloquium that says there's supposed to be a new agenda for calculus. Go to your chancellor, go to your president, go to your dean and say, "All these years you have been complaining like hell about the calculus instruction in our institution. How many calls have I received from you saying that you got a call from so-and-so's mother who said, 'Why does my boy have to be taught by a graduate student?' "

"Now's your chance to invest some money to make it better. I want so-and-so much released time for these two faculty members in the mathematics department so they can work on restructuring calculus for three years. I want so-and-so many funds to support a secretary who will be typing the new manuscripts that the students are going to read; and so forth." Administrators hear those kinds of requests from all other departments all the time. They hardly ever hear them from mathematics departments because mathematicians are too timid to ask. Now is your chance.

•

MICHAEL C. REED, co-author of *Methods of Modern Mathematical Physics,* is Chairman of the Mathematics Department at Duke University. His research is in the area of nonlinear harmonic analysis and in the application of mathematics to biology. He received a Ph.D. degree in mathematics from Stanford University.

Involvement in Calculus Learning

Linda Bradley Salamon

WASHINGTON UNIVERSITY

It is, I hope, appropriate for an English professor who's addressing a group of mathematicians to begin with an allusion to that eccentric figure who bridged both fields, Charles Dodgson (a.k.a. Lewis Carroll), to the effect that deans of arts and sciences like to believe three impossible things before breakfast .... We want to believe

- that our colleges can present a steadily improving, timely curriculum which will provide both students and faculty with continuous challenge, and
- that political "peace in the valley" can prevail among the various disciplines with which we work, with their very various intellectual styles and varying current successes, and
- that *both* those goals can be accomplished within our budgetary means, or with clearly foreseeable new resources.

The probability that the teaching of calculus can find a point of intersection in that three-way matrix is small enough to make Alice blink in Wonderland.

The good news from the dean's office is that one repeated assertion in the preparatory materials for this conference is false: on my campus, calculus is *not* the course about which most student complaints are registered; that dubious distinction belongs to chemistry. Nor do my colleagues in the sciences use performance in calculus, even implicitly, as a "weed-out" device, nor does our Medical School use it as a shorthand admissions indicator. Again, chemistry—particularly organic chemistry—admirably and accurately fulfills those roles.

Paranoia ill becomes you, though you have real enemies. The principal source of complaints about calculus, one where my Lone Ranger's peace-keeping skills most frequently intervene, is the "Engine School." I must admit, as an outside observer of the mathematical community but as an experienced student of pedagogy, that engineering educators have a point.

The objective of instruction in calculus ... is to bring each student to the most thorough and functional understanding of this sophisticated subject that she or he can achieve ... like it or not.

Because I cannot begin to consider whether calculus needs to cover partial fractions in order to prepare for a later encounter with Laplace transformations and other such *arcana*, I want to make four points from the perspective of a teaching colleague and educational administrator. They relate to classroom management and class size, to the use of the computer, to the role of mathematics in general education, and, of course, to money.

All my remarks assume that the objective of instruction in calculus—I'll qualify that by saying "at a research university," but I *really* mean anywhere—is to bring each student to the most thorough and functional understanding of this sophisticated subject that she or he can achieve ... like it or not. (Remember, please, that I hail from the only other discipline in our colleges where 80% or more of the beginning students are unwilling draftees; I know the consequences.)

The first implication to be drawn from a goal of bringing each student to his or her best achievement is obvious. We must take them as we find them both in ability and in preparation, from effectively near-zero to Advanced Placement.

Because mathematics is so linear and progressive a discipline, students' differing readiness at entrance dictates that at all but the most select institutions, there be several different calculus courses. I think they should vary *not* by the student's immediate use for calculus (as a biology or business major, say, rather than a protophysicist or engineer), but by the pace at which they move, the degree to which they pause over relevant precalculus topics before introducing new material, and conversely, the depth and sophistication of concepts they have time to include. The same textbook, after all—if one's not enslaved to its teaching manual—can be creatively utilized in quite different ways.

On a large campus, three or four different calculus courses—including an honors effort—might be underway. One of those can certainly be lean and lively; I doubt that all can. Selection should be thorough and informed, and prerequisites should be vigorously enforced, to the point of requiring preparatory "college algebra" or other euphemistic courses, if necessary.

What simply will not do is the model of a common syllabus for 1000 students, so that interchangeable Professor X can, on 15 minutes' notice, give a lecture on a particular topic to an anonymous mass, then whip back to his office and his Fourier transforms untouched by human minds. *Count on* your dean's complaining about that.

The second implication of seeking each student's achievement is what the NIE has taught us all to call "involvement in learning." In mathematics, this term that describes attempts at personal, internalized mastery surely means homework—homework that's required, evaluated, and included in the final grade.

In mathematics ... mastery surely means homework—homework that's required, evaluated, and included in the final grade.

Here's where my compulsive friends the engineers have a point, and you know it's true. Even in a minimal calculus course that's only teaching calculations—given the limited concentration and persistence of today's students—checked homework is needed; if you choose to concentrate on concepts and what the kids call "word problems," it's essential.

Now, does this imperative dictate the end of large lectures in favor of 30-student classes? Maybe, but I doubt it. "Help" sessions and drop-in math labs staffed by grad students go a long way toward giving students control over their own learning, as do 20 minutes for questions before or after each lecture.

All across our undergraduate curricula, moreover, the one thing that computers can reliably do better than we can is rote drilling in mechanics and routines. If the machine can do it for German adjective endings, with repetitions and branching that match the student's pace, surely they can do it for power series or what-have-you. (And that's leaving aside the intrinsic reasons for introducing students to computers and their algorithms.) As a dean, I'd rather pay for new terminals or PCs or networked micros than for new assistant professors—even supposing the department chair can find us talented candidates who can articulate clearly in English and relate comfortably to puzzled and insecure students.

As a dean, I'd rather pay for new terminals or PCs or networked micros than for new assistant professors.

About the issue of surrendering the teaching of the logic behind integrals and derivatives to some HP calculator that can score a B on tests without studying, I have no right to comment, but I hope that, as a profession, you won't succumb completely. If mathematics has a role to play in the general education of our students—in shaping their minds as informed and disciplined instruments—it is in this domain of conceptualization.

To deal with numbers in the real world, they need to know not the right formula to plug in to do the job of the moment but skills like estimation and approximation, concepts like scatter and risk, the meaning of graphic representation, the sheer likelihood of trial solutions—intuitive probability, if you will. Those ideas don't require a calculus course, to be sure, and I'd wish all mathematics departments would offer a clever course in finite mathematics for those students who won't take calculus.

Teach in the best way you can find to present the real power (and maybe even the beauty) of your discipline to an anxious or indifferent kid.

But for those whose *only* mathematics course will be calculus—a very great many—using computerized calculations that avoid fundamental ideas about numbers seriously deprives them, know it or not. What could replace the pencil sketching, the concrete visualization that I see our best math students doing? Honor the integrity of your discipline, please; teach what it's all about.

By the same token, with all due respect to the "user groups" represented on the previous panel, I hope you won't take the service function of calculus as paramount. Teach the topics they need covered, sure; but teach in the best way *you* can find to present the real power (and maybe even the beauty) of your discipline to an anxious or indifferent kid. We all know that the best means for meeting the objective with which I began—to help the student achieve the best understanding she can—is a willing, imaginative teacher who likes the material and will work hard at expounding it.

The tacit purpose for my invitation to be here today is not these opinions of mine, of course, but the question, "What will deans pay for?" or, more precisely, "Will deans pay for more mathematics faculty?" I think I know our tribe and its bronzed responses well enough to answer.

First, we do have those engineering and business deans riding like sheepherders into our peaceful valley, and for financial reasons we have to satisfy them. We'll defend our mathematicians to them if you give us the ammo *and* if what you do is defensible. If what they demand is more, differentiated sections and more evaluated homework (and it sometimes is), *you* should be making peace (and common cause) with them, not complaining about their students, or their demands.

Calculus is the second largest course after English composition on many campuses, as on mine, and we deans simply must attend to that brute fact.

Next, when mathematics department chairs ask us for more slots, of course they have to get in queue with the other two dozen or so department chairs, and available slots will be allocated on the basis of institutional priorities. The best rationale for attaining a high priority is unlikely to be the putative need to teach smaller classes *per se*, but if you devise genuine and compelling pedagogical reform that demonstrably requires additional staff used in imaginative and effective ways, we will certainly listen. Double your teaching staff? No. But relief for a worthy experiment, to continue if it succeeds? Highly probable. Calculus *is* the second largest course after English composition on many campuses, as on mine, and we deans simply must attend to that brute fact.

Demanding folks that we are, though, we'll also expect your candidates to be talented differential geometers or harmonic analysts, and that requirement raises different questions. I'm not a graduate dean, but my colleague who is both that and a mathematician reports mixed news about the pipeline. Are university mathematics departments producing enough new Ph.D.s well prepared to teach undergraduates to satisfy the ambitious goals of your proposed curriculum project, on a national scale? Guessing the answer to that, as an undergraduate dean I must look to our mathematics major program and its ability to retain students beyond calculus, indeed beyond graduation. In this partially closed system, perhaps we do indeed need to invest the resources that will bring us mathematicians for a new century.

LINDA BRADLEY SALAMON, Dean of the College of Arts and Sciences at Washington University in St. Louis, is co-author of the Association of American Colleges' *Integrity in the College Curriculum.* She is Past Chair of the AAC and a trustee of the College Board. A scholar of Elizabethan culture currently teaching Shakespeare, she received a Ph.D. degree from Bryn Mawr College.

Calculus in the Core of Liberal Education

S. Frederick Starr

OBERLIN COLLEGE

Despite being a college president, I still am a historian. I was asked the other day by a student perplexed over the mathematics competency requirement at Oberlin, "Why do it?" The specific question had to do with calculus. I suggested that if Newton could invent it in his sophomore year, he can study it in his.

Speaking as a historian, I would like to note a very curious exchange that took place between Newton and his friend John Locke which has very much to do, I think, with the subject at hand. When Newton had finished the *Principia,* realizing full well the importance of what he had done, he sent it over to Locke to get Locke's estimation of it. Locke looked it over, and couldn't decipher the math. So he sent it to a friend in Holland and asked him to check it out. The friend read it and said that the math was okay. Locke then read everything *except* the mathematics in the *Principia* and wrote very intelligently on it.

I was asked ... by a student [about] calculus. I suggested that if Newton could invent it in his sophomore year, he can study it in his.

Now, the issue here is whether Locke, who obviously had absolutely no contact with the *process* of mathematics that underlay the *Principia,* was able, as an educated person, to deal with that work or not. The assumption on which this conference rests, I would gather, is that Newton should have encouraged his friend Locke to study the mathematics necessary to deal with it because otherwise he would simply be dealing with the products of other people's thoughts and never be able to engage in the process. That really raises the first of five points I would like to lay before you here.

I don't get the sense, reading through the various papers that have been prepared, and hearing the discussion, that there is much agreement as to the basic purpose of the enterprise—the reform of calculus. Is calculus a service course? Is it a course that is providing techniques, methods, manipulation, and so forth? Or is it truly part of some core learning that an educated person should have dealt with? Is it really dealing with concepts, or can someone—as was asked earlier this afternoon—deal with the manipulations without genuine understanding?

As a pedagogue, that is an absolutely preposterous proposition to me. But obviously, if calculus isn't in the core, if it is instead a service for others, then one can get by with all kinds of mischief. It seems to me that great clarity on that point is required before anything else can proceed coherently.

My colleagues in mathematics at Oberlin have taken the rather uncompromising view that calculus has to do with thinking, with concepts, with the core of a liberal education. From that they proceed to deal with other questions. Discrete mathematics, for example, is being offered as a separate parallel course in the sophomore year. There is concern for the verbal dimension of thought, great concern in fact, among them. There is also a good deal of experimentation going on with computer algebra systems, both within the classroom and without.

But all this flows from some clarity on the basic purpose of the enterprise. You might be concerned whether this might lead to someone who has thought some fine thoughts but can't *do* anything. The answer, I think, is that the purpose of education isn't to teach you how to *do* something, but how to *be* anything.

The purpose of education isn't to teach you how to do something, but how to be anything.

Since the mathematicians at Oberlin are apparently enabling their students to go on into careers in science at a remarkable rate—to get their Nobel Prizes and to get their whole science program selected as first in the country among colleges—they're obviously doing something right. I would suggest that it has a lot to do with their very clear beginning point, namely, that calculus is part of the core of learning. It is not simply a "how-to" course.

The second proposition that I would like to lay before you has to do with the process of study. There was a very interesting body of research carried out by several investigators at Oberlin College in the last four years on the undergraduate preparation of scientists in the United States. They focused on a group of fifty liberal arts colleges that were the most productive of scientists and a similar number of universities.

David Davis-Van Atta and Sam Carrier, who did this research, documented, among other things, that there has been a precipitous decline in the the percentage of people entering careers in science and also an increase in the attrition rate of intended scientists as they enter college and proceed towards graduation.

Calculus is part of the core of learning. It is not simply a "how-to" course.

They showed, moreover, that the only institutions that are successfully bucking this national trend in the sciences are those that base their pedagogy on a kind of apprenticeship system—those in which the student is brought into the laboratory, in which there is a direct, hand-to-hand contact, not with graduate students, but with real professors.

If that is true, and there is so clear a correlation that I'm left with no doubt about it, then I think you've got to ask whether your discussions are taking this into account. Is mathematics open to that kind of direct apprenticeship-based engagement? Does the advent of the modern computer in fact provide a wonderful opportunity to go in that direction and to think of much more active forms of pedagogy than have been used in the past?

My third point is very different in character. I'm concerned that you might be replicating in a perverse way some of the negative features of a national discussion in which I've participated over some years regarding the teaching of foreign languages in the United States. In fact, one of the background papers drew the parallel and I was pleased to see at least that the relevance of this parallel was acknowledged.

The attempt to improve foreign language teaching culminated in a Presidential Commission a few years back. Frankly, the mountain didn't quite give birth to a mouse but something on the scale of a rat. It never really brought about the great transformation.

Because it is in the schools where your problem is being formulated ... if you don't bridge the geological fault separating you from the schools—there won't be any progress.

I think the great flaw in the Commission's approach, and in the thinking of the late seventies and early eighties, is the assumption that you can build a house from the roof down. I wonder if this flaw also might not be present here. Can you really hold conferences and talk seriously about calculus at the university level and not spend an equal amount of time on the secondary school mathematics curriculum? It seems inconceivable to me.

I would be most interested to know what has gone on since the MAA and the NCTM ten years ago cautioned high schools not to teach calculus unless they do it at a university level with university standards. It seems to me that this question has to be opened wide, not just for precalculus courses, but for the entire secondary school preparation, or you will not progress an inch. Because it is in the schools where your problem is being formulated, if it's not addressed at that level—if you don't bridge the geological fault separating you from the schools—there won't be any progress.

This suggests, by the way, that once you do cross that fault and deal with the whole process, then you can also consider calculus in several dimensions, progressively, in the course of teaching over time.

My fourth point is this—that calculus was not invented here. There are people elsewhere in the world who know how to teach it reasonably well. In fact, many demonstrably are doing a better job at it than we are. I think we're simply handicapping ourselves if we don't examine very carefully the pedagogical systems, pre-calculus and calculus, that obtain in those countries that are doing a good job.

You may say that such societies are different. They send a lower percentage on to college. Fine. Then demonstrate that that decisively negates the experience of country X. It seems to me the burden of proof lies on someone who would argue that international experience in this area is irrelevant.

Do a survey. You will find a phenomenal ignorance of just what calculus is all about.

And now, finally, a rather more practical matter, that takes me back also to the Locke-Newton link. I don't believe that calculus—or for that matter, mathematics in general—has a very strong constituency among faculties of our universities and colleges. All of us, all those outside the mathematically-based fields, are unfortunately too much in the position of John Locke. And we're not embarrassed about it, as Locke at least was.

The problem is very serious because if you are saying that mathematics, or specifically calculus, belongs in some core curriculum of what an educated person should know, you are saying that to a group of colleagues who themselves haven't taken calculus and who don't know what it is. Ask them. Do a survey. You will find a phenomenal ignorance of just what calculus is all about.

It seems to me that there is something very serious about this matter. What I would suggest is rather naively grand, but do-able. If you do proceed from the first principle that mathematics in general and calculus in particular should be part of the equipment of an educated person, then take time to offer an accessible calculus course for your colleagues in other departments.

Crack that problem. It's do-able. It's not the political dimension I'm concerned with here, but the intellectual dimension. Until this gap is bridged in at least one institution to prove that it is possible, we are talking about such remote worlds that, although people might as a matter of political bargaining give you your required hours in the classroom, or your piece of the budget, they won't really understand why they are doing so.

There are outposts in economics and various areas of the social sciences where your task will be easy. But most of the social scientists and nearly all those in the humanities are illiterates in mathematics and in calculus. Hence there is a need for teaching that is directed toward the professorial community itself simply as a means of making up, even at this late date, for the fatal neglect of mathematically-based learning in our primary schools, in our secondary schools, and in most of our colleges. Until the professors are mathematically literate, don't expect them to understand why students should be.

S. FREDERICK STARR, President of Oberlin College, is a specialist on Soviet Affairs, founding secretary of the Kennan Institute for Advanced Russian Studies at the Smithsonian Institution, the author of numerous books in his field, and a member of the Trilateral Commission. As a result of his initiative on the so-called Oberlin Reports on Undergraduate Science, the National Science Foundation and many private foundations have strengthened their involvement with mathematics and science at the undergraduate level. Dr. Starr received a Ph.D. degree in history from Princeton University.

RESPONSES

PARTICIPANT RESPONSES

Question What are the principal weaknesses in calculus instruction?

Answers

- No sense of adventure and discovery
- The students — poorly prepared, even more poorly motivated, and programmed not to think
- The instructors — often indifferent and inexperienced, with little commitment to teaching or to mathematics
- The confusion between education and training
- Stale mechanical manipulative mimicry
- How can algebra students solve a word problem if they can't do arithmetic?
- Its almost nobody's first priority
- Calculus serves too many masters. It is not a service course.
- It is dull and boring.

Mathematics as a Client Discipline

George E. Alberts

NATIONAL SECURITY AGENCY

As a representative of (arguably) the largest employer of mathematicians in this country, I have found the Calculus for a New Century colloquium remarkable. The huge turnout is a clear sign of growing concern and interest. The candor and quality of the discussion have been impressive. Although the National Security Agency also employs large numbers of engineers and computer scientists, and despite the temptation to comment on the full range of fascinating issues and ideas, these remarks will focus on the more narrow issue of the future health of the mathematics profession itself. In that regard, it seems worthwhile to view mathematics itself as its own "client discipline" in discussing reform and revitalization of the calculus.

Tom Tucker helped set the tone for candor by suggesting it might indeed be appropriate for the secondary schools to assume the responsibility for teaching calculus. In his "desert metaphor" the secondary level might be the appropriate "sinkhole" for the calculus. "Mainstream" calculus, which controls the pipeline from which the mathematics teaching faculty is replenished, could be taught at the secondary level. We may indeed "only need a few people who know the inner workings of calculus." I think not.

All of what I have heard described by clients as desirable seems essential for professional mathematicians themselves. As Gail Young wrote in his background paper, "the future of our subject depends on improving [calculus]." Calculus has frequently been described as the first real mathematics course (with the possible exception of a good Euclidean geometry course in high school). Although Anthony Ralston eloquently argues that calculus is dead, and discrete mathematics must take its rightful place as the core mathematical subject, I endorse Cathleen Morawetz' remark that trends in parallel computing suggest a convergence of the two subjects.

Complaints about lack of understanding, the need for more conceptual, intuitive, and at the same time more rigorous calculus echo a growing concern of our Agency's mathematics community—at a time when we need to hire more outstanding mathematicians, to do creative mathematics, the supply seems to be declining.

Mathematics' "clients" are said to be leading innovators, and might well be approaching the capability of teaching their own calculus. Professional mathematics must begin to learn from their clients. Ron Douglas suggested that attempts to reform calculus have been well-intentioned but short-lived "castles in the sand." Oberlin President Frederick Starr remarked on the short attention span of reformers. The present calculus curriculum, which Tom Tucker aptly described as unchanged in 100 years, is more a castle made of stone.

If the mathematics profession is to avoid ossification, its practitioners must take charge of the reform movement, and, while meeting the legitimate concerns of the other "clients," focus on reform of the calculus to rejuvenate American mathematics as a discipline first and a service second. The alternative may well be the calculus "super-nova" suggested as a possibility by Lynn Steen.

We do indeed need to resolve the basic purpose of all this activity, to reassert the significance of mathematics as core training for all educated people, to address the demanding full range of problems—not, primarily, because of our other constituencies, but because of ourselves. A lean and lively calculus, while serving the needs of those other constituents (which it can if they share in its revitalization) will at the same time do something more fundamentally important: attract and inspire successively better generations of American mathematicians.

GEORGE E. ALBERTS has served for twenty-two years as a professional mathematician at the National Security Agency, Fort George G. Meade, Maryland. He is presently an Agency executive and Chairman of its Mathematical Sciences Panel.

Calculus and the Computer in the 1990's

William E. Boyce

RENSSELAER POLYTECHNIC INSTITUTE

Whether or not a "crisis" exists, there is widespread agreement that improvements are possible in the calculus courses offered in many colleges and universities. Powerful inexpensive calculators and microcomputers, now widely available, provide numerous opportunities for courses adapted to the contemporary environment.

First, calculus should be taught so that numerical computation is seen as a natural consequence of using calculus. Most students who take calculus courses do so to improve their capacity to solve problems in other fields. Generally speaking, a numerical answer is sought, so eventually something must be computed in order to obtain it.

On the other hand, computation can also lead naturally to analysis. For example, computation carries with it the obligation to consider the accuracy of the results. Therefore it provides an opening to discuss such questions as estimation of errors or remainders, or speed of convergence of an infinite series or iterative process. Ideally, analysis and computation should appear as complementary and mutually reinforcing modes of problem solving, each used when appropriate, and each enhancing the power of the other.

Moreover, at every opportunity students should be encouraged to try to assess the reasonableness of their answers. A result that appears reasonable may not be correct, but one that appears unreasonable is almost certainly wrong (unless one's idea of what is reasonable is seriously deficient).

In addition to computing numerical approximations, other aspects of computing may also be important in calculus. One is the rapid and accurate generation of the graph of a function or an equation. To do this effectively it is essential to scale the problem appropriately. Beginners rarely, if ever, give adequate thought to this question, although it is crucial in obtaining a useful graph. Consequently, they often have difficulty in interpreting the graphs their computer screens depict.

In choosing a proper scale, it may help to do some of the things usually taught in the context of curve sketching, such as finding maxima and minima, checking for symmetry, and locating asymptotes. Thus, computing may make more clear the need to learn something about calculus and the behavior of functions, rather than merely resorting to trial-and-error computation.

The use of symbolic computational packages to reduce the need to perform tedious and repetitive algebraic procedures is highly attractive. However, the use of such packages also incurs a cost. One does not learn to use a sophisticated symbolic computation package (MACSYMA or Maple, for example) instantaneously. At least some of the time saved by using the package must be invested up front in learning how to use it.

There is also the "black box" question: should we permit students to use a symbolic computation package without some understanding of what the package is actually doing? My view is that in order to think constructively about the behavior of models of physical phenomena, one must have *some* specific information about *some* particular functions. It is probably not necessary to know how to evaluate $\int \sec^5 x\, dx$, but one should certainly know $\int \cos x\, dx$.

The boundary between what is essential knowledge and what is not may be unclear, but surely we should insist that students must learn to execute some procedures themselves, even while relying on a computer to handle the more complicated cases.

The use of a computer may not save much time in a calculus course, although it will give the course a somewhat different orientation. Assuming that it does save at least a few class days, what should we do with them? I suggest that a good use would be to attempt to foster better problem solving capability among our students.

The problems might be mathematical ones. Virtually every meeting of a calculus class offers the opportunity to expand on the day's assignment, to go a little off the prescribed path, and to explore interesting related material. If there is a little extra time in a course, this would be a good way to spend it. It might even help to attract to our discipline some of the many very good students who now see no attractive future in the serious study of mathematics.

Another possibility is to do more in the area of applications and mathematical modeling. I am not particularly enthusiastic about "realistic" applications. They must often be couched in terminology unfamiliar to many students and require too much time to describe the underlying problem.

It is better to use simple problems and models (even if "unrealistic") so that everyone can understand them.

Even simple problems can be embellished and modified so as to illustrate the ideas and principles of mathematical modeling. Given students' paucity of experience in mathematical modeling, even the simplest problems (other than the standard template problems) will prove challenging enough for almost all students.

Finally, what is needed as much as anything in teaching calculus is the proper attitude: a recognition that the computer is here to stay; that it offers insight into the phenomena of change; and that it can provide a springboard for the discussion of interesting mathematical questions, including many of those that are now part of standard calculus courses.

•

WILLIAM E. BOYCE is Professor of Mathematical Sciences at Rensselaer Polytechnic Institute. He is formerly Managing Editor of *SIAM Review* and Vice-President for Publications of SIAM. He is author (with Richard C. DiPrima) of *Elementary Differential Equations* and *Boundary Value Problems*.

The Role of the Calculator Industry

Michael Chrobak

TEXAS INSTRUMENTS

To make calculus instruction more applicable to real-world problems, educators must focus more on fundamental ideas, off-loading the mechanics of executing formulas and equations to modern computing aids. Embracing such tools—specifically, hand-held calculators—will bring about instructional changes needed to better prepare students for tasks in advanced education and the workplace.

Several issues must be addressed by educators and calculator manufacturers if this tool is to be incorporated effectively into calculus and lower-level mathematics programs.

First, calculator manufacturers must tailor their products, based on guidance from educators, to meet maintained education requirements. Close communication with the educational community must be, with new designs being based on the special requirements of the classroom. Sometimes this can be as simple as uncluttering a keyboard, thereby preserving and accentuating required functions.

The design of the TI-30 SLR+ calculator is one example where this collaboration between educators and industry has been effective. Specifically tailored to requirements at the high school level, this calculator was constructed with a hard shell case for increased durability. Solar power was chosen to eliminate the need for batteries. Large, brightly-colored keys were incorporated to identify mathematics function groups, to provide for ease-of-use, and to stimulate learning.

Second, proper use of calculators in the classroom requires textbooks developed with the calculator in mind. Indeed, some states already are making this a requirement. For instance, in California, publishers must now detail the use of calculators in their text, not as a supplement, but as part of each lesson. To achieve this, manufacturers, educators, and publishers must work together, incorporating effective use of calculators into instructional materials.

An example of planned technology emerging in the textbook industry can be found in a classroom calculator kit developed by TI and Addison-Wesley. The two companies have worked together to produce instructional material that incorporates the use of a hand-held arithmetic calculator for an elementary mathematics curriculum.

Third, the use of calculators in testing is another key issue. If students are taught with calculators, it follows that they should be tested with them as well. Today, testing materials incorporating the use of calculators do not exist. Manufacturers can assist in the creation of new testing materials, providing development support as well as inputs based on the design and operation of their products.

Currently, the Mathematical Association of America (MAA) has begun to reform mathematics education with a calculator-based placement test program. This effort is under the direction of the MAA's Committee on Placement Exams (COPE), managed by John Harvey of the University of Wisconsin at Madison, with TI providing funding since 1986.

Testing with calculators at the high school level will help to ensure that students have proficiency of calculator use by the time they reach college. Today, this is usually not the case.

While the calculator itself should never be the main focus in mathematics instruction, it should be viewed as an important tool to assist in the educational process. Calculator manufacturers should be included in future mathematics reform, ensuring that calculators for the classroom are designed appropriately, and that they integrate well into textbook formats and testing materials.

MICHAEL CHROBAK is Scientific Calculator Manager of Texas Instruments, Dallas, Texas. Previously, he served Texas Instruments as a Project Engineer and a Program Manager for financial and scientific calculators. He holds a B.S.E. from the New Jersey Institute of Technology, and an M.B.A. from Texas Technological University.

Calculus: Changes for the Better

Ronald M. Davis

NORTHERN VIRGINIA COMMUNITY COLLEGE

Calculus is the mathematical study of change. Yet, for the large part, calculus has changed little in the past thirty years. It has remained static in a continuously changing environment.

I strongly agree that our present calculus content and the way in which we teach it needs revision. Calculus needs to provide students with the abilities of reasoning, logic, and judgment. Calculus courses need to emphasize clear thinking and not merely symbol manipulation. There is an underlying fundamental importance of calculus for the sciences and technologies. We as teachers must comprehend this and must share this knowledge with our students.

As teachers of calculus we cannot ignore the vast array of available technologies that can enhance our efforts. We must incorporate calculators and computers as aids for our teaching and as tools for our students. These tools will not only simplify calculation and symbol manipulation, but will also require students and teachers to heighten their understanding of and insight into the concepts of calculus.

I am convinced that change will only be accomplished through a concerted and coordinated effort from industry, business, government, and education. For two-year college faculty, the effort will be exceptionally taxing. Teaching loads of fifteen hours or more often provide little time for curriculum development. Minimal computing support at two-year colleges restricts student and teacher access to computers.

Since the transfer of courses requires two-year college calculus courses to be accountable to four-year colleges, two-year college faculty will move slowly in making changes. Development of a dynamic calculus course will require a joint effort on the part of the mathematics faculties at two-year and four-year colleges.

A sense of isolation exists for the calculus teacher at many two-year colleges as their departments often have little or no travel funds. They have limited opportunities to meet with other calculus teachers from two-year and four-year colleges. Since calculus constitutes only 10% of the course load at two-year colleges, it cannot often be given the added attention that a revision would require.

I strongly believe that a mathematics course that involves a dynamic approach to calculus and portions of discrete mathematics needs to be developed and implemented in place of the present-day calculus. The teaching of such a course will require much preparation and retraining. I am, therefore, concerned that two-year colleges may be unable to commit the resources to prepare their faculty adequately to teach this dynamic course.

As a teacher I become invigorated with each opportunity to rethink course content and instructional approach. I am excited at the prospect of a new, dynamic calculus course.

RONALD M. DAVIS is Professor of Mathematics at Northern Community College, Alexandria Campus. He has served as Second Vice President of the Mathematical Association of America, and as Chair of the AMATYC-MAA Subcommittee on Curriculum at Two-Year Colleges. He received a Ph.D. in mathematics education from the University of Maryland.

Imperatives for High School Mathematics

Walter R. Dodge

NEW TRIER HIGH SCHOOL

The high schools' perspective of calculus is two-pronged. For some of our students calculus will be a course they take in high school. This course will consist of our ablest students taught in small classes by our most capable teachers. For others, calculus will be taken following high school. For these students the high school's objective is to prepare them in the best possible manner with the mathematical concepts and skills required for success in a college calculus program.

The universities' view of calculus is quite different from that of the secondary school. For them very often calculus is the lowest course offered, populated by students of varying abilities, and with teachers who are often not the ablest. In addition, quite often the class size is very large. Although this is a problem on a national scale, I do not believe it can be solved on a national basis. It *must* be solved within each local institution.

If we are truly to bring about calculus reform we must not only change the curriculum, but must also change drastically our delivery systems. If calculus is to become more conceptual and more intuitive, educational structures must allow this to occur: student-faculty interaction, group discussions, and laboratory experiences must be given top priority. A change in the content of the course without parallel change in the method and quality of instruction is doomed to failure.

One has to ask if there really is need for change in calculus? One very clear outcome of the Conference is that there *is* a compelling need to change the content as well as the conceptual nature of the course. The graphical, algebraic, and numerical capability of calculators and computers has definitely made many of the skills of a traditional calculus course very suspect, if not archaic.

Numerical methods of solution, implemented via the calculator or computer, open a whole new arena of calculus to the beginning calculus student. The graphical capabilities of both calculators and computers can enhance a student's understanding of the fundamental ideas of calculus. Therefore change should occur both for the improvement of the curriculum and the understanding of the student.

Exactly what parts of the traditional course can be eliminated is not clear. Just how much skill work is necessary and how much can be eliminated is a crucial question—often mentioned at the Conference—that must be answered.

The entire curriculum of the high school will be affected by these changes. If students are to be successful in college, the school curriculum will also have to become more conceptual and will need to incorporate technology. Skills will have to be developed in the use of both the calculator and computer. Approximation, estimation, and reasonableness of answers must be emphasized. Mathematics will become more "decimalized." Algorithmic reasoning will also become a high priority educational need.

The Conference has certainly stimulated thinking and planted the seeds for a change in the calculus. If this change is to come about, cooperative efforts between secondary school teachers and college faculty must be organized. Changes at one level have effects on the other, and the success of one will depend upon change in the other.

WALTER R. DODGE is a mathematics teacher at New Trier High School in Illinois. He has been a member of the AP Calculus Exam Committee, an AP Calculus Reader, Table Leader and Exam Leader. He is currently President-Elect of the Metropolitan Mathematics Club of Chicago.

An Effective Class Size For Calculus

John D. Fulton

CLEMSON UNIVERSITY

To create a Calculus for a New Century, more has to change in the calculus of today than course content. Calculus is not suited for mass education, yet calculus in many if not most colleges and universities is for the academic treasury the most productive "cash cow" in the university.

To be sure, English writing courses have more student enrollments than does calculus. But English writing classes are taught in sections of sizes twenty to twenty-five students. While writing courses generate more revenue for the university, they have been successfully portrayed to university administrators as being more labor intensive with respect to commitment of teaching staff. As a result, they cost more per section to teach.

In the September 1987 issue of *College English,* the National Council of Teachers of English published its recently adopted standards "No more than 20 students should be permitted in any writing class. Ideally, classes should be limited to 15."

The justification given by this Council for the class size is that

> ... the work load of English faculty [should] be reasonable enough to guarantee that every student receive the time and attention for genuine improvement. Faculty members must be given adequate time to fulfill their responsibility to their students, their departments, their institutions, their profession, the larger community, and to themselves. Without that time, they cannot teach effectively. Unless English teachers are given reasonable loads, students cannot make the progress the public demands.

Interchange "Mathematics" with "English" and a reasonable case for calculus is made as well.

Calculus for a New Century will be labor intensive with respect to teaching staff. In the early stages of development of a new calculus, considerable experimentation will be necessary. Guidelines for a new calculus likely will follow from such a body as CUPM. Text material incorporating the guidelines can be expected to follow. In the development period, experimental classes taught by faculty would seem of necessity to be small classes, as would the control classes with which they would be compared. Faculty teaching the classes would require released time from other duties to lead this development.

If calculus ultimately incorporates many of the ideas and recommendations discussed at the Colloquium, then it will be essential that the new calculus be considerably more labor intensive than the old. Also, it seems clear that a new calculus must rely more heavily upon experienced and committed faculty and less upon part-time faculty and graduate teaching assistants.

A new calculus must be taught with more attention to concepts than was the old. The assigning and grading of nonstandard problems as homework, more classroom discussion, board work, and class presentations by students, more word problems, and more questions requiring essay responses address the effective teaching of concepts, but do not lend themselves to a calculus class with thirty-five or more students per instructor.

If the new calculus uses technology effectively—as it must—expect it to be more labor intensive, not less. In addition to the concepts of change, limit, and summation of the old calculus, expect hand-held, micro- and mainframe computers to be used creatively in the new calculus to instill concepts such as approximation, estimation, error analysis, asymptotic behavior, and goodness of fit.

More frequent and longer discourse with students will be required to instill these numerical concepts. It has been alternately predicted over the years that radio, television, and now computers, videotapes, and electronic blackboards, would make classroom teaching obsolete. All of those predictions have been wrong. Likewise, we expect that the principal value of any cleverly devised expert system for the teaching of calculus will only be to enhance the effects of good classroom teaching.

The practitioners in our client disciplines—the engineers, the biological and physical scientists, the social scientists, the business and economic scientists—expect a lean and lively calculus to contain examples and problems for students which reflect applications in their disciplines. Faculty will have to lead in determining these applications and translating them into the teaching mode for the new calculus. This effort will not only be labor intensive, but quite likely it cannot be done by graduate teaching assistants. We expect applied problems to be nonroutine, troublesome to students of calculus, and requiring considerable discourse with knowledgeable teaching staff.

One goal of a Calculus for a New Century must be to assist more students, especially students from minority groups, through the filter which calculus represents and into the life stream of science, mathematics, and technology. The Professional Development Program for calculus students at the University of California, Berkeley (described in the background paper "Success for All" by Shirley Malcolm and Uri Treisman) seems to have had spectacular success for minority students with mathematics SAT scores in the 200 to 460 range. Its small group workshop methods seem transferable to all students. Its fifteen- to twenty-student recitation sections must, however, be regarded as labor-intensive. Moreover, it would seem that any successful program for significantly increasing the student success rate with calculus will be just as labor-intensive.

Calculus is with us now. A Calculus for a New Century will evolve from our present calculus, yet at too many colleges and universities an insufficient teaching staff is currently assigned to calculus. Its "cash cow" status with many university administrators is shameful.

Through student tuitions and in some cases through formula funding, calculus enrollments generate considerable funds for college and university coffers. With expenditures only for the teaching of large course sections, all too frequently taught by part-time faculty, with help from graduate teaching assistants, and with supplies limited to a little chalk, perhaps an overhead projector, and some transparencies, the net profit from calculus teaching can be considerable. Moreover, at many universities, graduate teaching assistants tend to pay their own salaries by paying their own tuitions and through formula funding generated by their own course registrations.

No calculus, old or new, can be effectively taught without sufficient teaching staff to allow regular and extensive feedback from students. Homework and quizzes must be graded line by line. Tests to effectively measure students' grasp of concepts in calculus should not be "multiple choice" tests. (Even Advanced Placement calculus exams have their essay portions.)

Like no other course at the freshmen-sophomore level, calculus concepts build with each successive class. A student who gets behind in the first few classes almost certainly will be lost for the entire course. Early and regular student feedback is thus essential for the course. Assessment of each student's grasp of concepts must begin early in the course and continue frequently thereafter. Conferences with students, often extensive, should be held regularly.

There is a need for teachers of calculus to instill in their students the ability to communicate scientifically, or mathematically. To at least some degree, then, calculus teachers are teaching writing. The circling of answers indicated as (a), (b), (c), or (d) on a multiple choice quiz will not stop a student from writing such mathematical garbage as $3 \times 4 = 12 - 1 = 11 + 4 = 15/3 = 5$, which just as frequently occurs on student papers as in calculator instruction books.

Only close scrutiny of student homework and quizzes, essay questions on quizzes, classroom discussion of concepts, and student classroom or office presentations can assist in meeting an effective mathematical communication objective for the teaching of calculus. The very nature of the calculus, new or old, suggests that it should be labor intensive.

If mathematics faculty have failed to communicate the need for effective class size for calculus to university administrators, it may well be a major cause of the inadequacy of the old calculus. We should not necessarily call for small sections, but for a sufficient teaching staff to be assigned to calculus to allow for regular feedback for students.

If we have allowed class size in calculus to increase beyond that which is pedagogically sound, then this Colloquium has given us a new opportunity, a new beginning. We must communicate the new enthusiasm for a new calculus—perhaps a lean and lively calculus—worthy of support by mathematics faculty, by faculty from client disciplines, and by academic administrators. Since the new calculus will evolve from present calculus, however, effective class size for calculus classes should be communicated as an imperative for the present.

•

JOHN D. FULTON is Professor of Mathematics and Head of the Mathematical Sciences Department at Clemson University. He is a member of JPBM Committee for Department Chairs, and Chair of the MAA *ad hoc* Committee on Accreditation. He received his Ph.D. in mathematics from North Carolina State University.

Don't!

Richard W. Hamming

NAVAL POSTGRADUATE SCHOOL

I have attended a number of similar conferences, devoted a lot of time to worrying about the calculus as it is currently taught, have thought long and hard about the topic, and decided that action is worth far more than endless talk. As a consequence, I have written a book that I thought would meet my standards of what the calculus course should be for the future. There are many other possible books besides the one I wrote, but at least it is something to go on rather than endless talk.

Listening to all the talk has caused me to compile a list of "don'ts:"

1. Don't try to optimize the calculus course. We are trying to provide a mathematical education, and the optimization of individual courses leaves too much to fall between them. We should view calculus as part of a student's total mathematical education, as a means to an end and not as an end in itself.
2. Don't look to the past. Our society has passed recently from a manufacturing society to a service society, and our teaching should be directed towards our students' future needs, not to our past activities.
3. Don't try writing a textbook by a committee. All too many of our texts are written by uninspired second and third-rate minds, and it is foolish to expect students to respond well to them. You need, more than anything else, an inspiring book. The particular contents are of less importance.
4. Don't think that you can move by small steps from where you are to where you want to be. We are at present at a local optimum, as can be seen by the fact that we have essentially only one text. The various books that are widely used so resemble each other that even the proofs are the same. Anyone who knows the least about the calculus theory of optimization must realize that any small step from the local optimum will be a degradation. To get to a better relative optimum you must move a large distance.
5. Don't think that discrete and continuous mathematics are separate topics. Any competent mathematician is well aware of the fact that the Riemann zeta function and prime numbers are closely related.
6. Don't think that calculus is only the development of tangent and area problems. It is also used widely to get new identities from old ones and to handle generating functions that arise just below the surface of combinatorial problems. The convergence of these generating functions is of no importance.
7. Don't neglect complex numbers as we now do; they are basic for the unification of various parts of mathematics as well as being essential in many areas.
8. Don't think that the old mechanical problems are of interest to the current crop of students—they are not! Both probability and statistics are of much more interest, are more useful, and can provide much more interesting problems. Furthermore, those engineers who claim that mechanics is more important to them than are probability and statistics are simply living in the past.
9. Don't think that money can buy the changes you want: you must use persuasion. The "New Math" is a perfect example of what not to do.
10. Don't present mathematics as a fixed, known thing for all eternity. Present the changing definitions that we actually have, the various attitudes towards mathematics, and the philosophy of mathematics. We are educating students, not just training them.
11. Don't emphasize the doing of algorithms. A friend of mine in Computer Science recently said to me: "Why think when you can program?" Adjusting this to the teaching of calculus: "Why teach the students to think when you can train them to follow algorithms?"
12. Don't start abstractly and move to the definite and concrete; rather start with the definite and exhibit the process of abstraction, extension, and generalization. They are the heart of mathematics. You cannot get motivation otherwise.
13. Don't teach huge classes of 500 students. Education

appears to require more personal contact. I hear that the large classes are for reasons of economy. I suspect they are also to let 19 professors escape teaching calculus. No wonder so few students major in mathematics!

14. Don't continue to act as if mathematics was the exclusive consumer of the calculus courses. If you want reality to respond, you must pay far, far more attention to the vast sea of user's needs.

RICHARD W. HAMMING is Adjunct Professor of Mathematics at the Naval Postgraduate School in Monterey, California. He worked at Los Alamos during the war, and from 1946 to 1976 at Bell Telephone Laboratories. He is active in the Association of Computing Machinery, IEEE, and various statistical societies. He received his Ph.D. in mathematics from the University of Illinois.

Evolution in the Teaching of Calculus

Bernard R. Hodgson

UNIVERSITÉ LAVAL

Plutôt la tête bien faite que bien pleine
Montaigne, *Essais*

One of the most remarkable features of this Colloquium is the huge number of attendees (above 600), much more, we are told, than the organizers originally expected. While such a level of participation may be interpreted as reflecting the extreme intensity of an actual "crisis in calculus," it can more simply be seen as an indication of a greater awareness among the mathematics community of the impossibility of keeping calculus teaching essentially unchanged, as it has been for so many decades.

The fact that computers, micro-computers, and even hand-held calculators are compelling changes in the way all mathematics, and especially calculus, is being taught, has been advocated over the years by various people. But time now finally seems to be ripe for real collective action to be initiated.

It might be instructive to recall briefly a few publications and meetings that have taken place since 1980 that have contributed to efforts to modify the way computer science is being perceived in influencing both mathematics and its teaching. Although by no means exhaustive, the following list nevertheless reveals the actual situation as an evolution which started slowly among a few enthusiasts and has progressed steadily so that now it concerns a great many people in various countries.

1980: Publication of the book *Mindstorms: Children, Computers and Powerful Ideas,* by Seymour Papert. The computer is presented as an "object-to-think-with."

1981: In a paper published in the *American Mathematical Monthly* (88 (1981) 472-485), Anthony Ralston argues for the consideration of a separate mathematics curriculum for computer science undergraduates, beginning with a discrete mathematics course rather than calculus.

The paper "Computer Algebra" (by Pavelle, *et al, Scientific American,* Dec. 1981) makes symbolic manipulation systems known to the general (scientific) public.

1982: A distant early-warning signal by Herbert Wilf: "The disk with a college education" (*Amer. Math. Monthly* 89 (1982) 4-8).

1983: Proceedings of a Sloan Foundation conference centered around Ralston's thesis on the balance between calculus and discrete mathematics: *The Future of College Mathematics* (Springer-Verlag).

1984: Some sessions of ICME-5 (Adelaide) devoted to the teaching of calculus and the effects of symbolic manipulation systems on the mathematics curriculum.

NCTM 1984 Yearbook on *Computers and Education.*

1985: NCTM 1985 Yearbook (*The Secondary School Mathematics Curriculum*) includes papers related to the issues raised by symbolic computation.

A symposium is organized by ICMI (The International Commission on Mathematical Instruction) in Strasbourg on the topic: *The Influence of Computers and Informatics on Mathematics and its*

Teaching (Proceedings published by Cambridge University Press, 1986).

1986: The Tulane Conference: *Toward a Lean and Lively Calculus.*

A symposium takes place in Tunisia on the influence of computer science on the teaching of mathematics as it relates especially to developing countries. (Proceedings published by the International Council for Mathematics in Developing Countries.)

1987: The International Federation for Information Processing organizes a working conference in Bulgaria on the subject: "Informatics and the Teaching of Mathematics."

This Colloquium: Calculus for a New Century.

1988: The programs of both the AMS-MAA annual meeting and ICME-6 meeting (Budapest) contain many activities related to computers and the teaching of mathematics (and especially of calculus).

A recent event has definitely served as a catalyst in accelerating the evolution suggested by the preceding milestones, namely the commercial availability of hand-held calculators with graphic or symbolic capabilities (like the Hewlett-Packard *28C,* the Casio *fx7000G,* or the Sharp *EL5200*). In spite of several inherent limitations and their often awkward usage, these devices have already had a considerable effect among teachers of mathematics: if many of these teachers still believed they could safely ignore Wilf's early-warning signal of 1982, a vast majority now realize that these calculators are here to stay, that they can only become cheaper and more powerful over the years, and that students will be expecting to use them.

In addition, symbolic manipulation systems, reserved to a small group of specialists only a decade ago, have now become sufficiently widespread that both mathematicians and users of mathematics are aware of their usefulness. These systems have created a situation where a mere technician, with little mathematical background, could work on problems which, because of their mathematical sophistication, would escape the expertise of today's typical engineer. What sense will this technician make of the "answer" (numeric, symbolic, or graphic) produced by the computer? How can such a technician appreciate the validity and limitations of the mathematical model being used?

This raises forcefully the difficult question of minimal competency in mathematics for engineers and other users of mathematics. One is reminded here of a similar situation actually occurring in statistics, where almost anyone can use commercially-available packages like SAS or SPSS to painlessly have the computer print pages and pages of output in all forms (tables, pie charts, bar graphs, etc.), often without having the interpretative abilities to thoughtfully use this information. (And here enters the consultant statistician!)

But things have evolved greatly over the last ten years in the teaching of statistics. Instead of concentrating on tricks for the calculation of various statistical parameters, teachers of statistics now stress the development of a sense of appreciation and judgment. Exploratory data analysis is quite typical of the shift in approach, where the calculator or the computer plays a central role in working with the data.

Although the inertia of the system is far greater in the case of calculus, a similar change *will* occur. It is compelled on us by the wide availability of calculators and numeric, symbolic, or graphical software which our customer departments are quite eager to use. Even if no clear relationship has yet been identified between, say, procedural skills developed by *lengthy* hand manipulations of algebraic expressions and the understanding of the underlying algebraic concepts, it seems beyond doubt that a shift will take place from purely computational to more complex interpretative abilities—in other words from calculation to meaning. More than ever, the adage of the *bonhomme* Montaigne prevails, and development of mathematics judgment greatly contributes to this *"tête bien faite."*

A lot of people in different places are now getting their feet wet in trying new approaches to calculus teaching. Attendance to this Colloquium indicates that this corresponds to a real need. We are now in a phase where experiments need to be performed, evaluated, and communicated to others. Identification of new curricula and production of related materials is a difficult and unrewarding task. But only such efforts can produce, as was wished by Robert M. White in his keynote address, a calculus that is no longer a filter but a *pump* in the scientific pipeline.

BERNARD R. HODGSON is Professor of Mathematics at Université Laval (Québec). Besides his research interests in mathematical logic and theoretical computer science, he is an active member of the Canadian Mathematics Education Study Group and regularly teaches courses for elementary school teachers. He received his Ph.D. degree from the Université de Montréal.

Collective Dreaming or Collaborative Planning?

Genevieve M. Knight

COPPIN STATE COLLEGE

The halls of the National Academy of Sciences echoed the voices of 600 plus persons who gathered to reflect on the current status and future direction of Calculus! Messages were loud and clear that today's profile of calculus compares poorly with the world of reality. If one moves away from the innovational uses of disks, symbolic algebra, and calculators not much is going on. The future complexion of calculus must mirror with zest the magnitude in which science and technology are changing.

The presentors spoke of urgency:

- For the calculus reform to consider the effort a national collaborative one (however, the effort must be generated at local levels and coordinated through networking).
- For collegiate mathematics to be less isolated and to link to other subjects, to professional users external to the classroom, and to secondary school mathematics.
- For an in-depth analysis of the practice of teaching calculus that focuses on purpose, content, methodology, and assessment—beyond two-hour tests and a final examination.
- For the keepers of calculus to "do something" about what is and and what should be.
- For the teaching arena to capture the beauty of the subject with meaningful instructional activities centered around ***students'*** understanding and needs.

After several speakers, I began to become concerned that too much time was being allocated to talking about "issues." Flooding my mind were the reflections gathered from reading the background papers, from personal interactions with colleagues, and from my own twenty plus years of teaching calculus. As a practitioner I was ready to engage in an intellectual mental brainstorm embracing reality with the host of talent in the meeting room. At this junction the voices representing NSF chimed softly, "We are listening to the mathematics sciences community and here is our proposal. Reach beyond the "talking stage" and initiate the process to develop this new calculus curriculum."

Lunch time was alive with groups of voices buzzing about the points of view presented during the morning session. As I circulated among the crowd listening to bits and pieces of conversations, I began to sense that the initial steps had commenced and the mathematics sciences community was generating the seeds for curricula reforms in calculus.

Speakers at the afternoon session artfully integrated the morning discussions to reflect their positions as "clients" and "administrators." The images of calculus we project to users and others are pictures that are fuzzy and out-of-focus. The Colloquium challenged mathematics professors to stop, to reflect, to regroup—and to enter into the world of technology and the 21st century.

The picture of calculus now takes on many different forms for the individuals who attended this conference. I leave with a commitment to share my notes with colleagues back home. In addition, I'll shepherd some fine-tune analysis by all departments whose students are required to take the calculus sequences. At least 30 different topics, issues, and concerns were voiced during the conference. I urge readers to generate from the following list questions to be answered by their mathematical science faculty:

- Purpose and sincerity of calculus reform
- Management of resource and materials
- Calculus textbooks
- Faculty
- Placement and assessment
- Discrete mathematics
- Methodology
- Composition of calculus topics
- Use of technology
- Demographic data
- Drop in supply of human resources

The collective dream can become a reality. Whatever we are now calling calculus will not survive in the new century. Mathematicians must assume the responsibility for what will replace it.

GENEVIEVE M. KNIGHT is Professor of Mathematics and Director of Mathematics Staff Development Programs, K-8, at Coppin State College. She received a Ph.D. degree from the University of Maryland at College Park. Her research interests are in non-traditional approaches to the teaching and learning of mathematics in an urban school district.

Calculus—A Call to Arms

Timothy O'Meara

UNIVERSITY OF NOTRE DAME

This colloquium will either be a watershed in mathematics education or a flop. Which it is depends entirely on the kind of follow-up that will be taken in generating interest in the mathematical community at large.

My initial reaction during the first few hours of the Colloquium was one of bewilderment and disappointment. Based on the announced title of the event, I subconsciously expected to be presented with a brand new calculus on a platter. In actual fact, I heard a recitation of the problems facing us in mathematics and science education, apocalyptic predictions for the year 2000, philosophical messages from our sponsors in science and engineering, and a call for mobilizing all sorts of forces in our society. Self-analysis and self-criticism were conspicuous by their absence.

My moment of truth came at the cocktail hour at the end of the first day. First, I realized that our real concern was with mathematics education at the college level, not with calculus. The word calculus was just a snappy way of putting it all together. More importantly, I realized that Calculus for a New Century is far from being a final product—it is still a movement. We are in process. The next step in the process must be an awakening, followed by an involvement of the entire mathematics community.

We are all aware of the crises in mathematics and science education in our country today: a severe shortage of mathematicians; few decent teachers in our schools; alarming drop-out and failure rates in many of our colleges; problems with the pipeline; dependency on foreign talent; a disgraceful inability to nurture mathematicians on our own soil and in our own culture. In fact, we are so accustomed to this recitation that we have been lulled into accepting it as immutable.

This was the starting point of the Colloquium. It will have to be repeated and repeated as the process continues. Those with a mercenary turn of mind will be interested to know that calculus accounts for half a billion dollars of business each year.

On the apocalyptic side, we were reminded that push-button calculus is just around the corner. So why teach students how to find one volume of revolution after the other. The calculator will surely tell us all. Will calculus become strictly utilitarian? Need its inner workings be known to any but the elite? If calculus is just a skill, should it be taught in our universities? I would hope not. Will discrete mathematics replace the continuous?

Mathematics will continue to flourish, but where will its creativity come from? From our economists, our biologists, our engineers? Will our mathematicians be sufficiently flexible and imaginative? I hope so, but I am not convinced.

Interestingly enough, little was said of the intrinsic beauty of mathematics. Epsilons and deltas were mentioned in whispers. Mathematics as part of liberal education for the new century was not mentioned at all. I find that alarming.

Mathematics education at the college level will have to be stripped of rote. Concepts will have to be emphasized. Research mathematicians will have to view teaching as an honorable part of their lives. And this will have to be transmitted by example to their doctoral students who will become the next generation of professors. All of us will have to convey our enthusiasm for our subject, not only to each other as we now do, but also to our deans who control the purse strings, and most of all to our students.

The next step in the process is up to us. If the surprisingly large turnout at the Colloquium is any indication, then I think we are on our way.

TIMOTHY O'MEARA is Kenna Professor of Mathematics and Provost of the University of Notre Dame. In his tenth year as Provost, he maintains his research interests in algebra and number theory. He has published three books in these areas; a fourth, *The Classical Groups and K-Theory*, coauthored with Alex Hahn, will be published by Springer-Verlag in 1988. He received his Ph.D. from Princeton University.

Surprises

William M. Priestley

UNIVERSITY OF THE SOUTH

On a "typical" final examination in a calculus course last year, a student who was given permission to push buttons intelligently on the latest hand-held calculator could easily have ensured himself of a passing grade. Yet the failure rate among college calculus courses is surprisingly high, and surprisingly few such courses incorporate any use of a computer. At the same time, a study of enrollment trends leads to a prediction that the next century will find more students studying calculus in secondary schools than in colleges and universities.

These unwelcome surprises from the conference "Calculus for a New Century" call for concerted effort to avoid a *fin de siècle* crisis in the teaching of calculus. For some of us who teach wrong mathematics in liberal arts colleges, however, a pleasant surprise came out of the conference as well. This was the response of a small but significant number of participants who endorsed as a partial remedy our most cherished forlorn hope and lost cause: the insistence upon clear writing in English by students of calculus.

These bewildered students have been so ill conditioned by repetitive drill problems whose answers are so transparent as to require no explanation that they think the proper response to every mathematics problem is to dispose of it as quickly as possible without one word of explanation to the intended reader of their work. The reader, of course, is their dedicated instructor (whose role will later in life be played by their exasperated boss) whom they confidently expect to reward them highly if the correct answer can be found scribbled anywhere at all in their chaotic ramblings.

The most overlooked shortcoming in the teaching of mathematics is the failure of teachers to insist that their students justify their answers—if not with complete sentences, at least with a few suggestive English phrases. In the case of an optimization problem in calculus, for example, it is surprising how much good is done by a teacher who demands that students

- Never put an "equals" sign between unequal expressions, and
- Pepper their computation with the proper use of a small glossary of words like *let, denote, if, then, so, because, attain,* and *when.*

Students understand the theory behind the technique of optimization if and only if they can carry out these demands.

But the list of worthwhile changes to bring to the calculus is never-ending. A dean and professor of English, who was surely sympathetic with the idea of writing, admonished conference participants *not* to try to teach writing. She had already heard about too many other things to do. Surely almost everyone agreed with her.

The major problem my students have at the outset of a calculus course, however, is that they don't (or can't) learn from reading the textbook. No one needs to do a statistical survey to know that this is true generally, and teachers who give tests that calculators can pass probably don't expect their students to be able to read mathematics.

How can this problem be solved? Recall how Newton solved a famous problem and discovered the calculus: he calculated the area beneath a curve between 0 and 1 by attacking instead the problem of finding the area between 0 and x. The technique of solving a hard problem by attacking in its place a still harder problem that ought to have been impossible is one of the most surprising methods learned in mathematics.

The solution to the hard problem of getting students to read the textbook is to attack instead the impossible problem of teaching them how to write. What would happen if all of us who instruct should insist to our students that we really expect them to learn to write? What would happen if we told our classes that a student cannot learn to think like a mathematician without learning to write like a mathematician? Some students might actually learn, grow up and study analysis, and attain the background needed to teach calculus to all those high school students of the next century.

As for the rest, they may never learn to write. But if they try to learn how to write, they will have the biggest surprise of their mathematical lives. They will learn how to read.

WILLIAM M. PRIESTLEY is Professor of Mathematics at the University of the South and author of a calculus textbook written for liberal arts students. He has published papers in analysis, participated in an NEH-sponsored summer seminar on Frege and the foundations of mathematics, and is interested in using the history of mathematics to promote the better teaching of mathematics. He received his Ph.D. degree in mathematics from Princeton University.

Calculus for a Purpose

Gilbert Strang

MASSACHUSETTS INSTITUTE OF TECHNOLOGY

The following comments are not so much about what was said in Washington, as about questions that were momentarily raised, and then dropped. Some of the unspoken or barely spoken needs are fundamental to success in the classroom—and the success of this whole initiative.

One is the student's need for a clear sense of purpose, and for a response that encourages more effort. Actually the instructor has the same need. Our questions to the panel representing the "client" disciplines might have been combined into a single question: What is calculus for? If the course is to be recognized as valuable and purposeful by students, that has to be answered. The afternoon audience was asking the right question!

It may seem astonishing that we who teach the subject don't know everything about that question. Of course it is not at all astonishing. First, there are too many answers. No one can be familiar with all the uses of calculus. But that is more than the students are asking, and more than they could be told. Students need to have *some clear purpose.* There is a second reason why we don't answer well—we are very much inside the subject, teaching it but not seeing it.

Calculus is a language, but what do we have to say in that language? There are many texts that explain the "grammar" of calculus—the rules of the language—but that is not the same as learning to speak it. We need ideas to express, things to say and do, or it is a dead language.

I was struck that in Tom Tucker's list of groups that have to contribute to a reform, authors were not mentioned. It was an oversight, but I think they are at least as important as publishers. I am convinced that the textbook itself is absolutely crucial in explaining the purpose as well as the rules. That is the job of the book, and I think it can be done.

I was fascinated by the observation that calculus courses reflect so little of the last 100 years—almost nothing since Riemann. Students cannot fail to see that in the biographies and to draw conclusions. But new ideas have developed, as well as many new applications. The instructor and the book are responsible for making time for new ideas in the classroom.

My last comments are in a different direction—about computers. It is one thing to believe, as we do, that they will come to have a tremendous part to play in calculus. It is quite another thing to see clearly what that part will be.

I can see one effect, which may not be central but will make a big difference. Lynn Steen mentioned a recent study that revealed that in more than half of the courses, homeworks are not graded (or even looked at). In other words, the student gets no response. That zero is worse than any grade. To work well without recognition is a lot to ask. We ask it because of pressures of time and of student numbers. Those are pressures that the computer is made for. I believe we will approach (slowly) a homework design in which the computer does the time-consuming part and we do the thought-producing part.

One difficulty is obvious, but not yet discussed. The computer is too quick. If we ask a direct question—a definite integral, or a system of linear equations—it answers immediately. The human part. is reduced to input of the problem. The student becomes—and knows it—the slowest and weakest link in the system. That is frustrating, not educating. This difficulty will be overcome, and I hope there will be publicity for successes (even partial successes) in finding the answer.

I hope the other difficulty will also be overcome—to make this course not a barrier but a door. We need first to see clearly where it leads.

GILBERT STRANG is Professor of Mathematics at MIT. He is the author of the textbooks *Linear Algebra and Its Applications* and *Introduction to Applied Mathematics.* An earlier book on the finite element method reflected his research in analysis and partial differential equations. His advanced degrees are from Oxford University and the University of California at Los Angeles.

Yes, Virginia ...

Sallie A. Watkins

AMERICAN INSTITUTE OF PHYSICS

Yes, Virginia, there is a Santa Claus. And yes, fledgling physicist, there is a gift in the making for you. Thoughtful, concerned persons are designing a new calculus course for your generation. It will be lean, lively, and true. I'm going to guess some of its other characteristics.

If you learn integration, it won't be to build character; you will use calculators and computers as natural tools in the course; you will learn to solve differential and difference equations; you will develop the power and know the satisfaction of mathematical proof; you will learn to make valid approximations, to do modeling; you will see where all of this is going, because there will be a steady supply of real-life applications; you will solve related problems as homework to be turned in for grading; you will see a symbiotic relationship between continuous and discrete mathematics in your course. Your calculus will be a richer course than today's, but your text will have a mass of less than 3.4 kilograms. There you have it, in a nutshell.

But there remains a burning question which will concern us both, Virginia: How will the lean and lively calculus affect the field of physics? We physicists talk about "the calculus-based introductory physics course"—and we mean it. Calculus is the mainstay of physics.

Only yesterday, the world we physicists were able to study was a perfect one. We neglected friction; we talked about physical phenomena in first approximation; we generally did violence to reality so as to arrive at solutions in closed form.

Take Newton's second law of motion. In point of fact, F does *not* equal ma. Nor does F equal $m\frac{d^2x}{dt^2} + b\frac{dx}{dt} + kx$—except in first approximation.

We live and move and have our being in a nonlinear world. Suddenly we have the mathematical capability of treating that world as it is. Shall we have a burning of the physics books and charge our physicists to begin the discipline anew?

Research in the learning theory of physics has shown that students are hampered by the baggage of a world view that insists on the existence of friction. Would these students have an easier time with a course in which they could feel more conceptually at home, even though the mathematics were messier?

But would something be lost if we were to redo physics to match the nonlinear world it studies? If Galileo and Newton had had the capability of handling the mathematics that we have today, would our physics be the beautiful, simple, clear structure that it is?

I submit that one aspect of the unique character of physics as a discipline is that it empowers its practitioners to abstract the simple out of the complex, to sense which features of complexity can be suppressed or ignored without loss of validity, and to construct sweeping generalizations.

Yes, students of the new lean and lively calculus will be equipped to deal with the nonlinear world as, indeed, it presents itself—but if they limit themselves (and us) to this level of information, we will find ourselves in a lose-lose situation.

Not everything that can be done should be done, Virginia.

SALLIE A. WATKINS is Senior Education Fellow of the American Institute of Physics, on leave from her position as Dean of the College of Science and Mathematics at the University of Southern Colorado. Her field of research is the history of physics. Dr. Watkins received a Ph.D. in physics from the Catholic University of America.

REPORTS

PARTICIPANT RESPONSES

Question What can be done to improve calculus instruction?

Answers

- When teaching is considered a worthwhile intellectual activity by the research community, [then] it will improve.
- Math departments should test students one year after they have finished calculus
- Commitment of our best universities to give good mathematicians time to work on materials.
- Remember the "good old days" of SMSG and UICSM?

General Comments

- Can we compete with rock videos?
- Change slowly.
- CAS is not the answer.

Calculus for Physical Sciences

First Discussion Session

Jack M. Wilson and Donald J. Albers

We began by considering the following agenda of issues:

- Should numerical techniques be taught as part of calculus?
- Should topics from linear algebra be included?
- Probability and statistics.
- What role for statements or proofs?
- Should mathematics teachers be required to have more of a background in physical science?
- Should discipline examples be included? If so, how?
- Role of drill and practice.
- Role of problem formulation and conception.
- Should there be formal instruction in problem solving techniques?
- Is there a special relationship between physics and calculus?
- Visualization of solutions. Graphics.
- Making physical sense of solutions.
- Relevance to the workplace.
- Should we do everything, or a few things well?
- What's in? What's out?
- The TA problem: language, time, training, motivation.
- Should calculus be taught as an umbrella course or partitioned into various courses for various majors?
- How do we handle large sections of calculus?
- Is (should) technology be the driving force behind the new calculus?

After initial discussion to consolidate these diverse issues, the group prioritized the major issues as follows:

A. Should we do everything in calculus or a few things well? What should be left out? What should be added?

B. What should be the role of proof?

C. Is (should) technology (be) the driving force behind the new calculus?

D. Visualization of solutions, graphics, and making physical sense of solutions.

E. Should numerical techniques be taught as part of calculus?

The absence of the other issues from this priority list means that either group members were in agreement on the issue or that they did not view the issue as controversial.

We first discussed what items should be excluded from calculus and what should be added. There was general agreement that we could not develop a detailed list in the allotted time, but could examine a short list of possibilities.

Several participants suggested that a zero-based budgeting approach be employed. There was consensus that we should reduce the amount of time spent on closed-form integration techniques. Some suggested that trigonometric substitutions provide students with a valuable review of trigonometry. Others argued that practice with integration methods help to teach pattern-recognition skills.

The suggestion that work with volumes of solids of revolution and the computation of centroids be eliminated did not get much support. Participants emphasized the need for applications such as those in which the integral is seen as a limit of sums.

Intuition, Calculation, and Proof

The elimination of epsilon-delta proofs in presenting limits was supported by all participants except the high school teachers, who said they must teach such ideas because of the presence of epsilon-delta questions on Advanced Placement examinations.

A reduction of time on derivative calculations was suggested by our group. The presence of calculators and computers that easily and quickly compute derivatives lends support to this call for reduction.

Our group also recommended the addition of more work with qualitative examples and exercises. For example, students might be asked to construct the graph of the derivative of a function given the graph of the function.

The group next considered the place of numerical techniques, including some elementary aspects of numerical analysis. There was general agreement that a careful selection of strategically-placed numerical methods should be included, but that they should not simply be grafted on to a standard text.

The question of the role of proof in a first course in calculus was sharply debated. A few argued that some formal proofs are essential to attract strong students to mathematics. A few mentioned that client disciplines often apply pressure on calculus instructors to minimize proofs. Some suggested that intuitive proofs be included. All agreed that concept building is a fundamental goal. To that end, it is essential to give clear, concise statements of theorems, and that changes in the hypotheses be explored to help motivate theorems.

Technology

The group felt that the question "Is calculus reform driven by the new technology?" in some way subsumed other questions on numerical techniques, on the statement-proof approach, and on developing a physical or graphical sense of solutions. This topic brought on an animated discussion, but there was a surprising consensus that calculus is being driven by the technology and that it may not be a bad thing. Although the use of computers in calculus teaching is very low, it was felt that the use of new technology by faculty is growing.

Several individuals noted that the computer allowed them to present old topics from a better and more easily understood perspective. There was also a discussion of the powerful graphic presentations now possible and the ability of the computer to explore qualitative features of complicated systems—a form of intuition building.

The availability of powerful symbolic manipulation programs, such as those provided by the HP-28C, will change the terrain of calculus instruction. Skills once important will now be available with the punch of a button. It will still be necessary to teach some of these skills, but how much and in what way remain open questions. This is one area that may free time for teaching other skills which now may be more important.

There was considerable discussion of the difficulty of retraining faculty. Students are coming to class expecting the faculty to have certain skills that they may not possess. This problem is as severe, or even more severe, at the college and university level as it is at the secondary level.

It was also noted that the computer has radically changed physics research and engineering practice, but that these changes have not yet been fully reflected in the teaching of these subjects.

Implementation

It was generally agreed that calculus courses are unlikely to change much until new textbooks and other instructional materials are produced. As a first step, the group recommends the formation of a blue-ribbon writing team made up of teachers of calculus from all levels. The team might have an advisory board composed of individuals from client disciplines in academe and industry. The first task of the writing team would be the creation of a draft syllabus for the new calculus. This syllabus would then be revised through review processes that might include meetings of focus groups at local, regional, and state levels, by professional organizations of teachers of mathematics, as well as at meetings of client discipline organizations. The revised syllabus could then serve as the basis for new calculus texts.

It is likely that several individual efforts toward the creation of new textbooks could use this syllabus as a place to start. Established authors might be influenced to include ideas from this syllabus.

It was also suggested that a sourcebook of applications from client disciplines be produced in order to heighten the appreciation of calculus by both students and teachers of calculus.

On the technology side, calculator manufacturers are urged to produce devices that are very user-friendly. Such efforts are likely to increase their acceptance by teachers and their use in calculus instruction.

•

JACK M. WILSON is Professor of Physics at the University of Maryland, College Park, and Executive Director of the American Association of Physics Teachers. He is also co-director of the Maryland University Project in Physics and Educational Technology and has published frequently in chemical physics, educational physics, computers in physics education, and public policy issues. He received his Ph.D. degree from Kent State University.

DONALD J. ALBERS is Associate Dean and Chairman of the Department of Mathematics and Computer Science at Menlo College. He has served as Editor of the *College Mathematics Journal* and as Chairman of the Survey Committee of the Conference Board of the Mathematical Sciences. He is co-editor of *Mathematical People*. He is currently Chairman of the Committee on Publications of the Mathematical Association of America.

Second Discussion Session

Ronald D. Archer and James S. Armstrong

This session, which consisted of a well-balanced mix of mathematicians and physical scientists, identified several major issues dealing with calculus content and instruction. Foremost among them was the need for continuing dialogue between mathematics faculty and physical science faculty.

The calculus course should be designed by mathematicians for mathematics majors, both pure and applied, and coordinated whenever possible with members of the physical science faculty. Whereas it is important for mathematicians to consider the needs of the scientist, conflicts should be resolved in a manner consistent with the goals of the calculus course.

Although all participants agreed that applications can be motivating, some cautioned against distorting calculus with an exaggerated emphasis on physical applications. Other participants expressed uneasiness with their own ability to discuss certain applications outside their own discipline, especially since their students often lack the necessary background to fully appreciate these applications.

Technology

Concerning the impact of technology on calculus instruction, there was strong consensus on several aspects of this issue:

- Technology is here and we must deal with it.
- Paper-and-pencil drill must precede student use of electronic devices.
- These devices should ultimately enhance students' understanding of the complexities and intricacies of the calculus.
- The proper balance between the use of electronic devices and traditional methods has yet to be determined.
- Caution must be exercised to avoid adding topics to an already over-crowded calculus curriculum just because technological advances allow teaching certain topics more quickly and efficiently.
- A need exists for a well-publicized national clearing house for calculus software.

Core Concepts

In thinking about a core for calculus courses, the group conceded that it was probably inappropriate to develop a curricular outline in one morning without adequate time for reflection. Nevertheless, there was widespread support for incorporating the following four major concepts as identified in the recent publication *Toward a Lean and Lively Calculus:*

- Change and stasis.
- Behavior at an instant.
- Behavior in the average.
- Approximation and error bounds.

There was also strong support for including formal proofs in the study of calculus, including epsilon and delta proofs. Most participants felt, however, that epsilon and delta proofs should be delayed until the second semester. Participants urged that it is vitally necessary to know your audience and their background before launching into these more sophisticated concepts.

Important pedagogical considerations related to the teaching of calculus were discussed at length. It was noted that the flavor of the calculus course should emphasize:

- The spirit and natural beauty of mathematics.
- The way mathematicians do mathematics.
- The intuitive dimension of mathematics.

For example, it is highly instructive for students to see the teacher develop the solution to a problem from scratch.

Textbooks

Much of the concern about calculus instruction centers on current textbooks. Weaknesses in current calculus textbooks include too much "plug-and-chug," too many topics, excessive use of highlighting and summarizing sections, and too many "template" word problems.

New textbooks, written to support calculus curriculum reforms, should minimize highlighting and, in fact, encourage students to read the mathematics. There is a strong feeling among mathematicians that the student must know how to read mathematics before they can write good mathematics. New textbooks should also include physical application problems carefully chosen from other physical science disciplines.

Homework should be assigned, collected, and graded at frequent intervals. Examinations and homework should reflect the course objectives. There appears to be significant faculty resistance to the use of common departmental examinations. Nevertheless, close supervision must be exercised over the preparation of examinations by less-experienced teachers.

Serious concern was expressed over the lack of emphasis on a cohesive mathematics-science curricular thread throughout the K-12 curriculum. This makes it extremely difficult for college faculties to use the natural connections between mathematics and the physical sciences in order to motivate the study of mathematics.

Good teaching must be rewarded in the same manner that good research is rewarded. The teaching of calculus can be exciting to both the teacher and the students, if we appropriately harness the new technologies, interact with our colleagues, and establish a reasonable core.

RONALD D. ARCHER is Professor of Chemistry at the University of Massachusetts, Amherst. He has served as head of his department and serves as Chair of the American Chemical Society Committee on Education. He is Chief Reader for Advanced Placement Chemistry for the Educational Testing Service. An active research chemist, Dr.

Archer received a Ph.D. degree from the University of Illinois.

JAMES S. ARMSTRONG is a Senior Examiner of Mathematics at the Educational Testing Service. His primary responsibility is the Advanced Placement Program in Calculus. Previously, he was an Associate Professor of Mathematics at the United States Military Academy at West Point, New York.

Calculus for Engineering Students

First Discussion Session

Donald E. Carlson and Denny Gulick

Among the 30 participants in our session nearly everyone spoke at one time or another, and showed the diversity in types of departments and academic institutions from which they came. The participants included high school teachers, college-university teachers, and a representative from a well-known publishing firm. Most were mathematicians; a few were engineers.

Syllabus

Most of the discussion centered on the topics of the calculus syllabus and the problems identified with students in calculus. Among the suggestions for inclusion in the calculus courses were:

- real-life applications and modeling,
- heavier use of approximation methods,
- computer graphics,
- symbolic manipulation, and
- discrete mathematics.

Real-life applications and models would tend to make more meaningful the concepts presented in the calculus. Not only are approximation methods important from the standpoint of all students of the calculus, but also they can be very fruitful in motivating certain aspects of calculus, especially if observational data provide the basis for these approximations.

Computer graphics can give not only a visual image of various concepts, but also a much more realistic image of functions such as exponentials and polynomials. In addition, symbolic manipulation, which is only now beginning to come into the classroom, could minimize the tedium of routine calculations.

Finally, we recognize the recent effort to incorporate discrete mathematics into the first years of college mathematics. In all likelihood discrete and continuous mathematics should be intermingled. How should it be effected?

If there are to be new items added to the calculus course, it is agreed that some topics in the already too-full syllabus would have to be eliminated. Although our session did not dwell on which topics should be eliminated, the question of including only some of the techniques of integration arose. Also, it was suggested that perhaps some geometric applications of the integral could be replaced by applications more indigenous to engineering and physics. Finally, should less class time be devoted to series?

At the same time certain participants emphasized that the fundamental concepts in calculus must remain in the course, along with at least a certain amount of drill work to develop manipulative skill. We must minimize the tedium of working problems where no thought at all is necessary. A basic question is the following: What would be an appropriate blend between geometrically-motivated concepts, definitions and theorems, relevant applications, and rigorous proofs?

Students

The second major topic centered on the calculus student: diverse backgrounds, work ethic, and retention of calculus concepts.

We are all aware that high school preparation in precalculus topics varies greatly. In addition, ever more students come to college having already encountered some calculus. Should colleges and universities meet the students "where they are," giving them a special calculus course? In mathematics, how should colleges and universities interface with the high school?

Several members of the session indicated that students nowadays do not seem to be highly motivated to study (calculus). Is that because the course is not packaged well? Or because the students have improper backgrounds? In order to have a real impact on the calculus curriculum, the "work ethic" of the calculus student will need to be changed.

Many students have difficulty retaining the methods

and the concepts of the calculus. Can effective retention be accomplished primarily through routine practice and by graded homework exercises? Should it be accomplished through exemplary applications? Probably the answer lies in a mix of the two. Again, before the calculus curriculum can be really successful, we will need to find an appropriate mix.

In conclusion, the participants noted that nothing that we suggest can have much effect on the calculus curriculum unless

- publishers are receptive to innovative texts and software;
- communication between mathematics departments and engineering and science departments is fostered;
- academic institutions demonstrate their concern for improved undergraduate education and reward efforts to that end.

We hope that the present symposium will help give new vigor to the calculus curriculum.

DONALD E. CARLSON is Professor of Theoretical and Applied Mechanics at the University of Illinois at Urbana-Champaign. He is active in mathematics education in Illinois and is educated in both engineering and mathematics. He received his Ph.D. degree in applied mathematics from Brown University.

DENNY GULICK is Professor of Mathematics at the University of Maryland. He has served as chairman of the undergraduate mathematics program at the University of Maryland and is co-author (with Robert Ellis) of a calculus text.

Second Discussion Session

Charles W. Haines and Phyllis O. Boutilier

The six major issues that were identified by our group are

1. How do we develop mathematical maturity so that students are better able to adapt to new situations in the future?
2. Personal computers and sophisticated hand calculators will soon be in the hands of many, if not, all our students. How do we make the best use of them?
3. Many of the engineering disciplines are seeking additional topics in calculus and changed emphasis in the first two years.
4. Student success rate in the first year is affected by algebra and trigonometric skills, life styles of the students, and faculty or teaching assistant enthusiasm for the course.
5. Suitable textbooks for some anticipated changes are not widely available.
6. And finally, where do we go from here?

The issue of mathematical maturity arose from a discussion of particular topics that might or might not be kept in a calculus course for the future, and from the realization that we cannot predict what will be needed. Thus, one of the primary aims of the mathematics curriculum, along with other parts of the curriculum, is to teach students how to learn by developing the fundamental skills and mathematical maturity it takes to learn on their own now and in the future. Some particular suggestions (not meant to be exhaustive) are:

- Get students comfortable with the concept of functions and families of functions.
- Encourage inferences from functions concerning their graphs and vice versa.
- Make much more use of the conceptual approach, which lies between pure skills and pure theory. Illustrate with geometrical proofs and convincing or plausible arguments. Pure theory probably is not desirable in an engineering calculus course.
- Allow time to linger over concepts, to consider consequences, etc.
- Do examples and then generalize.
- Emphasize relationship between symbols and ideas.
- Develop the ability in students to read and then do examples.
- Develop the material concerning topics such as series, differentials, optimization problems, substitutions, and others in light of the above.
- Develop two- and three-dimensional concepts thoroughly, then extend to n-dimensions.

Computers and Software

There was an overwhelming consensus that we must address the proper use of the newest generation of hand calculators and PC software in the calculus course, as they are already here and students will be using them even if we, as faculty, don't. Some criteria and issues for consideration when employing these technologies are:

- They can't be introduced as an add-on; they must be integrated properly into the course.
- When developing examples, make sure the proposer does the examples to make sure they work, as some current programs don't do what they say they do.

- Concepts must be developed and understood first, before being used on a calculator or a computer. This approach can be used to enhance and expand an understanding of a concept or procedure, but we can't reduce mathematics to a "black box approach."
- The emphasis should always be—are the answers reasonable?
- This technology can assist with placing more emphasis on geometrical concepts and intuition.
- The idea of challenging students to "crack" software was proposed, as it is an interesting way of having them learn the concept in order to "get around it."

The issue of repackaging the first two years of engineering mathematics was discussed only briefly, as there were only a few representatives from engineering. Some of the pressures for changing the first two years comes from the diverse needs of the different engineering curricula. Discrete mathematics, linear algebra, numerical concepts, probability and statistics were mentioned as candidates for incorporation into the first two years.

Questions and concerns these topics raise are:

1. What topics do we cut from the current courses?
2. Are the engineering disciplines willing to add another quarter or semester of mathematics?
3. Should there be a common first year calculus for all engineering disciplines, with the second year more tailored to each department's need?
4. Should some foreign engineering educational systems be studied to gain insight into other ways of approaching this problem? (This could also be expanded to cover other issues identified in this report.)
5. The use of the hand calculator and PC software also needs to be considered here in light of "repackaging."
6. "Lean" calculus may not be appropriate for the engineering disciplines.

Student Preparation

The issue of many students starting college underprepared in algebra or trigonometry drew many comments. Consensus seemed to be that pretesting and placement programs can be reasonable and effective. With more minority students and re-entry students being recruited for careers in engineering, placement programs for calculus and precalculus mathematics will be desirable.

Students should not be forced to repeat what they already know, as boredom will afflict the class, nor should they be in a calculus class where they are "down the drain" in the first two weeks due to lack of skills in algebra and trigonometry. Beyond carefully-designed placement programs, further reducing attrition in the calculus sequence must address other factors such as the students' study habits and motivation and faculty attitudes toward the course.

The large number of over-sized textbooks, most of which differ only by the number of exercises or by the color of the graphics, drew many comments. Most participants would like to see new lively texts but no consensus was reached regarding the content of these new texts. The mathematics community will be wary of adopting "non-standard" texts and publishers want reasonable expectation of sales before publishing a new text.

Follow-up Activities

Most of the participants put great emphasis on post-conference activities. Those activities should include:

- Local campus seminars by participants to inform and excite his or her colleagues in mathematics and engineering.
- District or regional workshops to exchange ideas and experiences.
- Information distributed nationally through as many vehicles as possible to inform the engineering community of this national initiative, as very few engineers attended this Colloquium. At the national level two societies come to mind, ASEE and SIAM. Other discipline-specific national societies should be informed. Involvement of engineering faculty is necessary for success.
- Curriculum development, experimentation, assessment, follow-up, feedback, i.e., a closed loop which if shared through a newsletter compiled and edited by a national committee could lead toward a better calculus course.
- A national committee to oversee and collect ideas, initiatives, experiments, etc., and to edit and disseminate the material to mathematics faculty and other involved persons including secondary school faculty throughout the country. This committee needs ample financial support over a five to ten year period.

CHARLES W. HAINES is Associate Dean and Professor of Mechanical Engineering at Rochester Institute of Technology. He teaches mechanical engineering and mathematics, has published in the areas of engineering analysis and differential equations, and is active in several professional societies. He received his Ph.D. degree from Rensselaer Polytechnic Institute.

PHYLLIS O. BOUTILIER is Professor of Mathematics at Michigan Technological University. She has served as Assistant Department Head and as Chair of the Mathematics

Division of the American Society for Engineering Education. Currently she is Director of Freshman Mathematics at Michigan Technological University. Boutilier received an M.S. degree in mathematics from Michigan Technological University.

Third Discussion Session

Carl A. Erdman and J.J. Malone

After discussing a wide variety of topics related to calculus instruction, the participants in our group decided—by majority vote—that there were four areas of major concern, which are presented below without regard to priority.

Student Background

Many students arrive at college not prepared to take calculus. Placement tests can help remedy some of the problem, but they are not a cure. Calculus courses should not be split into fast and slow tracks as a way of responding to student differences. It is preferable to have students take pre-calculus courses even though this may be resisted by students (and their parents).

If calculus is taught on the secondary level, it must follow a standard syllabus such as AP. The quality of this course is very important, since students may be turned on or off to mathematics from this experience. Indeed, the seventh and eighth grade experiences in mathematics may have already fixed student attitudes, a major problem which needs attention. Clearly, entry into a high school calculus course should not be at the expense of adequate exposure to algebraic skills. There is some suspicion of high school courses which are not similar to AP courses.

Syllabus Issues

The present calculus syllabus is overloaded; there is just too much material in the courses. However, there is no agreement as to what might be left out. We need some mechanism for deciding on what can be omitted. All of this is tied to a quantity-versus-quality agreement.

It is worth noting that only one-third of the participants felt that there was a strong need to change calculus in a dramatic fashion; many felt that we currently do a good job in calculus instruction. This may be a reflection of the group's makeup; all but one person (excluding the session leader) were calculus teachers, and they generally came from schools with good calculus programs. However, there were several comments concerning the need to have more time to deal with problem solving. A question was raised as to whether there was too much emphasis on training rather than on education.

There was a feeling that the engineering community wanted some discrete mathematics topics introduced, but probably did not want to give up much in the way of traditional topics. However, it wasn't clear that "discrete mathematics" meant the same thing to engineers as it means to mathematicians. There is a need here to define terms.

Technology

Those who saw a need for considerable revision in calculus often cited the need to incorporate new technologies as a motivating force. Computers can be used to develop intuition, to improve the pedagogy of the course, and to influence the choice of topics to be taught. These thoughts seemed to be broadly accepted.

There seemed to be agreement on the benefits of freshmen having (or having access to) a personal computer. However, there was no consensus as to whether the College should provide personal computer facilities or whether students should be required to buy them.

Goals and Implementation

What is the essence of calculus and how should we teach it? Certainly, it is not the repetitive working of routine problems. Some of the current calculus discussion should be directed toward a better definition of the objectives of calculus. The role of ideas or concepts as opposed to the role of techniques needs to be addressed. Some participants believed it was possible to teach concepts without using $\epsilon - \delta$ techniques.

Surveys of final examinations indicate there are few questions of the "state and prove" type. Is this a reflection of the goals of the course? It was noted that there is a need to stress understanding. But, when this is done, students often complain that the course is too theoretical. It appears that calculus instruction does not do a good job with either concepts or applications. An interesting question was posed as to whether engineers really care if their students have seen proofs in calculus.

It was strongly felt that there should *not* be separate sequences for each of the various engineering disciplines. This wasn't practical, and the course and educational objectives were better served with a variety of students in the course.

Other Issues

The issue of class size and the use of TA's was raised as an area of much less concern than were the previous four areas. It was noted that horror stories about non-English-speaking TA's abound and that some of them are true. But it was also said that some TA's are very good and that having a professor as instructor is no guarantee of good teaching. Some feelings were expressed that a valuable personal touch was lost with sections of large size.

Concerns with pressures produced by the engineering curriculum were also discussed. It was thought that engineering schools press for too hectic a pace; that they are forcing a compression in the mathematics presentation. One cause is the fact that most engineering schools will not openly admit that engineering is, in fact, no longer a four-year program.

Several participants suggested that a mathematics department resident in the engineering college made cooperation much easier and led to good curriculum planning. If the only mathematics department is in the arts and sciences college, campus politics can make cooperation much more difficult.

The final point was that engineering faculty needed to provide the mathematics faculty with examples of the kinds of applications they would like to see incorporated into the calculus.

CARL A. ERDMAN is Associate Dean of Engineering and Assistant Director of the Texas Engineering Experiment Station at Texas A&M University. He has been Head of Nuclear Engineering at Texas A&M, a faculty member at the University of Virginia, and a research engineer at Brookhaven National Laboratory. His Ph.D. is from the University of Illinois at Urbana.

J.J. MALONE is Sinclair Professor of Mathematics at Worcester Polytechnic Institute. He has also taught at Rockhurst College, University of Houston, and Texas A&M. His areas of research include group theory and near rings. He received a Ph.D. degree from Saint Louis University.

Calculus for the Life Sciences

William Bossert and William G. Chinn

Biologists complain that the mathematics curriculum is designed to meet the specifications of physicists and engineers and that our concerns and the needs of their students are ignored. Is this true? Are their needs for mathematical training really different and if so, how are they different and how could they be better met with changes in the mathematics curriculum? Several observations are frequently made by biologists.

Mathematics has been less central to the life sciences than the physical sciences. Perhaps this was true during the development of biology when description was more important than theory, but it is certainly not the case today. The vast majority of articles in some important biological journals from clinical medicine to ecology present mathematical models or the application of mathematical technique. The biology major in most colleges requires physics and physical chemistry courses that themselves need fundamental mathematical training. Many biology educators refuse to recognize the importance of mathematical training, however, and allot only two semesters in the crowded biology curriculum for it.

Within mathematics, calculus is less central to biology than are other subjects. Although there are many important applications of the calculus in all areas of biology, many fundamental biological research results—which should be included in college biology courses—depend on applications of modern algebra, particularly combinatorics, of probability, of statistics, and of linear algebra. (Some illustrations are given in the background paper of Simon Levin.) If we do not encourage young biologists to take more than two or three semesters of mathematics in college, should they all be devoted to the calculus?

For the life sciences, training in formulating problems is as important as training in solving them. Too often calculus courses present illustrative problems that are carefully selected to show off the technique. In biology the problems that students face even in elementary courses are either not well formulated or do not yield to simple techniques. Perhaps in engineering the best part of one's activity is applying well-studied models and their associated techniques to new situations, but that is not true for biology. Students must learn to abstract simpler formal descriptions from problems that they cannot solve, and need to be shown how to depend more on qualitative and numerical methods than on those traditionally taught in the calculus.

If the mathematical needs of the life sciences do differ in these ways, and others, what should be done about it? A first thought is to start from scratch and design a new course in college mathematics specially for biologists. In two or three semesters we could package all of the broader range of mathematics that biologists feel are needed, taught with a problem first and qualitative style.

This is not easy to do. Although some have tried it and several textbooks labeled "mathematics for the life sciences" are available, the efforts are not altogether successful. A simplistic critique holds that they are merely expositions of the same topics of traditional precalculus and calculus courses, perhaps reordered and with biological illustrations, and do not involve a new selection of topics and a "problem first then technique" attitude that might be required. Also, they regularly achieve breadth by simply being longer and placing more demands on our students.

There are some strong reasons for avoiding special discipline-oriented calculus courses. First, many students take this first course in college mathematics before their career directions are set. Many liberal arts colleges do not allow students to select a major until after the freshman year when the course would normally be taken. Second, the value of a mathematics course as a liberal arts experience is compromised if it is taught with a narrow disciplinary focus.

Third, and most important, there is a danger that the specialization of the calculus course will obscure the generality of the mathematical way of thinking that separate disciplines have come to mathematics to gain. It is useful for students to see that the mathematical modelling process and even some specific models transcend disciplinary boundaries.

Separate courses may be justified when resources permit, and when disciplinary education can be more efficiently begun in the calculus course. Students might also be usefully taught in separate sections that recognize their differences in previous experience and commitment to further training. Care must be taken, then, that the courses do not lose the conceptual and aesthetic essence of the mathematics.

It might be better to depend upon significant changes in the calculus curriculum, such as will surely come from this conference, rather than striking out on our own. If so, what changes would be particularly important to us that could also be desirable for other "client" fields and hence possible to be adopted? Some first thoughts are:

1. More illustrative examples from the life sciences. This obvious improvement is not a problem. Mathematics educators are regularly asking for teaching examples from applied fields.
2. More mathematical modelling. More time should be spent developing models, not just to apply known techniques for drill, but to introduce new topics which are required because of the model and not the reverse.
3. More linear algebra. Many applied fields deal with systems of differential equations which can be taught efficiently only after students can deal with matrices. This topic is also important to biology as the basis of multivariate statistics, which is too often left until late in the mathematical curriculum, or worse to computer packages.
4. More qualitative and numerical methods. The limited presentation of numerical methods in elementary mathematics courses is a puzzle to most biologists. Taylor's theorem should be exploited early and often in demonstrating numerical solutions of differential equations and optimization problems. Matrices are regularly taught by the presentation of important numerical algorithms. This might be another good reason for having some linear algebra in our calculus course: as a good example.
5. More applications of probability and statistics. Least squares rarely appear in calculus courses although it may be more appropriate there than in a good statistics course.

There is no disagreement on the value of more biological illustrations in calculus courses. We simply need to generate more catalogs of appropriate problems like the "Mathematical Models in Biology" complied by the MAA some years ago. These could enrich the calculus course greatly, since the life sciences at the current time depend on numerous realistic, yet simple models that may be more appropriate to elementary calculus courses than are those from physics or engineering.

There should be better communication between life sciences and mathematics faculty at a college to provide this input. Perhaps team teaching of the calculus course or specialty sections or adjunct courses in separate disciplines could present the biological relevance of the calculus. Convincing students of the relevance requires more than adding biology illustrations to a mathematics course. Biologists must increase the use of mathematics in their own courses.

In general, changes which broaden the range of topics and deal with real problems, perhaps in qualitative or numerical ways if they are the only ones available, would be desirable for biologists. In return we must understand that as mathematics becomes more central to the study and advancement of the life sciences, we cannot relegate it to the remains of the biology curriculum, with a priority lower than chemistry or physics.

One year of college mathematics is simply not sufficient, not for the study of organs, cells, organisms or populations and communities of organisms, and certainly not for the study and practice of medicine. Biologists will not be satisfied with the calculus curriculum, or the mathematical programs that biologists, to the dismay of many mathematicians, lump under the name of calculus, until enough time is allowed for it to be taught properly.

WILLIAM BOSSERT is Professor of Science, with teaching responsibilities in biology and applied sciences at Harvard University. His areas of research work include function of the mammalian kidney and rapid evolution in animal populations. In the past year he designed an advanced calculus course for biology concentrators. He received a Ph.D. degree in applied mathematics from Harvard University.

WILLIAM G. CHINN is Professor Emeritus of Mathematics at the City College of San Francisco. During 1973-1977 he served on the U.S. Commission on Mathematical Instruction and in 1981-1982 he was Second Vice President of the Mathematical Association of America.

Calculus for Business and Social Science Students

First Discussion Session

Dagobert L. Brito and Donald Y. Goldberg

Mathematics courses in American colleges and universities serve a diversity of students, programs, and interests and are provided by a diverse collection of institutions of higher education. Their traditional functions have been to provide a set of conceptual and computational skills and to serve as a screen for the allocation of scarce positions in various programs, both at the undergraduate and graduate level. Advances in computing technology as well as development of new mathematics have raised questions about the appropriate choice of concepts and skills which fulfill the first role. Changes in the composition of the undergraduate student body have raised questions about the second.

Diversity

Programs in business and economics vary widely to meet the needs of students who have a diversity of mathematical preparation and career aspirations. Some undergraduate business programs are open to all students; others find it necessary to limit admission. Many degree requirements include a calculus course; some require success in a calculus course (or sequence of courses) for admission to the program.

Graduate management school curricula, in the tradition of the "case method," have little or no need for formal mathematics. However, case method schools, and even a few law schools, may require a calculus course for admission. Other curricula are more quantitative in their focus and require a higher level of mathematical sophistication. These quantitative schools have more stringent mathematical requirements such as optimization theory and econometrics, which may use differential equations, linear algebra, and other advanced mathematics.

The demand for mathematical training by economics programs varies similarly. The traditional economics major may require a calculus course for the bachelors degree but very often calculus is not a prerequisite for any specific courses in the economics department. Students who plan to do graduate work in economics, or to enter a quantitatively oriented business school, are encouraged to study a substantial amount of mathematics, often equivalent to a strong mathematics minor or even a mathematics major.

Other undergraduate economics majors, for whom the degree is terminal or who plan to attend law school or a non-quantitative business school, find no need to take mathematics beyond the minimum required. A substantial number of students enroll in economics courses as part of their general education or as a requirement for another major; therefore, economics departments are reluctant to impose a calculus prerequisite for most courses. There is, however, a consensus that training in calculus is very useful.

In other areas of social science, some statistical methods courses are calculus-based; others are not. Other social science and business courses which use discrete mathematical techniques may require a prerequisite mathematics course, but not necessarily calculus.

Functions of Calculus Courses

One difficulty encountered by the discussion group was identifying the variety of interrelated functions of the calculus courses. One function, a systemic one, is as a screening mechanism for admission into degree programs. Other functions identified in the discussion are to teach particular calculation skills, to inculcate mathematical maturity, and to provide significant experience with mathematical models.

The screening function is often resented by instructors of business calculus or calculus for social science courses. Yet with this burden come obvious benefits to mathematics departments: large enrollments in the "short" calculus can be "bread and butter" for many departments. Use of calculus—or indeed of discrete mathematics, of principles of economics, or other courses—as an allocation mechanism is not necessarily inappropriate.

Participants in the discussion, principally from state and community colleges, recognized the screening function of their courses. Broader social questions of the appropriateness of using courses as specific screening devices and implications for gender and ethnic diversity in the professions were not raised in the discussion.

Basic calculus skills—much of the algebra and geometry prerequisite, manipulation of functions and simple differentiation and integration—were seen as essential to the calculus course itself. The advent of the supercalculators certainly has lessened the need for extensive technical competence, especially for students taking a "short" calculus course; yet the group agreed that hands-on pencil-and-paper experience with some calculus techniques is a necessary condition for students to understand the fundamental concepts. For quantitatively-oriented students in the social sciences, the traditional mainstream calculus course can be expected to provide in the future, as it has in the past, foundations for further study of mathematics.

The development of some mathematical maturity in students was identified by many participants as a key function of calculus courses in which management students and others enroll. One participant referred to the development of a "reading knowledge" of mathematics; as an example, understanding the notion of a differential equation, even without the techniques to solve one, was viewed as a worthy goal of an introductory course.

Another example was that a glimpse—even without mastery—of an important and subtle mathematical structure can aid the development of mathematical maturity. Understanding the nature of mathematical abstraction and developing confidence in using the concepts of functions were viewed by others as important steps in the ability to use mathematics profitably.

The one function of the calculus course—indeed of any introductory mathematics course—which was most enthusiastically endorsed by the group was the development of skills and understanding in mathematical modelling. It may be unclear whether "modelling" is an identifiable skill to be taught, but all agreed that it is essential for students to be able to represent particular problems—in business, economics, or other disciplines—in the language of mathematics. It was clear to all that the increasing power and availability of calculators and computers, whatever their implications for skills instruction, heighten the need for student capability in modelling.

Why Calculus?

One question frequently asked by students, social scientists, and many mathematicians, is "Why calculus?" As noted above, some business and social science programs require a calculus course for the degree but not as a prerequisite for any course. One participant, from a state college, helped develop a course tailored to the needs of her institution's business program: the demand was not for calculus but for a mastery of the notion of numerical functions, particularly exponentials and logarithms, sequences, matrix algebra, and introductions to linear programming and probability.

William Lucas, of the Claremont Colleges and NSF, rejected the notion of calculus as "the mathematics of change." Continuous change, yes—but discrete changes, especially those of human beings and human organizations may be modelled using new discrete mathematics, much of it being developed outside of mathematics departments.

Gordon Prichett of Babson described the Quantitative Methods course at his institution, an undergraduate business college. The course begins with a discrete approach to fundamental algebra skills in the context of linear modelling. Difference equations, differential equations, and some calculus techniques are considered; a follow-up course introduces some statistical concepts, using the integral. The Babson course exploits classroom computers, linear programming packages, and symbolic manipulations; another notable feature is the requirement of homework collected daily.

For the Record

The discussion group, with few exceptions, was composed of college and high school mathematics faculty. As "providers" rather than "users" of calculus courses, the group sensed a need to learn from social scientists what, in fact, their students require in their mathematics courses. Some unresolved questions and comments follow:

- Has there been significant research to determine which mathematics skills and concepts are used in business and social science programs?
- Will calculus reform efforts be directed—in term of grants—to social science calculus courses, or only to the mainstream course for physical science students? One participant suggested that grant funds be allocated to development areas on the basis of student enrollments. (It was noted, in this regard, that the mainstream calculus mass may be too large to move; change may be more likely on the periphery.)

The discussion group participants urged calculus reform leaders to acknowledge the importance of the non-physical science course in general education, in preparing students for non-science professions, and in educating many of the nation's future leaders.

DAGOBERT L. BRITO, Peterkin Professor of Political Economy at Rice University, has written and edited works on arms races and the theory of conflict, on public finance, and on the economics of epidemiology. Formerly he was Chairman and Professor of Economics at Tulane University. He received his Ph.D. from Rice University.

DONALD Y. GOLDBERG is Assistant Professor of Mathematics at Occidental College. His research interests include algebraic coding theory and combinatorics. He has participated in reading Advanced Placement Calculus examinations since 1980. He received his Ph.D. in mathematics at Dartmouth College.

Second Discussion Session

Gerald Egerer and Raymond J. Cannon

The discussants were primarily drawn from the field of mathematics. A variety of viewpoints were presented, reflecting the diverse nature of the participants' institutions and their personal experiences.

It quickly became clear that there is not *one* ideal calculus course for business and social science students, since the mathematics requirements at represented institutions varies from one to four semesters. There are also wide differences among student backgrounds, mathematical preparation, and motivation. However, participants did agree on the importance of a wide range of issues.

Mathematics departments need to establish more fruitful collaboration with other disciplines. Suggested points of departure include interdisciplinary seminars, joint teaching of courses, consultation regarding course prerequisites, and the revision of the content of courses of mutual interest.

Participants further agreed that mathematical techniques should be routinely and extensively used in economics courses so that the formal mathematical requirements become genuine prerequisites. This is because economics is becoming as mathematical as physics, as a cursory perusal of the major journals makes clear. The need for mathematics in business studies, while not so intense, remains nonetheless real, e.g., inventory control, production theory, operations research, statistics and probability.

It was generally felt that the mathematics community should not be overly defensive in its view of the need for a minimum level of mathematical sophistication on the part of business students. Some of the benefits accruing to these students as a result of such courses include increased analytical ability and hence greater acceptability by graduate schools or improved career opportunities. Furthermore mathematics courses can provide students with an important opportunity to construct, articulate, and interpret formal models in the context of their own discipline.

Nonetheless, there was a lack of consensus as to how best to serve the needs of these business students. Should, for example, their calculus courses begin with a review of needed algebraic techniques or with a discussion of rates of change, with the algebra to be covered as needed? Should the mathematics be presented in a concrete context (such as the theory of the firm) or should it be presented more "purely" so that its universality is stressed? (The value of UMAP modules as a source of applications was noted by several participants.)

Some concern was expressed about the general atmosphere in which learning takes place. While society is broadly appreciative of the results of technology, it is at best indifferent to the underlying basic scientific activity.

Discussion then turned to the effect which computers and calculators are likely to have on classroom instruction. Questions were raised as to whether "black box" technology (for integration, as an example) is an acceptable substitute for understanding analytical concepts and procedures. What is the relationship between developing computational skills on the one hand and conceptual understanding on the other?

Finally, concern was expressed that instructors, because they feel uneasy discussing applications outside their own field, might omit a certain amount of otherwise instructive and helpful material.

GERALD EGERER is Professor of Economics at Sonoma State University. He has worked both as a government and as a corporation economist in London, and thereafter as an academic economist at several universities in the United States. He received a doctorate from the University of Lyons, France.

RAYMOND J. CANNON is Professor of Mathematics at Baylor University. He has served on the faculty of the University of North Carolina at Chapel Hill, where he received an award for inspirational teaching of undergraduates. He is actively involved with the Advanced Placement program in calculus. He received a Ph.D. degree from Tulane University.

Calculus for Computing Science Students

Paul Young and Marjory Blumenthal

According to Hamblen [1], over 25,000 bachelors degrees were awarded in 1983 to students who majored in some form of computer science, information science, or computing technology. Hamblen estimates that this total will have reached over 50,000 students per year by 1988. This is about twice the combined total for mathematics, statistics, chemistry, and physics.

Of these 50,000 students, there are no reliable estimates of how many are in technically-oriented programs for computer science and computer engineering, which typically require calculus as a prerequisite or at least as a corequisite. Nevertheless, accrediting organizations for undergraduate programs in both computer science and computer engineering require calculus as a part of accredited programs.

In addition, a variety of recommendations, by various ACM committees have included calculus as part of the recommended course sequence for computer science students. Thus, it can be expected that in the immediate future there will continue to be be significant demand from computer science students for calculus as a "service" course.

In spite of this, it is sometimes difficult to pinpoint the exact calculus topics which are valuable for completion of the typical computer science program. Obviously students taking numerical analysis must be well-grounded in most of the topics covered in introductory calculus and linear algebra courses. But numerical analysis is not a required portion of all undergraduate (or even graduate) programs in computer science.

Students taking courses in simulation and performance evaluation must similarly be well-trained in the analytic methods required for series analysis, statistics, and queueing theory. But, again, performance evaluation and modeling courses are not required of all computer science undergraduates. Students taking courses in computer graphics and image analysis must be prepared to handle the underlying analytic techniques for analysis of two- and three-dimensional bodies in space. And generally, as computer science becomes more experimental, it can be expected that increasingly sophisticated statistical techniques will be employed in computer science, and these should depend on knowledge of the underlying analytical techniques.

While it is possible for many computer science undergraduates to pass through their entire undergraduate curriculum seeing little or no use of analytic techniques, and to use no such techniques when employed after graduation, academic computer scientists generally see enough analytic applications, and potential for more such applications, that they are reluctant to allow undergraduates to complete undergraduate programs in computing with no exposure to analytic methods.

Furthermore, students planning graduate careers in computer science are more likely to see the application of such techniques in simulation and modeling, in performance evaluation, in image processing and graphics, in statistics, and occasionally in analysis of algorithms. Hence, most of the better programs in computer science now require calculus. This requirement is reinforced by the common belief that calculus should be part of the universal culture for all scientists and engineers.

Finally, many computer scientists regard their discipline as requiring the same sort of mathematical skills and abilities as required for mathematics generally. Hence any course which enhances, or checks, students' "mathematical maturity" is often welcomed.

Still, it is safe to say that means of the analytic techniques (e.g., solving differential equations) taught in the current standard calculus sequence have little or no direct bearing on the core areas of computer science, including programming languages and compilers, software and operating systems, architecture and hardware, and perhaps even theory and analysis of algorithms. (In the latter area students occasionally use integration to place upper- or lower-bounds on the running times of algorithms, and sometimes use generating functions to produce solutions to recurrence relations which arise naturally in the analysis of algorithms.)

Purpose of Calculus

Given the above, it follows that computer scientists are *not* looking for particular collections of applications skills for computer science students taking calculus. Fundamental understanding of the underlying concepts is what is important for those students.

Thus, what is important in calculus for computer scientists is mastery of fundamental concepts, the ability to perform basic symbolic manipulations, and above all the ability to use analytic models in real applications, to know what analytic techniques are applicable, and understanding how to use them. "Plug and chug" for fast applications in some particular application domain is seldom, if ever, useful for computer science students in these programs.

This analysis does not imply that current approaches to teaching calculus are satisfactory for computer science students, but it does imply that what may be good for computer science students may well be good for quite a variety of other students as well:

1. Students should be taught, and must understand underlying concepts and intuitions.
2. Students should be given problems that involve modeling real-world problems analytically and solving them. Analytic, closed-form solutions should be stressed when appropriate, but numerical approximations should also be explored so that students gain an understanding both of the relationship of discrete, approximate solutions to continuous processes and of the notion of modeling large finite processes using continuous models. Integration, differentiation, and series summations all provide examples where the interplay between the discrete and the continuous should be apparent.
3. Series summations also provides an example where there is an opportunity to teach induction, recursion, and closed-form solutions of finite summations in the context of the calculus sequence. Recursion and its relation to induction is so fundamental to computer science that these topics must be taught in calculus courses designed to meet the needs of computer science students.

Using Computers

The panelists also discussed the role of programming and of computer aids in teaching calculus. It was generally believed that it was impractical, and indeed objectionable, to require programming ability as a prerequisite to calculus. It was felt not only that programming cannot reasonably be required of all students taking the course, but that in fact packaged mathematical software will often be the only computational technique that will ever be needed by many students. (Editorial comment: The panel did not discuss what could be achieved if an introductory computer science course were required as a prerequisite to the calculus.)

In view of the fact that calculus is seldom required as a prerequisite to any standard undergraduate computer science course except numerical analysis, this seems unfortunate. It is surely true that a calculus course taught to students who are proficient in programming could be designed so that it simultaneously enhanced the students' understanding of the relationship between analysis and computing and their understanding of basic mathematical principles.

While use of certain software packages (e.g., those which display convergence of rectilinear approximations to the area under a curve or convergence of tangential approximations to a derivative) may provide useful visual tools in helping all students understand analytical methods, such tools clearly do not address underlying issues connecting analysis to computer science.

Discrete Mathematics

Although calculus—not discrete mathematics—was the topic for panel discussion, it was clear from the general discussion that, with respect to computer science, the problem of teaching discrete mathematics for computer science majors was of more concern to the panelists than the problems of teaching calculus. Computer scientists typically believe that discrete mathematics, including mathematical logic at various levels, elementary set theory, at least introductory graph theory, and above all combinatorics at all levels are more important to computer science than the calculus course.

Those mathematics departments that have failed to offer such courses for computer science students have frequently found their computer science departments teaching these courses. Indeed, some computer scientists believe that such courses are better taught by computer scientists who are more familiar with the applications, or equally typically, many computer scientists believe that the more elementary topics often covered in discrete mathematics courses are best done directly in the context of the computing applications, where the underlying concepts (say of graphs, sets, and propositional logic) are both easily motivated and simply defined.

One view of elementary discrete mathematics supported by the panelists was that much of this material, e.g., elementary graph theory, propositional logic, elementary set theory, etc., might be better taught in high schools rather than in the universities, and might more reasonably be taught in the schools than the calculus. One way to encourage such a development might be for mathematicians to develop an advanced placement test in discrete mathematics.

While some panelists questioned the ability of the mathematics community to influence the schools and the Educational Testing Service, others pointed out that 30,000 university mathematicians having two professional societies with faculty who had to read advanced placement tests should be influential in getting such a reform instituted in the schools and the Educational Testing Service.

It was also believed that placing the more elementary parts of discrete mathematics in the schools, and emphasizing the relationship between discrete and continuous mathematics in the calculus sequence, (e.g., in the study of limits of sequences) would not satisfy computer scientists' need for training in more difficult topics in discrete mathematics, but there was little agreement on what was important.

One view expressed was that the appropriate discrete mathematics for mathematicians to teach for computer scientists is the analysis of (nonanalytic) algorithms. But computer scientists like to teach this themselves, regarding it as integral and basic to computer science. What is needed is help in essentially mathematical topics, including probability theory and statistics, combinatorics, and mathematical logic.

There was little belief among the panelists that this sort of discrete mathematics could be integrated into the calculus curriculum, but there was recognition that some of it, for example elements of statistics and perhaps generating functions, could at least be introduced in the calculus sequence, so that students would at least understand the applicability of analytical methods to these areas when they meet these topics later in their careers as computer scientists. In the long run, mathematicians should be alert to the possibility that discrete mathematics may replace calculus as a basic mathematical prerequisite for computer science students.

A Final Suggestion

The panel also discussed, not entirely successfully, the question of finding suitable computer science examples outside of numerical methods to integrate into the elementary calculus curriculum. The difficulty here, unlike for more traditional fields like physics, economics, and some fields of engineering, reflects both the panelists' inexperience with various subfields of computer science, and the fact that applications are scattered throughout various subfields of computer science which are often not part of the core of the discipline.

Nevertheless, there was general agreement that computer scientists, students from most other client disciplines, and indeed prospective mathematicians are not well served by calculus courses which do not successfully integrate real problems into the course. Word problems, problems from client disciplines, and problems that require students to think about the meaning of their solutions are all important in calculus courses, and generally not well treated in current courses.

To help with this problem, it was suggested that a problem bank with motivating problems from a variety of client disciplines could be developed. Problems could be tailored and classified by client discipline, by relevance to standard course topics, and by degree of difficulty. Ideally, a data base of such problems could be maintained and distributed "on-line" via computer network. Such a resource could even be continually updated. One way to launch such a data base would be through a special panel, perhaps federally-funded. The panel could generate potential problems and examples, discuss their suggested problems and examples with colleagues from the client disciplines, and revise them based on feedback from colleagues and students.

Reference

[1.] Hamblen, John. *Computer Manpower, Supply and Demand by States.* Quad Data Corporation (Tallahassee, FL 32316), 1984.

PAUL YOUNG is Chair of the Computer Science Department at the University of Washington in Seattle. He is Vice-Chair of the Computing Research Board and has served as Chair of the National Science Foundation's Advisory Subcommittee for Computer Science. He is an editor for several research journals and publishes regularly in theoretical computer science. He received his Ph.D. degree in mathematics from the Massachusetts Institute of Technology.

MARJORY BLUMENTHAL is the Staff Director for the NAS/NRC Computer Science Technology Board. She has researched and written about computer technology and applications. Formerly with General Electric and the Congressional Office of Technology Assessment, she did her graduate work in economics and policy analysis at Harvard.

Encouraging Success by Minority Students

Rogers J. Newman and Eileen L. Poiani

If minority students are to participate in Calculus for a New Century, then all those involved in and responsible for its instruction must be concerned about what is happening in the present decade.

The large and increasing minority population (in this report, "minority" is defined to be Black, Hispanic, or Native American) in the United States offers a critical national resource for the mathematical sciences. Demographic data show that by the late 1990's, one-third of American elementary school students will be of Hispanic origin.

At the same time, the proportion of minority students entering and persisting in higher education is slipping. According to data from the American Council on Education Office of Minority Concerns, between 1980 and 1984, Black undergraduate enrollment declined by 4%, while Black enrollment at the graduate level fell by 12%. During this period, Hispanic enrollment in higher education increased by 12% while American Indians/Alaskan natives experienced a 1% decline. More than half (54%) of the Hispanics enrolled in higher education attended two-year institutions. At the same time, the proportion of Black and Hispanic full-time collegiate faculty remained about the same (4% and less than 2%, respectively).

Among the measures of "success" in mathematics as perceived by the public is performance on the Scholastic Aptitude Test and the American College Testing Program. Students from most minority groups showed improved scores in 1987 when compared with those in 1985.

Fundamental Issues

To launch a discussion of encouraging success in mathematics by minority students, the following issues were raised:

1. There is no noticeable difference in the mathematical performance of minority students when compared with majority students with comparable mathematical background. The differences are caused by disparities in background preparation.
2. Many teachers have low level of expectations for minority student achievement.
3. Teachers, counselors and school administrators provide inadequate encouragement for minority students to pursue solid mathematics courses, such as Algebra I and II, as well as additional senior-level mathematics courses.
4. Small numbers of minority students aspire to careers in mathematics or related fields, both on the undergraduate and the graduate levels. Indeed, there has been a decline in enrollment of minority students in the calculus, and consequently among mathematics majors and in all other programs leading to professional careers in the mathematical sciences. There is a real need to attract high achieving minority students to mathematics-related majors.
5. Teachers need to focus on minority student strengths and to emphasize the need for academic discipline. Students must be encouraged to work hard. Most low income minority students do not have the family support system or tradition of exposure to higher education necessary to promote success.
6. Students must be able to read well in order to make full use of textbooks and notes.
7. We need to identify mathematically-talented students at an early age and nurture them through elementary and secondary school.

Pre-College Preparation

The group discussed several areas in which work should be done to address these issues. We began with a list of activities to improve pre-college mathematical preparation of minority students:

- Develop rapport with school systems for the purpose of encouraging more minority students to take at least two years of algebra and at least one advanced mathematics course in the senior year.

- Provide enrichment opportunities, especially for students from low income backgrounds, including cultural programs and field trips to science centers, technical industries, and college campuses.
- Develop summer enrichment programs for minority students by collaborative initiatives among colleges and universities, school systems, and funding agencies.
- Share pedagogical ideas about teaching mathematics in a way that makes it more inspiring and appealing and makes the image of mathematics more positive.
- Use television and related media to promote mathematics. For example, perhaps Bill Cosby could be persuaded to include a mathematician, or at least people using mathematics, in the scripts of this programs.
- Mathematics professional organizations should cooperate to support initiatives which identify and nurture talented minority students in grades K-12.

Collegiate Students

The major part of our discussion concerned collegiate-level mathematics. Everyone agreed that it is essential to maintain standards of quality and high level of expectations for all students in mathematics courses. In addition, it is especially important to identify and make use of minority student strengths while teaching mathematics.

When students have difficulty in mathematics, it is sometimes (but not always) caused by poor preparation. Here are some suggestions for encouraging success in calculus among students who enter college with poor preparation in mathematics:

- Provide the opportunity for personal attention so that instructors get to know each student as an individual and can follow his or her progress in coursework. This is easier to achieve in small class settings; it will require innovative approaches if the large lecture-recitation model cannot be modified.
- Enforce regular class attendance to complement the continuity of the calculus content. Motivation must be reinforced among poorly-prepared students.
- If "Desk-Top Calculus" is to become the norm, financial resources will be needed to enable low income students, many of whom are minority students, to acquire personal computers.

When students with good preparation in mathematics fall behind in calculus, it is usually for lack of motivation. Here are some suggestions for ways to promote success among such students:

- Expose students to reasons why calculus in its present and future forms serves as a critical entree into majors that lead to higher paying technical and professional careers.
- Integrate the resources of the personal computer with the teaching and learning of calculus.
- Invite guest lecturers (e.g., successful alumni/alumnae, or colleagues from client disciplines) to address the usefulness of calculus in other fields. (Note particularly the resources of the BAM and WAM programs of the Mathematical Association of America.)
- Provide a network of "mathematics mentors" to help motivate those who need support and to guide those already committed to the mathematical pipeline.
- Develop interactive approaches to the teaching and learning of calculus.

Adult Learners

Adult learners form an increasing percentage of those studying calculus. They bring to the classroom an intellectual maturity and curiosity which demands more depth of understanding about the relationship between mathematics and the real world than most textbooks provide. New materials must be developed to meet this growing and important need.

The percentage of adult learners is especially high in two-year colleges, a group of institutions that was not specially addressed at the Colloquium. Perhaps it should have been. Many graduates of two-year colleges often do not pursue higher education, so calculus becomes their last mathematics course. The needs of these institutions and their students require consideration well beyond the brief time that we were able to provide.

Increasing the involvement and success of minority students in calculus for non-mathematics major programs as well as for the mathematics major should be a national priority. Otherwise our nation will miss the opportunity to make full use of the potential of all of its students.

To encourage continued discussion of these important issues, our group recommends a national conference of educators and mathematicians at all levels to discuss issues surrounding the encouragement of minority success in mathematics.

ROGERS J. NEWMAN is Professor of Mathematics at Southern University at Baton Rouge, where he served as department chair for twelve years. He is President of the National Association of Mathematicians and served as Dean of the College of Sciences and Humanities at Alabama State

University. He received a Ph.D. degree from the University of Michigan.

EILEEN L. POIANI is Professor of Mathematics at Saint Peter's College and Assistant to the President for Planning. A former Governor of the New Jersey Section of the Mathematical Association of America, she is founding National Director of the Women and Mathematics (WAM) program. She is currently President of Pi Mu Epsilon and chairs the U.S. Commission on Mathematical Instruction of the National Research Council. Dr. Poiani received her Ph.D. from Rutgers University.

Role of Teaching Assistants

Bettye Anne Case and Allan C. Cochran

The discussion group was concerned with the use of teaching assistants (TA's) in a "lean and lively" calculus. Unsure of the exact nature of such a course, we made an attempt to consider some of the major issues of calculus instruction by TA's and other non-regular faculty, and to project possible effects of change.

Realities of the current situation, like a bad back, will continue to plague us whether or not we change the content or teaching of calculus. These grim realities include a lack of money, poorly prepared and scarce faculty, inadequate facilities, inadequate preparation of students, tension between research, teaching and service demands, and more.

Much of the discussion of this group reflected the problems expressed in *Teaching Assistants and Part-time Instructors: A Challenge* [1]. The Foreword to that work began the discussion and seems appropriate here:

> The dilemma of the beginning professor in our publish-fast-or-perish academic world is whether to devote time almost entirely to research or to put effort in teaching. Graduate teaching assistants must walk, even more, a thin line to acquit their teaching duties effectively and responsibly and enjoy teaching while efficiently pursuing studies and research. Discovering how to help them is our goal.
>
> In most mathematics departments there is a serious effort to help graduate students teach effectively, but the number of regular faculty members involved in this effort is necessarily small. Those involved faculty members may not find much agreement among their colleagues on these matters, and they find even less information on helpful activities. Compounding the problem is the current diversity of college teachers who are not in the professorial ranks. Both graduate departments with graduate teaching assistants and two- and four-year college departments may have part-time and temporary teachers. In graduate departments they teach the same level courses as graduate teaching assistants and are drawn from graduate students in other disciplines, undergraduates, moonlighters who are employed in government, industry, high schools or other colleges, and those who are sometimes called "gypsy scholars."

The role of teaching assistants (and other non-regular faculty) in the teaching of calculus (lean and lively or not) is of serious concern. A recent survey of colleges conducted by the Mathematical Association of America Survey Committee shows the percentage distribution by type of faculty for all single-instructor sections [2]:

	Mainstream			Non-Mainstream	
Faculty Type:	I	II	III	I	II
FT Professor	70%	73%	82%	47%	45%
FT Instructor	9%	14%	10%	13%	12%
Part-time	6%	4%	3%	13%	20%
Teach. Assist.	15%	9%	5%	25%	23%

More teaching by non-professorial faculty is indicated when the survey population is restricted to departments having graduate programs [3]:

Individual Calculus Classes (≤ 65) Taught By:

	Tenure Track	Other Fulltime: PhD	Non-PhD	TA	Pt-time Instr.
GP1-2	46%	2%	5%	41%	6%
GP 3	61%	1%	7%	23%	7%
MA-TA	74%	2%	11%	3%	10%
MA	73%	6%	8%	0%	13%

The categories GP1-2 and GP 3 stand for institutions with doctoral programs classified according to [4]. MA-TA stands for masters-granting departments with teaching assistants, while MA stands for masters-granting departments without teaching assistants.

Many of the views expressed in this discussion group reflect a wide range of constraints on the participant schools, including many shared problems:

- It might be easier to effect change with teaching assistants than across the "hard-core" regular faculty. At the least, TA's would not present unsolvable problems. One reason is that only the best teaching assistants teach calculus.
- The exploitation of "gypsy scholar" and "permanent visitor" instructors is deplorable on both ethical and humanistic grounds. No one had realistic solutions for this problem.
- There is tension between academic freedom and "what should be taught" which may be heightened by changes in the calculus.
- There will be a lot of expense in training people for a new mode. A TA present in our group described current computer applications in his calculus class; his supervising professor pointed out that this TA only teaches one section to allow time for preparation. A 50% drop in productivity would be difficult (impossible) to handle in many departments.
- There were no good ideas for how to retrain faculty.
- Hand-held calculators are feasible, but real computer experiences for all calculus students would not be feasible in a short or medium time frame at most schools.
- The need for gradual changes, for pilot sections and programs, was repeatedly mentioned. (Examples: Syracuse tried computer use in calculus in honors sections first; Clemson has three sections of calculus using the HP28C—furnished this one time at reduced cost.)
- The need for sharing and keeping resource files from successful classroom experiences was mentioned as a great help in producing good instruction.
- A large amount of time is needed for preparation, and there is a need for lead time in preparing demonstrations.

Discussion of departmental "coping devices" produced a variety of anecdotes and opinions: classes over size 60 (but under 100) having a grader but no recitations were often felt, given a suitable room, to work well. Lecture-recitation mode—100 to 700 students—was considered as a last resort by most participants. However, one member of the group related a case where a change to smaller sections brought more complaints and less enrollment. Subsequently, a change back to the lecturer-recitation mode was made.

Another "coping device," uniform departmental tests and syllabi, were blamed as the cause of "chug-and-plug" courses. Some felt that poor student performances over the years forced an easier course. The suggestion was made that, in fact, the major coping device—the common syllabus—did not go far enough in prescribing word problems and simple proofs.

The single recurrent theme of this discussion group was that teaching by teaching assistants and part-time instructors, although a very significant problem at schools with graduate programs, would not be *the*, or even *a*, major hindrance to changes in the teaching or content of calculus. A generally positive attitude was held concerning teaching by these people, along with concern for the dual demands of being a student and being a teacher.

Reference

[1] Case, Bettye Anne, *et al.* "Teaching Assistants and Part-time Instructors." MAA Notes, 1987.

[2] Anderson, R.D. and Loftsgaarden, Don O. "Preliminary Results from a Special Calculus Survey." MAA Survey Committee.

[3] Case, Bettye Anne, *et al.* Analysis of partial returns from the "1987 Survey of the Committee on Teaching Assistants and Part-time Instructors." Mathematical Association of America, to appear.

[4] Rung, Donald. "CEEP Data Reports: New Classification of Graduate Departments." *Notices of Amer. Math. Soc.* 30:4 (1983) 392-393.

BETTYE ANNE CASE is Associate Professor of Mathematics at Florida State University where she directs the teaching assistants. She chairs the CUPM Subcommittee on the Undergraduate Major and is Director of the Project on Teaching Assistants and Part-time Instructors of the Mathematical Association of America. She received a Ph.D. degree in mathematics from the University of Alabama.

ALLAN C. COCHRAN is Professor and Vice Chairman of Mathematical Sciences at the University of Arkansas at Fayetteville, where he directs the instructional program. A Past Chairman of the Oklahoma-Arkansas Section of the Mathematical Association of America, he received a Ph.D. degree in mathematics from the University of Oklahoma.

Computer Algebra Systems

First Discussion Session

John W. Kenelly and Robert C. Eslinger

The discussion group participants represented a variety of experiences with computer algebra systems (CAS). They ranged from individuals with no first-hand personal experience to authors of some of the current systems. At least one-third of the participants were currently engaged in curricular activities using CAS. (A list of the projects represented in the group is included at the end of this report.)

Amidst a lively discussion, the group reached agreement on several major issues. These included recognition of several forces for change in the approach to calculus instruction. Future students will be entering college with increasing experience with technology, in particular with graphing calculators and microcomputers. With this experience they will tend to see much of the current instructional material as being out-of-date, and to some extend irrelevant in today's environment.

Pressure for change will be exerted, many believe, by faculty in allied departments. They will see the content of the current calculus course as inappropriate for their students who operate in a computer-oriented atmosphere. In addition to these technological forces, the group felt we need to be sensitive to the pedagogical implications of research in cognitive psychology.

Many in the group were concerned about skepticism of their mathematical colleagues toward the use of CAS in calculus instruction. It was noted that the mathematical community was, in general, very conservative in their choice of textbooks and reluctant to supplement textual material with their own notes or assignments. This reluctance is incongruent with the accepted need for experimental text material to support change in calculus instruction. Unfortunately, there was no agreement on the mechanisms to develop these materials.

This discussion group benefitted especially from the contributions of several calculus text authors. In particular, the authors noted that the content of current texts reflects market pressures of conservative faculty interests in including specific topics. In this context, the group agreed that they could not select specific algebraic skills, for example, that would be rendered obsolete by the use of CAS.

Black Box Syndrome

There were also issues on which participants voiced disagreement. A primary point of dispute could be characterized as the "black box syndrome." Some participants were concerned that students would use mechanical systems without understanding underlying concepts. Others argued that these systems allow students to concentrate on higher-level reasoning instead of on lower-level manipulative skills.

For example, on-going projects were described in which solutions to word problems were broken down into appropriate steps by students using CAS. It was also pointed out that these systems facilitate student exploration, thereby leading to discovery learning.

Disagreement also arose over the complexity of the user interface with CAS. Some felt that the difficulty in learning these systems was sufficient to inhibit their use by many faculty and students. It was argued that faculty were reluctant to abandon traditional approaches to teaching for uncertain results with a system that required a substantial investment in time in order to master. There was also concern that students' mathematical maturity upon entering calculus was inadequate to use these systems productively. In contrast, participants who were using CAS indicated that initial training in its use was not a significant barrier.

In summary, participants in this CAS discussion group concluded that the mathematical community is in the experimental phase of developing effective use of CAS in the teaching of calculus. Some expressed frustration with the lack of a clear statement either identifying problems with current practice in calculus instruction or outlining objectives for change. However, the following major issues were identified by the group:

- Student preparation for calculus.
- Balance between conceptual understanding and calculation.
- Relevance of calculus to allied fields.
- Utilization of current technologies.
- Resistance to change.
- Evaluation of experimental methodologies.

CAS Projects

Tryg Ager, Stanford University: Building an on-line calculus instructional system which uses REDUCE as an "algebra engine" and interactive graphics as an expository device. The project is directed toward precollege calculus, but covers all material in the AP calculus syllabus.

George Andrews (with Michael Henle), Oberlin College: Supplementary materials for calculus sections using Maple. Each exercise is designed to fit into the traditional curriculum and provide a deeper understanding of some aspect of the calculus.

Dwayne Cameron, Old Rochester Regional School District: Using muMath in honors algebra II, precalculus, and calculus classes to a limited degree.

J. Douglas Child, Rollins College: Macintosh Interface to Maple. Experimental lab course to better understand (and improve) student problem-solving processes.

Joel Cohen, University of Denver: Teaching a junior-senior level course in symbol manipulation for applied mathematicians, physicists, and engineers, using a number of computer algebra systems; also experimenting some with Maple in calculus.

Abdollah Darai, Western Illinois University: Using MACSYMA for the first time in a first-year calculus course.

Franklin Demana (with Bert Waits), Ohio State University: Developing a precalculus text that makes significant use of graphing calculators or computer-based graphics.

John S. Devitt, University of Saskatchewan: Using Maple as an electronic blackboard.

Harley Flanders, University of Michigan: Developing software specifically for teaching calculus (and precalculus). Includes symbolic manipulation, plane and space graphics. Classroom testing.

Richard P. Goblirsch, College of St. Thomas: Have offered calculus with numerical emphasis using computers; have begun experiments with SMP.

Edward L. Green, Virginia Polytechnic Institute: Employ IBM PC in calculus sequence; have partially developed graphics packages.

Don Hancock, Pepperdine University: Developing material for a calculus course using MACSYMA. The curriculum will emphasize the interplay between discrete and continuous ideas.

Alan Heckenbach, Iowa State University: Using Flanders' Microcalc for Calculus; using muMath in Master of School Mathematics Program.

M. Kathleen Heid, Pennsylvania State University: Projects using symbol manipulation programs:

1. Algebra with Computers. Continuing and pending NSF projects using muMath and programming in high school algebra; examines the numeric, graphical, and symbol-manipulative connections.
2. Computer-based general mathematics, using muMath for 10th and 11th grade students.
3. Applied calculus course using muMath and graphical programs to refocus course on applications and concepts. (Report in upcoming issue of *Journal for Research in Mathematics Education.*)
4. Algebra I with muMath, focuses on word problems using the symbol manipulator to perform routine procedures prior to student mastery.

James F. Hurley, University of Connecticut: Computer laboratory course using TrueBasic to graph and do numerical experimentation directly tied to concepts of calculus, and to motivate them.

John Kenelly, Clemson University:

1. Freshman calculus section covering regular departmental syllabus. Students are issued individual Hewlett-Packard HP 28C calculators.
2. State grant for 160 secondary mathematics teachers to receive and be trained in the use of Sharp EL5200 super-scientific graphic calculator. Study of curriculum implications of graphing calculator technology are part of the course.

Andrew Sterrett, Denison University: Using Maple in two sections of calculus.

JOHN W. KENELLY is Alumni Professor of Mathematical Sciences at Clemson University. He has been Chair of the College Boards' Advanced Placement Calculus Committee and he now chairs their Academic Affairs Council and is Director of the Advanced Placement Reading for the Educational Testing Service. A member of the Board of Governors of the Mathematical Association of America, he also chairs the MAA Committee on Placement Examinations. He received his Ph.D. degree from the University of Florida.

ROBERT C. ESLINGER is Associate Professor of Mathematics at Hendrix College. Until this year he served as Chairman of the Department of Mathematics and Head of the Division of Natural Sciences. He was Governor of the Oklahoma-Arkansas Section of the Mathematical Association of America and currently serves as a national Councillor of Pi Mu Epsilon. He received a Ph.D. in mathematics from Emory University.

Second Discussion Session

Paul Zorn and Steven S. Viktora

The discussion section began with a simulated demonstration of the computer algebra system SMP. Next, a large number of participants described computer-oriented calculus projects—some involving CAS—at a variety of institutions, large and small.

Computer Algebra Systems (CAS) already affect the teaching of calculus and will likely play an even greater role in the future. The defining feature of CAS is a symbol manipulator. Most systems do arithmetic, operations with polynomials, linear algebra, differential equations, calculus, abstract algebra, and graphs. The group discussed how CAS are used now and what effect they may have on the teaching of calculus in the future.

A surprisingly large number of colleges and universities already use computers to aid in teaching calculus. Most approaches seem to fall into one of two categories. Some institutions use what might be called special purpose packages. Individual programs, whether locally or commercially produced, are typically used only for certain topics, not for the whole calculus course.

Other institutions use "full service" CAS programs for a wide range of tasks. A common characteristic of these courses is an emphasis on problem solving. Exploratory projects, realistic applications, and modeling are stressed.

After participants described computer-oriented projects at their own institutions, discussion turned to more general issues: problems, strategies, and prospects for the future. Four main themes and issues emerged:

1. *The need to define—and agree upon—goals and objectives for calculus instruction.* No clear consensus exists on what calculus is or should be. Perfect agreement is impossible, but it should be possible to agree upon a small finite number of acceptable calculus courses.

Among the questions: Which particular traditional calculus skills and topics are really essential? More generally, what balance should be struck between routine algorithm performance—differentiation, antidifferentiation, etc.—and "higher-order" activities—problem-solving, mathematical experimentation, and understanding of concepts? Shall we proceed conservatively (tinkering with the present standard course) or radically (designing a new course from the ground up)?

There was no systematic effort to answer all of these questions. It was agreed, however, that future calculus courses *will* differ significantly, not just in details, from present courses. Developments in computing will force such changes.

2. *How will computer algebra systems and other forms of computing drive, or be driven by, changes in calculus instruction?* Discussion centered on both curricular and pedagogical matters. Among curricular effects of computing, the following were cited:

- Computers handle *routine* operations; hence, time can be spent on better things.
- More realistic applications are possible when computations are cheap.
- Approximation and error analysis are important and useful dimensions of calculus, but they are computationally expensive. Cheap, easy computing solves this problem.
- Some traditional topics and methods arose historically from the high (human) cost of computation. Cheap machine computation renders them obsolete.
- Computing *may* save time in the curriculum. On the other hand, present calculus syllabi are crowded—it would be a mistake simply to add more material.
- Whether we like it or not, the calculus curriculum will change. Hand-held machines, if nothing else, will force this; too many traditional exercises become inane when performed by computer.

The effects on calculus pedagogy of modern computing are just beginning to be felt, and so are hard to predict confidently. They might include:

- A more active, experimental attitude of students toward mathematics, supported by less painful manipulations. Conjectures can be made and understood as part of the process of mathematical proof.
- With computers, mathematical objects can be represented graphically and numerically as well as algebraically. This should lead to a deeper, more flexible understanding of "function," for example.
- Computers permit a larger sheer number of examples and exercises to be worked out. This could speed students' development of mathematical intuition.
- Computers could support more *qualitative* reasoning in mathematics. E.g., students could see concretely that polynomials grow more slowly than exponential functions.
- With computing to handle details, students can carry out multi-step problems without foundering in calculations. This should foster more effective problem-solving.

3. *New approaches to calculus require new materials: books, software, problem sets, etc.* Who will write them, and with what rewards? How will re-invention of the wheel be avoided? Many participants reported finding standard text and exercise materials unsuitable for computer-aided courses. Writing suitable material is difficult and time-consuming. It was agreed that cooperation is essential in developing and sharing problem materials. NSF was mentioned as a possible source of support for distributing materials outside the usual commercial channels. Meetings of people involved in computer projects would also help spread the word, and build morale. The *Computer Algebra Systems in Education Newsletter,* based at Colby College, is a start.

4. *It is widely agreed that calculus courses should become more conceptual. Clearly, computing can help support this goal. But what problems do such changes raise, and how will we solve them?* CAS's can do much of what we traditionally teach. If CAS's do handle such tasks, what will students do? Will their algebra skills atrophy, or perhaps, aided by better motivation and more varied experience, even improve? How will more conceptual matters be *tested?* Can a more conceptual calculus be taught successfully in large, bureaucratic settings?

No agreement emerged on such questions. It was suggested, though, that we mathematicians should not apologize for asking for more resources to solve the problems attendant upon improving calculus instruction, any more than our natural science colleagues do for requiring *their* version of laboratory resources.

The group was unable to agree on, or even discuss at length, every issue raised. Some open questions raised during the discussion follow:

1. How will the calculus syllabus be affected by CAS? What topics should stay, and which should go? What should be added?
2. How will changes in college calculus affect the whole precollege curriculum, not just the Advanced Placement program?
3. What is the "right" machine (if there is one)?
4. What makes a "friendly" CAS? How do we ensure that CAS are easy for students to use?
5. Should programming be taught? If so, how much?
6. Where do hand-held calculators fit in? Present calculators are not yet as powerful as CAS, but they are relatively inexpensive and convenient.

PAUL ZORN, Associate Professor of Mathematics at St. Olaf College, is currently a visiting professor at Purdue University. He is co-leader of a project, supported by the National Science Foundation and the Department of Education, to integrate numerical, graphical, and symbolic computing into mathematics. Dr. Zorn received a Ph.D. degree from the University of Washington.

STEVEN S. VIKTORA is Chair of the Mathematics Department at Kenwood Academy, Chicago. Previously he taught in teacher-training colleges in Ghana for five years. He served as President of the Metropolitan Mathematics Club of Chicago, and is currently an author for the University of Chicago School Mathematics Project. He received an M.A.T. in mathematics from the University of Chicago.

Objectives, Teaching, and Assessment

First Discussion Session

Alphonse Buccino & George Rosenstein, Jr.

This group unanimously agreed that calculus teaching needs improvement. Opinions varied on the existence and degree to crisis. Despite the differences there was clear enthusiasm regarding the need for the "Calculus for a New Century" program.

Objectives

The objectives of the calculus sequence must be reformulated within the context of the objectives for the undergraduate mathematics major and of undergraduate education generally. Within this framework, we identified several categories of objectives, including:

- Transcendent purposes: communicating the power, excitement and beauty of calculus.
- Understanding: calculus as a tool for modeling reality.
- Skill: symbolic manipulation.
- Development of faculties: geometric intuition and deductive skill.
- Rite of passage: calculus as a talent filter for mathematics majors and students in other disciplines.
- Generic skills: writing, reading, and note-taking.

There was strong agreement that the higher-order cognitive objectives of insight and understanding should be emphasized.

These higher-order objectives are strongly related to skill development and manipulation. In any reexamination of the objectives of calculus, the relationship between objectives at different cognitive levels must be emphasized. Additionally, the responsibility of mathematics and mathematicians toward the other disciplines that utilize calculus must be considered and clarified. Finally, calculus courses need to be fit into a context that includes goals for majors and for general education, as well as the preparation of students for calculus.

Teaching

The participants agreed that whatever objectives might emerge for a particular course, students ought to be cognizant of them. Daily interaction (e.g., quizzes, "board problems," graded homework) provide valuable information for the student about the objectives of the courses and the teacher's expectations, and for the teacher about the progress of the class and of individuals. Feedback in this form must not be merely routine drill.

Instructors are often too distant from their students and often are unaware of serious problems or significant successes individuals may be experiencing. Consequently, a climate of sensitive awareness with substantial feedback should characterize calculus classes. Regular accountability and close scanning to assess status and progress should occur.

There was some discussion of the role of textbooks in teaching calculus. To a large extent, books define both the content and style of presentation in our standard courses. In particular, textbook problems seem to stress drill and template applications, often relegating "interesting" problems, should they occur, to the end of a list of forty or fifty problems.

Although courses appear to depend on the textbook, it was noted that class time is frequently spent repeating the material in the text. Many believed that this repetition is peculiar to mathematics. Some participants reported that students would read the book if the instructor made his or her expectations clear and endeavored to enforce them. Participants generally agreed that teaching students to read the text was worthwhile.

Writing was another generic skill that teachers believed was worthy of class time. Writing for clarity enhanced students' higher-order thinking skills, forced students to consider their results, and increased students' appreciation for the problems of writing for the benefit of others. Participants mentioned journals and summaries as post-learning activities and brief "write everything you know about ..." exercises as introductions to topics.

Assessment

In discussing assessment, the group focused on testing to assess student progress and achievement. There is a substantial need for consciousness raising, faculty development, and technical assistance in the construction and administration of good tests based on current knowledge about assessment.

However, there is also a substantial need for research and development on assessment and testing to advance the state-of-the-art, especially with reference to the higher-order cognitive objectives.

Although some members of the group questioned the accuracy of the profile of present tests as 90% manipulation and template problems, there was general agreement that tests frequently do not ask students to perform higher-order conceptual tasks. With some prodding, the group produced several kinds of questions that seemed to require more than routine skills.

Assessment of transcendent objectives probably could not be fit into a grading scale, but should be part of the evaluation of any course. These objectives do not fit most teaching evaluation procedures either, nor, apparently do some other higher-level objectives. Several members of the group expressed the view that these evaluations thereby interfered with true teaching effectiveness.

What To Do

One major omission from the Colloquium program was discussion of research and development on the teaching and learning of calculus. Development of a program providing support for coordinated research projects and activities is essential. In addition, the value of natural experiments that continually occur to improve calculus teaching and learning can be enhanced through support of such things as clearing houses, communication, reporting, and travel. Research and development should include investigation of experiences and practices in other nations.

Dialogue with colleagues in other disciplines should occur. This can be a grass-roots effort involving such simple things as brown-bag lunches. Efforts pyramiding upward to an integrated and synthesized perspective on calculus and its relation to other disciplines are also needed.

Course and classroom testing is a major problem for all college teaching, and not just calculus. Instructors simply are not trained in even the rudiments of test construction and administration. Each campus should address this problem and provide necessary technical assistance to improve the situation.

Participants saw advanced technology as an opportunity for re-examining the objectives of the calculus. Because advanced calculators and recent computer software trivialize many of the computational skills on which many courses seem to be based, we will soon be forced to confront the inadequacies of these courses.

ALPHONSE BUCCINO is Dean of the College of Education at the University of Georgia. Previously he served in several management positions in the Directorate of Science Education at the National Science Foundation, and as Chairman of the Mathematics Department at DePaul University. He received a Ph.D. in mathematics from the University of Chicago.

GEORGE M. ROSENSTEIN, JR., is Professor of Mathematics at Franklin and Marshall College. He received his Ph.D. from Duke University in topology. His current research interest is in the history of American mathematics. He is presently the Chief Reader for the Advanced Placement Examination in mathematics.

Second Discussion Session

Philip C. Curtis, Jr. and Robert A. Northcutt

Our group discussed these three topics from two perspectives: proposed changes in orientation of the calculus course from one devoted primarily to technique to one devoted to conceptual understanding, and increasing utilization of computing technology in instruction both in calculus and in preparatory subjects.

There was general agreement that a change of emphasis in the teaching of calculus is a desirable goal—a change from a technique-oriented course with applications to one where the major emphasis was on understanding major concepts. There was no consensus, however, on the extent to which this new goal could be separated from the current goal of technical mastery.

If emphasis on concepts is to be achieved, however, there are several necessary ingredients which are not now present. First, there has to be general agreement on the part of the teaching staff that this change is desirable and that individual efforts in this direction should form an important part of a faculty member's professional life. Teaching techniques would then need much more attention and support than is now the case.

Secondly, the orientation of the textbooks must change to include much more of an emphasis in both text and exercises on the understanding of concepts and their applications to problems. Central to this success will be high-quality feedback from the instructor, teaching assistant, and homework reader. Understanding and communication of ideas on the part of the student will not be achieved without this feedback.

The similarity to the problems encountered in the effective teaching of writing is inescapable here. Although technique-oriented questions will probably not disappear from examinations, concept-oriented questions must form a more important part than is now the case. Both students and faculty should be well aware that this is the orientation of the course and of its assessment.

It is clear that this change in orientation will not occur quickly; it will be difficult and mistakes will be made. It should be realized from the outset that if this change is to be successful it will be an evolutionary development rather than a revolutionary one.

It is inescapable that technology in the form of increased computing power will play an increasingly important role in the teaching of calculus. It already does to some extent and this role will increase rapidly. The challenge will be to manage the role of technology in a way that deepens the mathematical understanding and capabilities of students rather than just replace what they now do by paper and pencil.

The opportunities are many. Understanding the function concept should be greatly enhanced by the graphical capabilities of the new calculators. Derivatives as rates of change and areas as limits of sums can be easily visualized where this was only imperfectly realized before. More realistic problems can be confronted where parameters are not carefully chosen so that solutions can be given in closed form.

The role calculators will play in the elimination of symbolic calculation at this point is not as clear. Algebra is the language in which mathematical problems are posed. Correctly stating the problem, organizing it in such a way that it can be attacked, and finding a route to a solution are always the more difficult parts of problem solving. It is certainly conventional wisdom that at this juncture technical mastery forms an important part of insight. The challenge will be to utilize the technology in such a way that will preserve and deepen the insight, but lessen the burden of tedious algebraic and numerical calculation.

Our challenge will be to manage the changes proposed, rather than to let the many forces outside mathematics dictate the directions. Historically, we have not done a good job in either the implementation of new directions or the necessary follow-up. "New math" is the classic example. The necessity to carefully plan, realistically implement, and train teachers are major obstacles.

Training faculty through better apprentice/mentor programs, giving appropriate rewards for successful teaching, and increasing recognition of the professional aspect of calculus will help. Bad practices in class presentation, assessment, and materials must be eliminated. The mathematics profession must take an active role in this management process. Calculus is a part of the over-all fabric of mathematics and as such, instruction prior to calculus and after calculus must reflect an awareness of what is happening in the calculus, and why.

Many students do not now develop an overview essential to understanding calculus. Technology should allow instructors to broaden objectives in this area. The function concept, change, qualitative behavior, and global insights can become more accessible for students. Problems in background review, remediation, and practice can be addressed using calculators and computers, but this alone will not solve the problems of apathy, motivation, or attitudes. Students also must bear responsibility for the process of their learning. Good teaching requires both an active teacher and an active student.

The introduction of technology into the calculus classroom will not be devoid of problems. Technology has a good side in its utilization, but misuse can polarize many of our present faculty, instill a false expectation in problem solving, and may give rise to a host of materials so specialized as to make faculty resistant to innovation.

Finally, there should be a conscientious effort on the part of the mathematics profession to involve secondary school teachers in the design and in the implementation of any proposed changes. It should be the goal of all concerned to narrow the spectrum of preparation of students for calculus to an acceptable and accessible range. In the future, an increasing number of the better students will be receiving their single variable calculus instruction in the secondary schools. Consequently, a clearer understanding of the desired goals of calculus instruction is essential at all levels.

PHILIP C. CURTIS, JR. is Professor of Mathematics at the University of California at Los Angeles. He has been active for several years in the University of California's state-wide programs on admissions and articulation with the schools. A former Fulbright fellow, Dr. Curtis held several visiting faculty positions and served as chair of his department. He received a Ph.D. degree from Yale University.

ROBERT NORTHCUTT is Professor of Mathematics at Southwest Texas State University. Formerly Chairman of his department, he is the Texas Section Governor of the Mathematical Association of America. He received his Ph.D. degree from the University of Texas at Austin.

Third Discussion Session

Lida K. Barrett and Elizabeth J. Teles

It is no secret that calculus, as it is now taught, is not understood or appreciated by most students. The mathematical power and beauty of calculus that mathematicians know and relish cannot be delivered to students in the context of our present calculus course. Our group identified this problem as *the* crucial issue in calculus.

There were basically two parts to our discussion. In the beginning part, the group identified five issues of importance. The first centered around the fact that students do not understand or appreciate calculus, but see it as a selection of rules and problems which they must memorize in order to be allowed to proceed. Calculus is therefore seen as a filter, or a rite of passage. According to Robert White, it is time for this to end: calculus must become a pump, not a filter. In order for it to become this pump, mathematicians must market calculus as *they* know it both to students and to client disciplines.

The second issue is that the content of calculus is seen as rigidly controlled by a large number of forces. Among these are client disciplines, accreditation boards, departmental expectations, and textbooks. Accurate current information on the needs of client disciplines might yield a different perspective, perhaps even support for more teaching of concepts and less manipulations. The actual specifications of accreditation boards should be determined. Faculty instead of the textbook should determine the syllabus and the course.

A third constraint is the quality of instruction: teaching by increasing numbers of part-time faculty and graduate students; the inability of the faculty to communicate with students; and the insistence of most universities that all tenured faculty carry on research. It was generally agreed that institutional rewards of rank and salary, in most settings, are greater for mathematical research than for quality teaching, curriculum development, or research on teaching.

In addition, international teaching assistants assigned to recitation sections often fail to communicate with students, not only because of their accents but also because of their inability to comprehend the underlying meaning of the questions. Lastly, even though calculus teachers often give lip service to the problem-solving approach, it is often not the technique actually used in the classroom.

Articulation among all levels of mathematics is a very important issue since success in calculus depends on the student's previous mathematical experience. In order that calculus be taught for understanding and appreciation, precalculus and other high school courses must support this endeavor. Of particular importance is articulation among high schools, two-year colleges, and four-year colleges and universities on the calculus course itself.

Finally, the role of technology in the Calculus for a New Century was explored. It was readily agreed that technology should play an important role but that its use should be carefully constrained. There is just as much danger in the improper use of technology as there is in ignoring it altogether. Technology must be a facilitator rather than an appendage.

To address these issues will require changes first of all in the reward structure for faculty, supported by administrations and departments. Changes will also be needed in teaching methods brought about in part by the new technology. Other new methods could include greater use of a historical perspective, more student motivation for problems by ensuring that they understand the question before they are asked to respond, and the teaching of differentiation and integration rather than the manipulative skills of differentiating and integrating. Furthermore, long-term commitment to teaching must be encouraged. The primary resource needed—faculty time—must be made available and faculty effort must be rewarded.

In the second half of the session, the group explored objectives, teaching, and assessment. Many suggestions were discussed in the short time available, and consensus was reached in many areas.

Students must be taught how to think about new problems as well as how to solve template problems. An alliance must be built with client disciplines to aid not only in the development of courses but in the development of support for courses. Teaching style and presentation of material must convey enthusiasm for knowledge and an appreciation of the beauty of calculus.

To support better teaching, the group as a whole recommended a newsletter-type publication that would contain course outlines, teaching techniques, reports of experiments, and examples of new types of problems (e.g., open-ended problems and problems that use multiple techniques).

Of great importance to good teaching is a well-organized assessment program. Assessment was seen as needed in a variety of ways—assessment of students as they enter courses (i.e., placement, high enough as well as low enough), assessment of curriculum, assessment of faculty and teaching, and assessment of the accomplishments of the course. Whether a course meets its objectives is often difficult to determine. New courses, however, must be assessed by new standards, and not by standards established for old courses.

The primary objective must be marketing of the new calculus as a gateway to future study in the majority of other disciplines. Some skills needed in client disciplines are changing, contact with these disciplines must be established in order to appropriately reflect their changes in our courses. However, it is up to mathematicians to convince others that a lean and lively calculus will meet their needs. It is important at this time that mathematicians become proactive rather than just reactive.

LIDA K. BARRETT is Dean of the College of Arts and Sciences at Mississippi State University. Previously, she was Associate Provost at Northern Illinois University and head of the Department of Mathematics at the University of Tennessee. President-Elect of the Mathematical Association of America, Dr. Barrett received a Ph.D. from the University of Pennsylvania.

ELIZABETH J. TELES is Associate Professor of Mathematics at Montgomery College, Takoma Park. She is Chairman of Maryland-Virginia-DC Section of the Mathematical Association of America and currently is completing a Ph.D. in Mathematics Education at the University of Maryland. She has published papers on assessment testing, on use of computers in mathematics, and on gifted middle school students.

ISSUES

PARTICIPANT RESPONSES

<u>Question</u> What are the principal weaknesses in calculus instruction?

<u>Answers</u>

- The low priority given by the "mathematical community" ... One doesn't get tenure, promotion, or prestige by being concerned about freshman mathematics.
- "Black box" instruction where the student is the "black box".
- The attempt to make it all things to all people
- Too many students taking calculus
- Collusion between instructors and students to cover up what they don't know
- Impossibly diverse audience
- I hate current textbooks!
- The curriculum — outdated, obsolete
- The BC level AP exam — a senile, cumbersome mass whose time has long since passed ...

Calculus Reform: Is It Needed? Is It Possible?

Gina Bari Kolata

New York Times

Calculus, the "gateway to all areas of science and engineering," in the National Science Foundation's words, and the mathematics course most dreaded by hundreds of thousands of undergraduate students, is changing. Mathematicians agree that because of advances in computer technology, calculus cannot remain the same course that it has always been. And there is a growing feeling within the mathematics and science community that the course is desperately in need of revitalization. So the real question is whether calculus will change haphazardly or whether its alterations can be planned and its subject matter and teaching invigorated.

The National Academy of Sciences and the National Science Foundation hope to initiate a national debate over how calculus can and should be changed. In order to start the debate, the National Science Foundation would like to invest about $2 million a year for the next several years in conferences, workshops, and demonstration projects. The program began with a Colloquium on the future of calculus teaching, held in Washington D.C. on October 28-29, 1987. The colloquium is supported by the Sloan Foundation and sponsored by the National Research Council of the National Academy and the Mathematics Association of America.

An Unprecedented Effort

The new effort to revitalize calculus is almost unprecedented in educational circles. It would affect mathematics departments, all of which teach calculus. In fact, calculus is the overwhelmingly dominant mathematics course taught in colleges. It would affect the more than half a million students who enroll in calculus courses each year. It would affect textbook publishers, who invest substantial resources developing and promoting their calculus books, and it would affect other science departments who use the high failure rate in calculus courses as a means to eliminate weaker students and who frequently design their own courses around what students should have learned in calculus.

The revitalization of calculus, however, is most emphatically not another instance of a "new math"—the ill-fated attempt to change the teaching of mathematics at the elementary level. Bernard Madison, a mathematician from the University of Arkansas and calculus project director for the National Research Council, says that this effort will be carefully coordinated and will involve everyone whose life will be changed by a new calculus course.

Mathematicians and educators have learned their lesson from the new math.

Mathematicians and educators have learned their lesson from the new math. Changes cannot be imposed on people and they have to occur gradually. "This is not just a curriculum issue," says Madison. The entire infrastructure of calculus must be changed. "We have to lay the groundwork for the project and give it the kind of visibility and prestige that is unquestionable. This is a political and social process."

Is Calculus Irrelevant?

The movement to reinvigorate calculus began several years ago with challenges to the very existence of calculus. Some mathematicians said that calculus is irrelevant for today's students. It should be replaced by discrete mathematics, the sort used in computer science.

But others argued strongly against this extreme view, saying that calculus is of central importance, although it may no longer be the course it should be. Several meetings were held, including one sponsored by the Sloan Foundation that took place at Tulane University in January of 1986. The Tulane conference resulted in a now well-known collection of papers called "Toward a Lean and Lively Calculus" and published by The Mathematical Association of America.

At the same time, the National Research Council through its Board on Mathematical Sciences and its Mathematical Sciences Education Board was examining the state of university mathematics in general, and finding it wanting. Madison explains: "I've been looking at various aspects of mathematics instruction since 1980, and I knew there was something wrong. Mathematicians in the early days thought they were just being mistreated and not given enough money." Fewer and fewer students were selecting mathematics as a major and even fewer were going on to get Ph.D.'s in mathematics. Yet more students than ever before were taking mathematics courses because other departments required them. If mathematics courses were truly exciting, more students might be lured to the department.

"We have a faculty that is relatively inactive as producing scholars. Most are primarily teachers," said Madison. Many are uninspired by their subject matter and fail to inspire their students. The National Research Council is now putting together an agenda to reinvigorate all of college and university mathematics and it considers the calculus project a special—and crucially important—case.

"We have a faculty that is relatively inactive as producing scholars. Most are primarily teachers."

Last fall, the NSF requested funds from Congress to support a revitalization of calculus and now the agency expects that Congress will appropriate those funds. Reformers believe that the groundwork is laid for a change. But no one expects that change will be easy and no one expects that the impact of a new calculus course will be confined to mathematics departments. Calculus is not just an isolated course. It is essential for other sciences, including physics, chemistry, biology, computer science, and engineering. Many colleges and universities require that students majoring in business, psychology, and other social sciences take it as well.

A Dominant Course

Of all the mathematics courses offered at universities, calculus is the most well known and most often taken. The best data on enrollment in college and university calculus courses derives from a survey conducted every five years under the auspices of the Conference Board of the Mathematical Sciences. According to Richard Anderson, a mathematician at Louisiana State University who directed the 1985 survey, more than 600,000 students took a calculus course at a college or university in the fall of 1985—the most recent data. If anything, even more students are taking it today. More than 40,000 instructors teach calculus. "It's an absolutely dominant course," Anderson says.

For many students, their grade in calculus will determine whether they go on to study mathematics or science. But, at least in large universities, fewer than half of all students who enroll in an introductory calculus course complete it with a grade of C or above. And many students who pass do so only after repeating the course. Ronald C. Douglas, a mathematician and dean of physical sciences at the State University of New York at Stony Brook, says that at his university as many as 20 to 25 percent of students in introductory calculus courses are taking the course for the second, third, or even fourth time.

Many students who pass do so only after repeating the course.

Because calculus is considered a difficult course, it frequently is used as a "filter"—other departments use it as a way to eliminate of less-than-stellar students. "Math departments often complain that physics or chemistry departments don't want to kick out the students so they figure they'll just send the students over to the math department. That will weed them out," says Ronald Graham, director of mathematics and statistics research at AT&T Bell Laboratories in Murray Hill, New Jersey. Any changes in calculus, then, will change departments' abilities to use the course in this way.

It is even said that medical schools use calculus grades to distinguish among their applicants. Douglas says that he has often heard from students who moan that if their grade in calculus is not changed to an A, their dreams of becoming a doctor will be for naught.

Those leading the movement to revitalize calculus are often asked, however, why they want the course to change and what, in fact, they mean by change. Calculus has been around for hundreds of years—it began with Isaac Newton and has been developed by some of the greatest mathematicians that ever lived. It describes such fundamental processes as motion in physics and diffusion in biology. Science students through the centuries have studied it. It is "a monument to the intellect," according to John Osborn of the University of Maryland. What is there to change?

"What a subject is as part of the discipline is often quite different from what it is as a part of education."

First of all, says Kenneth Hoffman of MIT, it is in a sense misleading to talk of changing calculus. "Each branch of mathematics is whatever it is and each branch is to some extent unchanging," Hoffman explains. Calculus is calculus and no one imagines changing the mathematics itself. However, Hoffman continues, "what a subject is as part of the discipline is often quite different from what it is as a part of education. What calculus is as a part of mathematics is quite different from what we present to 19-year-old kids. We have come to realize that the way we look at calculus from a disciplinary point of view is not necessarily the way we should look at it when we teach it."

So the real question for mathematicians and educators is: What, if anything, is wrong with the way calculus is being taught and what can be done to improve it?

Reasons for Change

The first point that the advocates of a new calculus make is that what is being taught as calculus today bears little resemblance to the course 20 years ago. One reason for the change is that the group of students taking calculus is different and the course has been adapted in response.

During the past 20 to 30 years, college education has become more commonplace. Now the masses, rather than just the wealthy and privileged, go to college, and their high school mathematics education frequently is woefully inadequate. To accommodate these students, calculus courses were made less demanding. No longer are students asked to do proofs, for example, only to work out simple problems that are exactly like worked-out examples in their textbooks.

Different departments at colleges and universities had their own reasons for wanting students to take calculus, and everyone has a say in the course material. "Every user of calculus got a word in and calculus became taught so that the average student could learn it," says Lynn Steen of St. Olaf College in Northfield, Minnesota. "There was a major change in the philosophy of textbooks that no one had planned."

"We no longer ask students to understand."

"We no longer ask students to understand," says Douglas. "Now it is manipulation, pure and simple," meaning that students are just plugging numbers or symbols into formulas. The calculus tests reflect this. "What you test is what the students learn," Douglas points out. "In fact, we have a great deal of difficulty using class time on anything else besides what will be on the tests and we have abandoned testing anything but manipulation."

Reducing Rote Learning

Nearly everyone who has thought about an agenda for a revised calculus concludes that the course must include a renewed emphasis on mathematical concepts and understanding. "When you use calculus, you can't imagine that anyone will ever give you a problem like one on an exam," Douglas says. "What you might encounter will be a conceptual underpinning or you might encounter a problem where the whole purpose would be to turn it into the kind of problem that occurs on an exam." Any student who just learned by rote to plug into formulas would be lost in the real world.

"But," Douglas continues, "it's even worse than that. We now have computers and even hand calculators that will solve these calculus problems. There are now hand calculators that would get a B in most calculus courses. What we're teaching is not only the wrong thing—in that it is not what students will use—what we're teaching is obsolete. It is like spending all your time in elementary school adding and subtracting and never being told what addition and subtraction are for."

Mathematicians who are working to revitalize calculus feel quite strongly that the rote learning must go and that students should learn to rely on computers and hand calculators to do routine calculations. In place of the time now spent working out problems by hand, students should learn concepts—what mathematics is all about.

Neglect of Teaching

But, the revisionists argue, more than just the subject matter of calculus must change. "There are links that we cannot deny and cannot ignore between what is taught, who teaches, and who is taught to," says Bernard Madison, of the National Research Council. And calculus teaching is also in drastic need of reform.

The crisis in calculus teaching began in the late 1960s, according to Douglas. "For a number of reasons, the resources, effort, and money put into teaching calculus have fallen over the past 15 years," he says. During the 1970s mathematics departments found themselves squeezed for funds at the same time as mathematics enrollments increased dramatically. Students enrolled in these courses in greater numbers, they were more likely to want to major in the physical, biological, or social sciences, and they were more likely to have to take calculus.

So, to deal with the influx of calculus students, universities and colleges increased the number of students in their classes. Now it is common for calculus classes to be taught in lecture sections of 250 or more students.

Once or twice a week, the students break up into recitation sections, of size 30 or 40 in many schools, where a teaching assistant goes over the work with them. "In some places, the recitation sections are now larger than the classes used to be," Douglas remarks.

"The resources, effort, and money put into teaching calculus have fallen over the past 15 years."

One of the problems with this method is that much of the teaching of calculus is relegated to teaching assistants. These graduate students may not be particularly interested in the course and may have difficulty communicating even if they try. As many as half of all teaching assistants are not U.S. citizens and they frequently have only a rudimentary command of English.

Teaching assistants, says Graham, often "are not natural teachers. Many are not even math majors." In other cases mathematics departments press undergraduate students into service. "I have seen sophomores teaching to freshmen," says Graham. But hard-pressed mathematics departments that have far too few faculty to handle the hoards that take calculus courses feel they have no choice but to use teaching assistants.

Even when a bona-fide Ph.D. mathematician teaches calculus, the course has been made so instructor-proof by textbook writers that there is almost no opportunity for the mathematician to introduce any concepts that might reflect the beauty and excitement of the field. "At big universities, they standardize things and you teach by a syllabus," says Robert Ellis of the University of Minnesota. "You don't need a mathematician to do this. It requires at most 5 minutes preparation to teach a class, and usually a mathematician can do it off the top of his head."

Calculus requires that you know the material from the entire high school mathematics curriculum ... that's not true for most students.

All too often, the mathematicians who teach calculus regard it as a boring burden. Typically, says Madison, "we will go to the classroom, teach section 3.4, go home, and mow the yard." Teaching calculus has become routine and mindless for the professors.

Although other departments have had some success with impersonal large lecture sections combined with smaller sections taught by teaching assistants, "at least in calculus, this did not work," Douglas says. Calculus is different—it cannot be taught in an impersonal environment, according to Douglas. "Suppose you come into another sort of course, say freshman psychology," Douglas says. "You start from scratch, you don't build on anything. And you don't need to have mastered what you learned one week to go on to the next week's material. But calculus requires that you know the material from the entire high school mathematics curriculum. Already, that's not true for most students. And the other problem with calculus is not just that it builds but that you have to keep up. If you did not do well on one test, the chances are you will not do well on the next. There is no other course on the freshman-sophomore level like it."

Personal Involvement

To do well in calculus, most students need the personal involvement of a professor or teaching assistant, according to Douglas. "They need to have homework assigned and they need to have homework collected and returned. If you are going to teach calculus well, you have to have someone reading papers, grading them, and returning them. That doesn't save money."

The way it is now, all too many students have no personal involvement. Homework may not be assigned, and even if it is assigned, it often is not collected. Students let the course slide, thinking they will learn the material at the last minute before an exam. "There is no feedback," Douglas says. "When students get the sense that no one is personally interested in what they are doing in the course, they put very little effort in. They put in what they think is the minimum work necessary, and, of course, they often judge wrong."

The revitalized calculus ... "will be much more difficult to teach and will require much better prepared teachers."

A new calculus, which emphasizes understanding and concepts and in which no student can slip by learning to plug into formulas by rote, will require more money for teaching staff and it also will require that the staff put more effort into teaching and more time into giving students feedback. The revitalized calculus, Madison points out, "will be much more difficult to teach and will require much better prepared teachers."

Yet some say that the new calculus might be effectively taught even with the resources now available and even with the cost-cutting large lecture sections. "There are ways to teach better in large sections," says Donald Bushaw, who is vice-provost for instruction at Washington State University. "It's just a matter of practical advice, making the best use of the lecture and small group method. There is a good deal of conventional wisdom that is not being followed everywhere." Douglas points out that if calculus were better taught, fewer students would have to repeat it and this would also save money.

Textbook Stagnation

Still another issue for the calculus reformers is to encourage the publication of a new kind of textbook. The current crop of textbooks would be all but useless for the new course. But developing new textbooks means going against an entrenched industry that seems very content to go on the way it always has. Fourteen new calculus books are scheduled to be published in the fall of 1988, according to Jeremiah Lyons, an editor at W.H. Freeman and Company, and none are innovative.

Lyons explains that the textbooks have evolved to satisfy so many constituencies that publishers have gone beyond the bounds of reason. "The books published now are really our own fault," Lyons says. "We have gone well beyond an adequate number of exercises and worked-out examples." In addition to what many feel are bloated and expensive books, the textbook publishers supply supplementary material to the instructors at no additional cost. This includes a solutions manual that consists of worked-out solutions to every problem in the text and computerized test banks—a computerized compendium of as many as 3000 questions so that instructors can easily give weekly quizzes, without having to make them up, and even have the quizzes printed for them.

Educators "talk a great game of innovation, but if we move one standard deviation from the mean, they don't use our books."

For publishers, innovation involves risk. Educators, says Lyons, "talk a great game of innovation, but if we move one standard deviation from the mean, they don't use our books." If publishers were to experiment with a new sort of calculus text, they would be wary of investing the usual amount of money in it. For example, says Lyons, to take a chance on a new sort of text, publishers would want to forego the usual 1100 pages of text by providing far fewer worked-out examples and exercises. They would like to save money on illustrations—the typical calculus book now has a $75,000 art budget—by using computer-generated art, simplifying the illustration program, and not using a second color. And they would like to get rid of the $50,000 to $100,000 budget for all supplementary materials for instructors. "If we lower the cost and reduce the basic investment, we can gamble more," Lyons says.

Links to Science Curricula

Yet even if the textbooks change, there is still the problem of the responses of other departments to a new calculus. Any changes in calculus will have a ripple effect in other sciences, particularly physics and engineering. Courses will have to be revamped and re-thought, which further complicates efforts to change the mathematics curriculum.

In physics, says Edward Redish, a physicist at the University of Maryland, "the physics curriculum is tied very, very closely to math." Physics students usually take calculus at the same time as they take introductory physics and the two courses are coordinated so that as students learn a technique in calculus, they use it in physics.

Finally, there must be some way to evaluate whether a revitalized calculus is, in fact, more effective. Mathematicians say that it is even difficult to evaluate the effectiveness of the cookbook, standardized calculus that is being taught today. No one has ever done a national survey to determine such basic things as failure rates, pass rates, or even how many hours a week students spend in calculus classes. "We need that kind of data," says Anderson, for without it it would be impossible to even think of comparing a new calculus to the old.

Among the issues that NSF and the calculus reformers would like to see debated on a national level are what sorts of changes in calculus would best suit other science departments and how those changes could be accomplished. So far, those who have thought about the issues have reached no consensus.

Conflicting Advice

Even when scientists from other departments do want to see calculus changed, they do not always agree on what changes should be made to better serve their majors. Redish would like to see more stress placed on approximation theory and on numerical solutions to differential equations. He would like to see a great deal more emphasis on methods of checking qualitatively to see whether a number that comes out of a computer is approximately what would be expected.

Biologist Simon Levin of Cornell University would like to see more emphasis on qualitative analysis and less on computations; he would especially like to see his students introduced to partial differential equations, particularly as they describe diffusion.

Gordon Prichett, dean of the faculty at Babson College, a business college just outside of Boston, says that although business students routinely take calculus, the course for them is, strictly speaking, unnecessary. So, he says, "if we are going to teach calculus in the business curriculum at all, it would be for breadth and for problem-solving ability."

Advocates of the Status Quo

Redish, Levin, and Prichett, however, want to see calculus changed. Not everyone does. Some argue that the course is a classic that has withstood the test of time and that is crucially important for the other sciences. Most physicists, according to Redish, are happy with the status quo. Redish believes that this system, in which physics courses are developed around the current calculus courses, distorts physics and cheats students of a feeling for what physics, to say nothing of mathematics, is all about. But, Redish cautions, "I am not typical. I'm out here at the edge trying to pull my immense department. Most don't want to think about change. 'If it ain't broke, don't fix it' is an attitude I see a lot."

Most don't want to think about change.

Anderson, who has presented the idea of revitalizing calculus to forums of engineers and other scientists, says that the engineers and scientists frequently were uninterested in change. "They wanted students to have the same math that they had had," Anderson says. "They thought of it as good training and they would say, 'What was good enough for me is good enough for my students'."

"What was good enough for me is good enough for my students."

Still others just do not want to be bothered. Calculus, they say, is a bread-and-butter course for mathematics departments—it is one of those courses that keep mathematics departments in business. But it is not really worth the time and trouble to radically change it. And besides, any student who cannot pass calculus as it is currently taught ought to re-think his or her plans to become a scientist anyway.

Others are concerned that if a committee starts tinkering with calculus, the course is likely to become worse—less useful and less meaty. Ellis of the University of Minnesota thinks that it would be impossible to really teach mathematics—as opposed to routine formulas—to poorly-prepared and poorly-motivated students. The new calculus would be a much harder course than the course that is taught now, and it is not clear what could be done about the even higher failure rates that might result. Ellis also notes that a true change in calculus will have to have widespread support. "You can't impose this from above. It has to be up to the professors," he says.

The new calculus would be a much harder course than the course that is taught now, and it is not clear what could be done about the even higher failure rates that might result.

Nelson Markley, the chairman of the University of Maryland's mathematics department, worries that if too much is done with a computer or hand calculator, the students will never really learn what the calculations mean. "I tend to think a great deal about the analogies between mathematics and music," he says. "You can't learn to play the piano by going to recitals. You can err in the direction of going too far from actual calculations." So far, the revitalization of calculus "is not something that I'm enthusiastic about," Markley says.

But, Madison and others point out, change is coming anyway with the increasing sophistication of computers and calculators that do calculus problems. "The truth is that things are going to change," Madison remarks. "It is just a matter of whether we want to control that change and make it happen in a positive way or whether we will let it happen haphazardly. The only excuse for the argument for no change is that controlling the change is impossible. I've never accepted that."

•

GINA BARI KOLATA is a science writer for *The New York Times*. Formerly, for more than a decade, she covered biology and mathematics for *Science*, the official journal of the American Association for the Advancement of Science.

Recent Innovations in Calculus Instruction

Barry A. Cipra
ST. OLAF COLLEGE

Good teaching doesn't come easily, and good teachers are rarely satisfied with the job they've done in the classroom. Even if calculus were not the linchpin of college mathematics, instructors would probably keep on tinkering with the course. But because of its prominence in the undergraduate curriculum, calculus is the focus of much educational innovation. This brief article will describe some recent efforts to improve the way that calculus is taught. Since an exhaustive survey would require several volumes, we only sample a few of many themes.

Let the Computer Do It

Much of the current innovation in calculus instruction is centered around use of computers. As computers have become cheaper, smaller, friendlier, and also more powerful, many people have begun exploring their possible applications to the calculus curriculum. The computer as tedium-reliever, as expert draftsman, as super blackboard, and even as teacher, are some of the possibilities being looked at.

There is widespread agreement that calculus courses tend to be excessively "technique-" and "skill-oriented," with corresponding agreement that calculus ought to be a more "concept- and "application-oriented" subject. "I would say the problem is that most calculus instruction focuses on computational details," says John Hosack of Colby College. That is, Hosack explains, a student can pass the course by carrying out standard algorithms with little understanding of ideas. "We think that calculus should reorient itself toward concepts and applications."

The hypothesis at Colby and at a number of other colleges is that Computer Algebra Systems, such as MACSYMA, Maple, or SMP, will render obsolete the computational emphasis of traditional calculus instruction. A typical Computer Algebra System can carry out all the routine steps of an algebra or calculus problem, including formal differentiation and integration—precisely the skills that are at the core of the current curriculum. Most of these systems also house numerical equation-solving routines and superb graphics capabilities, two more topics traditionally treated in calculus. There is little doubt but that a Computer Algebra System could do very well on a typical calculus exam; it would certainly make fewer mistakes than the students. Indeed, Herbert Wilf wrote in 1982 of one system (muMath), calling it "the disk with the college education."

A Computer Algebra System could do very well on a typical calculus exam; it would certainly make fewer mistakes than the students.

Colby College, the University of Waterloo, Oberlin College, St. Olaf College, Denison University, Harvey Mudd College, and Rollins College are among the schools that have received grants from the Alfred P. Sloan Foundation or the Fund for the Improvement of Post-Secondary Education (FIPSE) to experiment with Computer Algebra Systems in calculus. Some projects involve software development. Doug Child of Rollins College, for instance, is adapting Maple for use on the Macintosh and writing an interface for what he calls an "interactive textbook." But the main thrust is toward rethinking the calculus curriculum to take advantage of these powerful programs.

Two replacements have been suggested for the traditional computational emphasis in calculus. One is "exploratory computation," in which the student looks at a number of related examples, such as graphs of successive Taylor polynomial approximations to the sine function. The computer does all the unpleasant stuff, the point being to illustrate vividly some concept or—more ambitiously—to have the students discover concepts for themselves. According to David Smith of Duke University, by the time students had seen the 19th-degree approximation to sin x, they were asking if you couldn't just let the degree go to infinity. He adds that sequences and series went very smoothly thereafter.

Every teacher's dream is to have students who will play with the homework, modifying problems to see how the answers change, and posing the dangerous question *What if?* But this goal gets lost in the press of routine exercises. Moreover, translating that ideal into practice is not an easy thing to do. According to Paul Zorn of St. Olaf College, it's hard to convey what you mean by "experimental math," especially when students have preconceptions about the type of problems they are supposed to do. Zorn suggests easing the students into the practice. "One way to do it is to give progressively more open-ended problems." Some instructors are now optimistic that, with machines shouldering the algebraic burden, students will be more receptive to the idea of open-ended problems.

Every teacher's dream is to have students who will play ... modifying problems ... and posing the dangerous question What If?

The other suggested alternative to the traditional emphasis on technique is the inclusion of more realistic and complex applications—problems that are not artificially tailored for simplicity (textbook arc-length problems are frequently derided) but that require a number of separate steps for their complete solution. Bruce Char and Peter Ponzo of the University of Waterloo cite the development of such multi-step, multi-concept problems as a major component of a Sloan grant there. The idea, Ponzo says, is to test the student's ability to put together a "long-winded" solution to a problem. The computer will handle the mundane calculations; the student will concentrate on setting up the problem and deciding on the sequence of formal operations that will lead to the solution.

One of the material goals of the Waterloo project is to compile problems into a "symbolic calculation workbook." Ponzo and Char both acknowledge that this has turned out to be rather more difficult than they had anticipated. Ponzo originally thought they would get hundreds of problems, but says "finding problems of that ilk ain't easy." The goal now is several dozen, of which an initial batch of five was published last April.

Another goal of the Waterloo project is to create computerized tutorials in calculus and to develop "user-friendly" software that will allow individual instructors to create their own tutorials in a nonprogramming, text-editing environment. The Waterloo project uses Maple, which was developed at the University of Waterloo in the early 1980s. ("Maple" is not an acronym; it stems from Canada's national symbol, the Maple leaf.) Maple is considered one of the easiest of the Computer Algebra Systems to learn, but Char, one of the group who created Maple, says that future work will look at making Maple even more user-friendly. "You still have to go through at least an hour of training in order to use Maple," he says, noting that this is a barrier to some calculus teachers who feel they don't have the time.

Complexity and limited access are two factors limiting the use of Computer Algebra Systems in calculus instruction. Some faculty are themselves reluctant to learn the special grammar and vocabulary of the systems. Others are concerned with the amount of class time required to instruct students in use of the systems; an hour spent on computer syntax is an hour not spent on the integral.

"I don't even know how to log onto a computer," says Alvin White, at Harvey Mudd College, who has other, more serious objections as well. "The more computer power we have, the less the students know what they're doing .... The infatuation with computers moves the student further and further from thinking and creating. The promise we were given by the calculator people is that we can spend more time on the underlying ideas. But in my experience, the time is spent on showing them more buttons to push."

"The infatuation with computers moves the student further and further from thinking and creating."

Proponents of the Computer Algebra Systems, however, claim that explaining the systems does not require an inordinate amount of class time; students tend to pick up what they need to know either from handouts or from each other. Paul Zorn, who has taught an SMP-based calculus course for several semesters, says that he spends "maybe half a lecture right at the start" giving an overview of the system. A big surprise, he adds, has been "how little frustration students seem to experience using the computer." The paradigm of giving a command and getting an answer comes very quickly. "It doesn't seem to be a big distraction."

Wade Ellis, Jr. of West Valley College in San Jose echoes this observation. Ellis and a colleague, Ed Lodi, wrote a computer activity book called *Calculus Illustrated,* which they used last spring at West Valley College with a group of 15 students. Ellis describes the student reaction as being more or less "isn't this what we're supposed to do?"—a change from the "why do we have to use computers?" reaction of five years ago. "Now everyone knows they have to use computers," Ellis says.

A BASIC Disagreement

For most proponents of Computer Algebra Systems, an important ease-of-use feature has been the removal of programming as a requirement for use of the systems. Herb Greenberg of the University of Denver, which has developed its own instructional software called "Calctool," says that students should not be aware of programming, only of the mathematical applications. "We are not teaching programming, and the students are not doing programming," Greenberg says of the program there.

George Andrews of Oberlin College notes that about 20 years ago he participated in developing an experimental text called *Calculus: A Computer-Oriented Presentation,* which he tried in class but "bailed out" when he found that "the programming got in the way." David Smith of Duke University adds that enrollment at Duke in a supplementary "calculus and the computer" course, which involves programming, is way down, from 100 approximately five years ago to a recent class of 12. Smith feels that the dwindling enrollment may be due in part to students' reluctance to take on extra work.

Jim Baumgartner of Dartmouth College disagrees vehemently with this anti-programming outlook. He disputes the idea that programming detracts from the ideas of the course. "If you choose the problems correctly, they're simple, they're short, and they teach the essence of programming." Dartmouth emphasizes programming from the very beginning. "What we're doing is building a base for further down the line," Baumgartner says, adding that later courses assume that students can program in *TrueBASIC.* (BASIC is particularly popular at Dartmouth.)

Many students (Baumgartner estimates 70%) come to Dartmouth with Macintoshes, and these are networked throughout the campus. The mathematics department gives each student a *TrueBASIC* disk with short demo programs, the idea being to give them something simple that they can modify, together with problems and solutions.

James Hurley of the University of Connecticut also reports success with *TrueBASIC* programming in calculus there. "We give them something they can use later on," Hurley says. The programming component, however, is done only in special sections of "enhanced" calculus, which include a one-hour computer lab each week. Hurley adds that a big reason for their success is the use of *TrueBASIC,* to which they switched in 1985. The Computer Science department, he says, had complained about their use of old versions of BASIC, which often encourage "sloppy logic" and poor programming habits.

The programming/anti-programming "dispute" is largely a comparison of apples and oranges; presumably there is room for both in the calculus diet. Nevertheless, ease of use is a major selling point of the Computer Algebra Systems. "If I had an extra hour per week," says Paul Zorn, "I'd still use SMP rather than a language like *TrueBASIC,* and do something else."

The Super Calculator

Access may turn out to be a more serious limitation than complexity, at least until Sun work stations show up on discount at K-Mart. Complexity is primarily a software problem, and thus amenable to programming improvements; but access is essentially an economic problem having to do with the cost and number of the machines required to run the software. According to John Harvey of the University of Wisconsin, smaller colleges have an advantage in this regard over the larger state schools: accommodating a few hundred students on a Computer Algebra System is economically feasible; accommodating several thousand students is not.

Harvey, who is project director for a Texas Instruments-funded grant to develop calculator-based placement exams for the MAA, believes that powerful pocket calculators will be the dominant innovation at large schools. It's not unreasonable, he says, to ask a student to buy a $100 calculator for three semesters of calculus. Harvey anticipates having experimental sections of calculator-based calculus at Wisconsin by the fall of 1988.

Powerful pocket calculators will be the dominant innovation at large schools.

John Kenelly of Clemson University is a strong advocate of the new generation of pocket calculators, particularly the Casio 7000, which primarily does graphics, the Sharp 5200, and the Hewlitt Packard HP 28C, which does graphics and some symbolic manipulation. Kenelly, who has been called "the 28C salesman of the math community," had the HP 28C on secret loan from Hewlitt Packard for six months prior to its release in January, 1987. In the summer of 1987, Kenelly taught a three-week course to 16 teachers of advanced-placement high school calculus, funded by South Carolina. Participants were each given a Casio 7000 calculator, the purpose of the course being for the teachers to write curricula for use of the calculator in their AP courses.

Clemson is embarking on a pilot program to begin using the HP 28C in freshman and sophomore engineering mathematics courses, from first-semester calculus through differential equations, matrix algebra, and engineering statistics. Hewlitt Packard is loaning Clemson some one hundred 28C's, which the mathematics department will in turn issue each semester to students who are signed up for the sections that will experiment with use of the calculator.

In 1987-88, six faculty members will teach one course each, "playing" with the 28C at appropriate "moments of opportunity," according to Kenelly. These same six hope to spend the summer of 1988 rewriting the syllabi for their courses to incorporate the 28C in a substantial way, and then teach experimental sections from these syllabi the following year. The summer of 1989 is then projected for special early training sessions for TA's and other faculty members. In the fall of 1989, all sections of engineering calculus will be using the 28C.

Kenelly observes that the faculty is excited by the project: "We have a colleague or two who think we're going to bring up a bunch of button-pushing deadheads, but the bulk of the community is behind it."

Kenelly adds that Clemson also has a grant for the coming year to train local high school teachers—he anticipates a total of 160 participants—in curricular uses of the Sharp 5200 graphics calculator. (This is Kenelly's third calculator, and he predicts—safely enough—that the electronics industry will keep coming out with more. The 5200 currently sells for about $100, and Kenelly foresees the price coming down to around $50.) Kenelly describes the 5200 as "very friendly," especially in its matrix capabilities; he says it will invert a 10x10 matrix in about 15 seconds. It also has a "solve"" key for finding zeros of functions, which Kenelly considers a "must" these days.

The Super Blackboard

In a curiously opposite extreme from the economically small pocket calculator, the computer as "super blackboard" is another technological innovation that holds promise for large universities as well as small colleges. Much of what an instructor writes in chalk on a blackboard—especially graphs—could just as easily be done on a computer screen, if only the screen were large enough to be seen by more than the first two rows of students. But special equipment—either new attachments that feed the computer screen's output into an overhead projector or large monitors posted around the classroom—can overcome that handicap. Using such displays an instructor can, for instance, draw an instant and accurate graph of the sine function on, say, the interval $[-\pi, \pi]$, and superimpose graphs of the first several Taylor approximations.

David Smith and David Kraines of Duke University are among the proponents of the computer as a super blackboard. With funds from Duke and the Pew Memorial Trust, they have equipped two classrooms with computers and display equipment (monitors). Smith and Kraines have written about their project in an article for *The College Mathematics Journal.* Their article is partly about the technical aspects of installation and display (and also about security from theft), but it also addresses the question of getting people to use the equipment.

"Ease of operation is the key to more use of computers in teaching."

"Ease of operation is the key to more use of computers in teaching," they write. "Only a few instructors will make any substantial effort to plan a computer demonstration, especially if they must set up the computer .... We have not hounded our colleagues as aggressively as we might have, but several have found uses for the classroom computers, sometimes in ways that might not have occurred to us."

Herb Greenberg reports that the University of Denver is planning to try out the Kodak Datashow, a device that transfers a computer screen display to an overhead projector. John Kenelly adds that Hewlitt Packard is working on a system to feed the infra-red signal from the HP 28C into the Kodak projector.

The Super Tutor

While the "super blackboard" uses the computer as a prop for the classroom instructor, a project under way at the Institute for Mathematical Studies in the Social Sciences at Stanford University is looking to replace the instructor altogether! A group headed by Tryg Ager under the supervision of Patrick Suppes began in 1985 to develop an intelligent tutor that will give interactive lessons in calculus by checking the validity of a student's reasoning as he or she works through the steps of a complicated problem or proof. The goal of the project is not actually to do away with the teacher, but rather to design a course "that could be taught with supervision by a teacher who is mainly playing the role of instructional manager," according to an interim report.

Ager argues that a student's time with the computer is spent more efficiently than in traditional instruction. "You don't have the sort of low-intensity activity like sitting in lectures, and the frustrating activity of doing homework" with the long turn-around time between lectures, he says, adding that the computer doesn't get tired of giving feedback.

The project, which is funded by NSF's Office of Applications of Advanced Technology, supplements the Computer Algebra System REDUCE with a system of "equational derivations" dubbed EQD. "The idea that students construct complete solutions to complicated exercises dominates our pedagogical model," according to the interim report. The report cites the use of REDUCE and EQD as one of three themes, the other two being the development of interactive instructional graphics and the development of an interactive theorem-proving system. The latter is based on work of Suppes' group on proof logic, which was tried out earlier in an advanced course in axiomatic set theory. Ager calls it "a terribly difficult problem" in artificial intelligence.

A student's time with the computer is spent more efficiently than in traditional instruction.

The project is aimed at eventual application in high-school calculus instruction; Ager says they would like to begin tests next Fall (1988). The reason for targeting high schools, Ager explains, is that the majority of high schools (Ager estimates 75%) have only a handful of students ready to take calculus, too few to meet school-board requirements for minimum class size.

Rural high schools especially are often too small to offer calculus. Tom Tucker, of Colgate University, participated in another solution to this problem, involving 13 students in five rural schools in upstate New York: Tucker taught them calculus over the telephone! The class, which "met" one hour a week in 1985/86, was sponsored by the state of New York, which provided the conference line and some fancy electronic writing tablets. Each location could speak and write at any time, except that Tucker's electronic pen took precedence.

Tucker says that several other states are mandating similar programs. South Carolina, for instance, is requiring high schools to offer advanced placement in at least one subject. John Kenelly says that South Carolina is considering the same system that Tucker used in New York, but adds that they are looking down the road toward even more sophisticated, computer-based communication systems.

Should High Schools Teach Calculus?

The quality of high-school calculus instruction—and mathematics instruction generally—is of concern to many at the college level. It is traditional for college professors to bemoan the poor preparation of their students, but the problem in mathematics seems to be acute. Indeed, *The Underachieving Curriculum* is the title of the American report of the Second International Mathematics Study, which gathered data from eighth- and twelfth-grade level classes in 20 countries around the world. According to the report, "the mathematical yield of U.S. schools may be rated as among the lowest of any advanced industrialized country taking part in the Study."

One of the culprits pointed to by the Study is the excessive repetition of material from one math course to the next: "I didn't learn much this year that I didn't already know from last year. Math is my favorite class, but we just did a lot of the same stuff we did last year," a fourth-grader named Andy is quoted as saying.

The U.S. pattern is based on a valid educational theory that Jerome Bruner calls the "spiral curriculum," in which topics are introduced early and then revisited repeatedly in progressively more complex forms. However, the Study points out that practice has not lived up to theory: "The logic of the spiral curriculum has degenerated into a spiral of almost constant radius—a curriculum that goes around in circles."

"The logic of the spiral curriculum has degenerated into a spiral of almost constant radius—a curriculum that goes around in circles."

The problem in high schools is pertinent to college calculus courses, because many students in high schools that do offer calculus are getting what John Kenelly tersely refers to as "crap calculus," in the expectation that these students will take "real calculus" in college. Schools are teaching "trashy" calculus, Kenelly says, and kids are skipping fourth-year "real" math to take it, so that mother and father can brag about it down at the country club.

Don Small, in a paper prepared for a January 1986 calculus workshop at Tulane University, writes, "The lack of high standards and emphasis on understanding dangerously misleads students into thinking they know more than they really do. In this case, not only is the excitement [of learning calculus] taken away, but an unfounded feeling of subject mastery is fostered that can lead to serious problems in college calculus courses."

Indeed, the problem is considered serious enough that a joint letter endorsed by the MAA and the NCTM was sent out in 1986 to secondary school mathematics teachers nationwide, urging that calculus in high school be treated as a college level course and only be offered to students who have a full four years' preparation in algebra, geometry, trigonometry, and coordinate geometry. The recommendation favored by the MAA and NCTM is that high school calculus students take university-level calculus, with the expectation of placing out of the comparable college course.

Advanced-placement high school students actually learn calculus better than college students, even after "ability level" is factored out.

According to Kenelly, who has a long involvement with the calculus AP exam, advanced-placement high school students actually learn calculus better than college students, even after "ability level" is factored out. John Harvey points out one possible explanation: high school students get nearly 180 hours of instruction, in smaller classes, and oftentimes with more careful instruction.

Should Colleges Teach Calculus?

While it is easy and in some ways satisfying to point the finger of blame at high schools, there is concern about the quality of calculus instruction at the college level as well. Large lectures taught by disinterested faculty, and classes taught by inexperienced TAs with minimal supervision, are among the problems cited. Many critics feel that the computer will become part of the problem rather than a solution. And there is nearly universal disgust with the current gargantuan textbooks.

"Getting the faculty back in calculus is the most significant thing that should be done these days."

"Getting the faculty back in calculus is the most significant thing that should be done these days," says Robert Blumenthal of St. Louis University. Blumenthal took over as lower-division supervisor in the mathematics department at St. Louis University in 1986. The department was faced with declining enrollments in upper-level courses, due, it was determined, to poor teaching in the lower-division courses. Looked at more closely, the department found most of the complaints concerned non-permanent faculty—TAs and part-time instructors. Blumenthal's solution: have only full-time faculty teach calculus.

"You need a department that feels calculus instruction is important and deserves the attention of the faculty. It's our bread and butter," Blumenthal says.

In addition to moving away from TAs and part-time faculty, Blumenthal eliminated common finals in calculus, which used to be multiple choice. Blumenthal was also unhappy with the textbook being used at the time. He feels that current calculus texts are too long and too wordy, and offers an interesting explanation: "These texts are written more to correct the deficiencies of the instructor than to help the student." Consequently he was "delighted" to find that Addison-Wesley has reissued the "classic" second edition of Thomas, which originally appeared in the 1950s. His department changed to Thomas last year.

There is nearly universal disgust with the current gargantuan textbooks.

Morton Brown of the University of Michigan is adamant on the issue of classroom size. The mathematics department at Michigan tried large lectures for several years, but abandoned them in 1985 in favor of smaller classes of 30-35 students that meet four times a week with one instructor. "In math it's more like a language," Brown says. "You have to get in there and find out what the student is doing." Most of Michigan's 40-50 sections of calculus each semester are taught by TAs, and most of the rest are taught by younger faculty, but Brown feels this is preferable to the lecture-recitation format. "We think large lectures provide very poor teaching," he says.

Peter Ponzo describes the opposite experience at the University of Waterloo. About 700 students at Waterloo take first-year calculus in lecture sections of 120-140 students that meet three times a week. These sections are supplemented by a two-hour "problem lab" of 30-40 students which meets with the professor and some graduate students. Ponzo says they make a conscious effort to put the best teachers into calculus. The department used to have small classes, some taught by graduate students, but about ten years ago it was decided that graduate students are not good teachers for first-year students. Graduate assistants, Ponzo says, "like to impress students with how much they know. They tend to make things far more complicated than they are."

Should TAs Teach Calculus?

The role and effectiveness of graduate teaching assistants is the focus of a study by the MAA Committee on Teaching Assistants and Part-Time Instructors, chaired by Bettye Anne Case of Florida State University. With support from FIPSE, the committee surveyed nearly

500 departments with graduate programs to ascertain patterns of present practice and issues. A preliminary report was published in 1987, and the committee plans to publish a "Resource Manual" in 1988 which will contain additional data along with models of training programs for TAs and part-time instructors.

The first survey found that the majority of doctoral-granting departments offer training programs, whereas the majority of master's-granting programs do not. Members of the committee (all of whom are from doctoral-granting departments) consider training and supervision of TAs to be of paramount importance.

Committee member Thomas Banchoff of Brown University writes, "If the TA gets the impression that the professors themselves, especially those in charge of the courses TAs teach, care about teaching effectively, then they are likely to develop well themselves and pass on their experience to those coming after them. On the other hand, if the faculty is perceived as uninterested in teaching or in working with students, it is this attitude that the students will take to their own TA jobs and pass on to the new graduate students."

The training of foreign TAs is of particular concern. Several schools have tightened their language proficiency requirements. At The Ohio State University, for instance, international students must be certified as having oral communication skills in English. This is done by the University, but the mathematics department then makes its own, independent judgment on whether to place the student in a classroom or hold him or her out for the first year as a grader.

Beginning in 1985, students are given four quarters to become certified, and based on experience in 1986/87, the department now requires that international students come in the summer as a precondition for support. According to Harry Allen, former chair of graduate studies for the mathematics department at OSU, all but one of the international students who came in the summer were certified by the end of the following spring, whereas of those who came in the fall, only a couple were certified.

Allen feels that the program at OSU has been highly successful, and points to a "drastic decline" in the number of complaints going to the undergraduate chairman's office over the past few years. Moreover, the foreign TAs themselves are positive about the training, in part, Allen says, because "the people who are running it are doing it the right way."

Should *Anyone* Teach Calculus?

Most college mathematics teachers learn their trade by the example of others and through their own experience; few have any background in educational theory, and most have little interest in it. Stephen Monk, of the University of Washington, is one of the few. Monk describes himself as a Piagetian, after the Swiss psychologist Jean Piaget, who is well known for his studies of the intellectual development of young children. Monk was moved by Piaget's stress on students' learning through their own activity. Students of calculus must proceed from their own intuition, he says, adding that calculus, which is usually taught as a formal subject, can be understood in the way that arithmetic is understood.

Consequently, Monk is keenly interested in the group dynamic of the classroom and the role—or function, as he discriminates it—of the teacher. "I was mildly flattered when I was asked to teach a large lecture course," he writes in an article in *Learning in Groups*. "The request indicated that I had joined a circle of competent teachers in my department who could give the clear explanations that a large lecture demands; who are sufficiently well organized, patient, and self-assured to get the subject across to a large, diverse audience; and who are free of the idiosyncrasies of style that are charming in a teacher with a few students but disastrous in a teacher with many .... Looking back to that time, I view my attempt to express my teacherly impulses as a lecturer in a large course as roughly parallel to attempts to express sensitivity for the less fortunate of our society as a county jailer."

"Collective work is a key ingredient to intellectual growth."

Monk has worked with psychologist Donald Finkel, of the Evergreen State College, on the idea of using small "learning groups" in the classroom with the teacher distributing worksheets and then acting as a roving "helper" rather than as "the expert." "The evidence that collective work is a key ingredient to intellectual growth surrounds us," they write. "Yet to judge by the typical college course, most teachers do not believe that it is either appropriate or possible to foster these important processes in the classroom." They call this attitude in which the teacher shoulders all responsibility in the classroom the "Atlas complex."

David Smith, of Duke University, has also embarked on another daring innovation in his calculus classes: writing assignments. "Failure to read and analyze instructions prevents students from getting started on a problem, and their ability to understand a solution process is related to their ability to explain in English what they have done," he writes in an article in *SIAM News* (March 1986). Smith now requires, on at least one problem from each assignment, that students explain in complete sentences how they set up the problem, the sequence of operations, and their interpretation of the result. His exams, which are wide-open, take-home tests (he even allows collaboration!), require writing on almost all problems. Typical instructions: "Evaluate each of the following six limits, if possible. State in a sentence or two what your technique is. If the limit fails to exist, say why."

A few of the students at Duke were really hostile to the idea of writing, according to Smith. They did it to get the grade, but thought he was being unfair. Most, however, were intrigued with the idea. Smith describes their course evaluations as being cautiously positive. Faculty were also interested but somewhat nervous about the time commitments and their own competence in grading students' work. Smith acknowledges that he was able to handle the time commitments (upwards of one hour per paper) in part because he had fewer students last year. However, he also points out that a writing center workshop at Duke helped him learn what to look for. One piece of advice he found helpful: Look for key features in a paper; if these aren't there, don't try to read it, ask for a rewrite!

The Rise (and Fall) of Discrete Math

While calculus is traditionally considered *the* mathematics course for first-year students, discrete mathematics has been proposed as an *avant-garde* alternative. Many of the modern usages of mathematics, particularly in computer science, are based not on the analysis of continuous phenomena, but rather on combinatorial and recursive principles. Engineers need calculus right away, computer scientists don't, the proponents of discrete mathematics argue. Why not teach people what they need to know?

The answer may be that it just doesn't work very well. Two schools that tried putting discrete mathematics into the first year, Dartmouth College and the University of Denver, have taken it back out.

In 1983 the mathematics department at Dartmouth College proposed a new curriculum in which a four-quarter sequence of calculus through differential equations was reduced to three, and a new course in discrete mathematics, which could be taken at any time after the first calculus course, was offered. "As it turned out, it was a disaster," says Jim Baumgartner.

The main problems were with the discrete mathematics course and the second-quarter multivariate calculus course. The discrete mathematics course hadn't been placed in the sequence, and it therefore attracted a tremendous variety of students, with anywhere from one to five courses of mathematics in their backgrounds. It was also hard to know what to teach, Baumgartner says, so the course varied widely from quarter to quarter, depending on who taught it. He adds that the Computer Science Department at Dartmouth dropped the course as a requirement, which also hurt. Finally, the Mathematics Department was hoping that the discrete mathematics course would make things easier in upper-level courses by introducing proofs, induction, and so forth, but it didn't seem to be doing that. Discrete mathematics at Dartmouth, Baumgartner summarizes, was a "total failure."

Multivariate calculus in the second quarter was also a mistake, according to Baumgartner. There was a "sophistication problem," he says—the students just weren't picking it up. (He notes, however, that advanced students, who took it as their first course at Dartmouth, did pretty well.) Students disliked the multivariate course intensely (more than usual, Baumgartner says), and the majority disliked the discrete mathematics course and found it quite difficult.

The third-quarter Differential Equations course was the most successful, according to Baumgartner. Its only weak point was the introduction of series here rather than in a prior calculus course. In the new program infinite series will move back to calculus and the differential equations course will get more of a linear algebra slant.

Dartmouth's new four-quarter mathematics sequence is largely a return to the original program: two quarters of calculus, including series and a chunk of matrix algebra, a quarter of multivariate calculus with linear algebra, and a quarter of differential equations. Discrete mathematics is being dropped discreetly.

At the University of Denver, Herb Greenberg and Ron Prather, who is now at Trinity College, used a Sloan grant to experiment with a combination of calculus and discrete mathematics. Their idea was to teach a "discrete structures and calculus" course in the first quarter, differential calculus in the second quarter, and integral calculus in the third quarter. The experiment began in 1983, and was abandoned two years later. The department, which combines mathematics and computer science, now offers a three-quarter calculus sequence and one quarter of discrete mathematics.

"Student evaluations were negative and somewhat surprising," Greenberg says. Students did not find the discrete structures course at all easy (Greenberg retitled the course "Destruct Creatures"), and, more importantly, they did not see the relevance of it. Even the computer science majors, Greenberg says, said it had nothing to do with the computer science they were studying, and called the course "a complete waste of time."

Greenberg was also disappointed by the lack of "carry-over" from discrete mathematics to calculus; for instance, students seemed to have no idea that proof by induction was a useful technique outside of the first quarter. The reason may simply be students' lack of experience in high school (and earlier) with non-computational mathematics. "They weren't mature enough to understand it," Greenberg says.

Enthusiasm vs. Realism

Many other individuals around the nation are doing innovative work in calculus instruction. Their efforts are in large part self-rewarding: they impart to their students their own excitement at doing and learning calculus. "The success of innovations depends predominantly on the enthusiasm of the person doing it," says Frank Morgan, now of Williams College. Programs and funding to improve calculus instruction will certainly increase in the coming years, and new ideas will continue to be discussed.

Morgan, however, seasons his enthusiasm with a concluding note of realism: "Any system you come up with, after awhile the students figure out how to get through it with the least amount of work."

BARRY A. CIPRA is a mathematics writer, teacher, and editor. The author of several research and expository papers in number theory, and of the calculus supplement *Misteaks,* he has taught at Ohio State University, M.I.T., and St. Olaf College. He received his Ph.D. degree in mathematics from the University of Maryland.

Some Individuals Involved with Calculus Projects:

TRYG AGER, Institute for Mathematical Studies in the Social Sciences, Ventura Hall, Stanford University, Stanford, California 94305.

HARRY ALLEN, Department of Mathematics, Ohio State University, Columbus Ohio 43210.

GEORGE ANDREWS, Department of Mathematics, Oberlin College, Oberlin, Ohio 44074.

TOM BANCHOFF, Department of Mathematics, Brown University, Providence, Rhode Island 02912.

JIM BAUMGARTNER, Department of Mathematics, Dartmouth College, Hanover, New Hampshire 03755.

ROBERT BLUMENTHAL, Department of Mathematics, St. Louis University, St. Louis, Missouri 63103.

MORTON BROWN, Department of Mathematics, University of Michigan, Ann Arbor, Michigan 48109.

STAVROS BUSENBERG, Mathematics Department, Harvey Mudd College, Claremont, California 91711.

BETTYE ANNE CASE, Department of Mathematics, Florida State University, Tallahassee, Florida 32306.

BRUCE CHAR, Computer Science Department, University of Waterloo, Waterloo, Ontario, Canada N2L 3G1.

DOUG CHILD, Mathematical Sciences Department, Rollins College, Winter Park, Florida 32789.

WADE ELLIS, Department of Mathematics, West Valley College, San Jose, California 95070.

HERB GREENBERG, Department of Mathematics, University of Denver, Denver, Colorado 80208.

JOHN HARVEY, Department of Mathematics, University of Wisconsin, Madison, Wisconsin 53706.

JOHN HOSACK, Mathematics Department, Colby College, Waterville, Maine 04901.

JAMES HURLEY, Department of Mathematics, University of Connecticut, Storrs, Connecticut 06268.

ZAVEN KARIAN, Department of Mathematical Sciences, Denison University, Granville, Ohio 43023.

JOHN KENELLY, Mathematics Department, Clemson University, Clemson, South Carolina 29631.

STEPHEN MONK, Department of Mathematics, University of Washington, Seattle, Washington 98195.

FRANK MORGAN, Department of Mathematical Sciences, Williams College, Williamstown, Massachusetts 01267.

PETER PONZO, Department of Mathematics, University of Waterloo, Waterloo, Ontario, Canada N2L 3G1.

DONALD SMALL, Mathematics Department, Colby College, Waterville, Maine 04901.

DAVID SMITH, Department of Mathematics, Duke University, Durham, North Carolina 27706.

ANDREW STERRETT, Department of Mathematical Sciences, Denison University, Granville, Ohio 43023.

TOM TUCKER, Department of Mathematics, Colgate University, Hamilton, New York 13346.

ALVIN WHITE, Mathematics Department, Harvey Mudd College, Claremont, California 91711.

PAUL ZORN, Department of Mathematics, St. Olaf College, Northfield, Minnesota 55057.

Calculus for Engineering

Kaye D. Lathrop

STANFORD LINEAR ACCELERATOR CENTER

The engineering with which I am most concerned is research and development engineering, usually conducted by individuals with at least a master's degree. Research and development problems usually require extensions of existing technology, and indeed, part of the research and development effort is directed toward finding new ways of solving engineering problems.

The individuals who conduct this research must be extremely competent applied mathematicians. They must be able to derive the equations that describe the systems with which they are working and they must be able to extract solutions from these equations with enough ease to permit parameter surveys and sensitivity analyses. If the equations they derive are not those for which standard solutions are available or for which usual numerical methods apply, they must be capable either of developing suitable approximate solution techniques or of approximating the equations to be solved to render them tractable. These people will have programming support, access to supercomputers and access to experts in appropriate disciplines, but they will often become the expert in the particular class of problem being addressed.

For most students who expect to do graduate work in the physical sciences, the conventional college calculus course is not sufficient.

For the people described above, calculus is a fundamental part of their mathematical training. For these people, and indeed for most students who expect to do graduate work in the physical sciences, the conventional college calculus course is not sufficient. Even the special four quarter (or three semester) courses sometimes offered for those majoring in engineering or the physical sciences are at the same time neither broad enough nor deep enough.

Chickens and Eggs

In addition to this problem of insufficiency, calculus suffers from a chicken and egg problem. Applied physics and most engineering is best taught using calculus and differential equations as a known subject, but most freshmen students don't speak these languages. At the same time calculus is not learned except by working examples. Since the best examples are those which motivate the student, these are most likely to be those selected from the students' field. To some extent this latter need is recognized by the offering of special courses tailored to disciplines—calculus for business majors or calculus for mathematicians. Although such courses do not resolve the paradox of needing to know calculus to learn the subject, they do lead to the proliferation of first level classes, some of which might as well be called "calculus for those who don't want to take calculus."

Calculus should be part of a curriculum of computational mathematics.

The real need is for a calculus curriculum to provide early a solid base of understanding and practical tools for subsequent engineering and physical science courses, and to provide later another coverage of calculus, both deeper and broader. I happen to believe that such a repetition of subject matter also greatly increases proficiency and understanding.

Another form of the chicken and egg problem occurs within mathematics itself. There is a tendency in some treatments of calculus to emphasize the proving of theorems and the pathologies of functions. Especially for engineering students, this more abstract information is best reserved for second courses in calculus. Even then this type of information is best illustrated by examining it in the context of examples relating the significance of the result obtained to practical applications.

Problems with Problem Sets

In addition to coordinating the pace of students' mathematical development with their engineering or physical sciences development, it would be most desirable to train them simultaneously in both areas. One way this might be done is by teaching calculus using problem sets and illustrative examples from engineering and physical sciences applications. This is not a novel idea. There has been a trend in this direction, and I think it should be accelerated.

It is clearly more work to prepare such educational examples and care must be taken to choose problems and illustrations appropriate to either basic or advanced levels of the students, but the payoff should be better motivation and clearer understanding of both subjects. Older texts seem to use predominately abstract examples of functions, and to spend inordinate amounts of time integrating and differentiating special functions, finding areas and volumes and so on, with little regard for the relevance of these acts. This state of affairs could be improved.

Every student in calculus should be deeply exposed to the derivation and use of numerical approximation methods.

In a related vein, many calculus texts use problems that can be solved by following recipes, and typical homework exercises involve a couple of hours of repetitively following the recipes. In general, more time should be devoted to making the ideas clear and less time devoted to rote learning.

In particular, calculus, engineering, and physics courses should spend proportionately more time on the topics of problem formulation and problem solving. Too often students' only exposure to the thinking that goes into the process of formulating and then solving a problem is the verbalization of the process by the instructor in the classroom. When one has mastered the solution of a particular class of problems, particularly if this mastery was gained some long time ago, there is a tendency to present the problem formulation and solution as a smooth, seamless, effortless process, forgetting the struggle, the trial and error, and the dead ends that are pursued before success is obtained.

As students progress to more advanced courses, proportionately more time should be spent on the techniques of problem solving. Examples in texts should include case histories discussing the reasoning that goes into a variety of approaches. Classroom lectures should include periods when experts are confronted with new (even unsolved) problems and asked to attempt a solution and describe what they are thinking as they do so. This sort of thing can be a close approach to the actual research process.

Using Computers

Many treatments of calculus, especially in books that have been revised over a period of years, scarcely recognize the use of computers. This is a serious omission on at least two counts. Present day computer programs provide extremely powerful symbol manipulation, including the capability of integrating and differentiating. The student should become aware very early of the capabilities of these programs.

More fundamentally, every student in calculus should be deeply exposed to the derivation and use of numerical approximation methods. In the fields of engineering and physical sciences, almost all solutions are obtained by numerical methods. The successful professional will use computational evaluations far more often than the techniques of classical calculus. The understanding of calculus is extremely important and the language of calculus is a useful shorthand, but once the applicable equations are derived, the greatest challenge becomes the determining of numerical solutions.

That is not to say that all the classical analysis that is taught is of limited use. Indeed, the validation of numerical approximations—that is, the experimentation necessary to show that a computer code produces correct solutions—is a discipline in its own right that usually involves comparison of evaluations of analytic solutions, where such are available, with solutions produced by numerical approximations.

The successful professional will use computational evaluations far more often than the techniques of classical calculus.

Both analysis and approximation are useful, but the curriculum should combine them synergistically rather than treat them disjointly. In my view, calculus should be part of a curriculum of computational mathematics, with the initial course emphasizing numerical differentiation and integration and with subsequent courses covering first ordinary differential equations and then partial differential equations.

Geometry

For research and development in engineering, clear visualization of the problem geometry and a clear understanding of the problem solution as a function of key problem parameters are imperative. I believe the calculus curriculum should emphasize analytic geometry and combine it early with the use of computer graphics displays. CAD systems modeling three-dimensional solids are commonly available to the engineer. Students should become familiar with these powerful tools that can describe all geometric aspects of the actual system and permit viewing from any desired vantage point.

As students begin to generate numerical solutions, they should at the same time become familiar with the tools available to display these solutions. They should also be trained in the techniques of processing masses of data to extract significant information. Visual detection of anomalies is an extremely important and powerful tool, both to detect errors in solution techniques and to discover new effects.

As computers make possible the solution of more and more complex problems, the importance of properly sifting the output becomes greater and greater. In high energy physics, for instance, terabits of experimental data are repeatedly mined with sophisticated tools to extract new physical effects. Monte Carlo evaluations of theoretical models gives equal amounts of data to be similarly sifted.

Differential Equations

The more advanced part of the calculus curriculum should concentrate on solutions of the most important classes of the differential equations of engineering and the physical sciences. In the last twenty years an enormous body of knowledge has been accumulated about these equations and about their practical means of solution.

In my view, a coherent treatment of solution techniques does not exist, even though "black box" computer algorithms exist for ordinary differential equations and are beginning to emerge for partial differential equations. Use of these solution techniques is extremely important in much research and development in engineering and the physical sciences. Hence students should be trained in these areas, being given a combination of classical analytical methods and numerical methods.

Summary

My suggestions for calculus for engineering and physical sciences can be summarized concisely: more mathematical training should be offered the undergraduate, and more of that training should be computer oriented. In particular, I'd recommend that universities

- Make the calculus curriculum broader (numerical methods) and deeper (two, perhaps three full years through partial differential equations).
- Closely integrate the calculus curriculum with the engineering and physical sciences curriculums.
- Emphasize computational mathematics as well as conventional analysis.

KAYE D. LATHROP is Associate Laboratory Director and Head of the Technical Division of the Stanford Linear Accelerator Center in Stanford, California. He has served as Associate Director for Engineering Sciences for the Los Alamos National Laboratory, and as Chairman of the Mathematics and Computation Division of the American Nuclear Society.

The Coming Revolution in Physics Instruction

Edward F. Redish

UNIVERSITY OF MARYLAND

The current introductory physics curriculum has been highly stable for almost thirty years and is nearly uniform throughout the country. It is closely linked with the introductory calculus sequence in that the ordering and content of the physics courses are strongly determined (sometimes inappropriately) by the students' mathematical skills, and in that the calculus course frequently uses examples from physics.

Despite its apparent stability, there are indications that three revolutions are beginning to make an impact on the curriculum: the explosion of new knowledge and materials in physics, new insights into the process of learning gleaned from studies in cognitive psychology, and the power and widespread availability of the computer.

I suggest that these changes, with the computer acting as a lever on the first two, are making inevitable a revolution in the way we teach physics. These changes will require associated changes in the calculus sequence which will have to carry the burden of introducing a wider variety of mathematical techniques to scientists and engineers than is currently the case. These new techniques include numerical and approximation techniques, the study of discrete equations, and pathological functions.

Physics and Calculus

Physics and mathematics have been close colleagues throughout their history. Newton's development of the theory of mechanics was intimately tied to his invention of the calculus. The understanding of the theory of electromagnetism was achieved when Maxwell completed his set of partial differential equations. The mathematical studies of transformation theory and tensor analysis at the end of the nineteenth century, when applied to Maxwell's equations, led Einstein to develop his special and general theories of relativity. This close interplay continues today in a variety of forefront research, including the development of string theory and the theory of chaos.

Explosion of new knowledge, ... new insights into the process of learning, ... and the power ... of the computer ... are making inevitable a revolution in the way we teach physics.

The close connection between physics and mathematics is nowhere more obvious than in the structure of the undergraduate physics majors' curriculum. The standard curriculum covers a variety of subjects including:

- mechanics
- electricity and magnetism
- thermodynamics and statistical mechanics
- modern physics and quantum mechanics.

These subjects form the foundation of fundamental material on which more specialized subjects, such as condensed matter physics, atomic and molecular physics, nuclear physics, plasma physics, and particle physics, are built.

The standard approach to these subjects is a spiral. They are first covered in an introductory manner in a survey course in the freshman and sophomore year, redone in individual courses in the junior and senior years, and finally covered one more time in graduate school. Both the spiral character and the specific ordering of topics in the physics curriculum is primarily controlled by the mathematical sophistication the student is assumed to have.

The control imposed by mathematics is very rigid and even influences the detailed order of presentation in individual courses. For example, the first semester of physics is usually taken with first year calculus as a corequisite. In teaching mechanics, nearly every textbook author introduces the subject by considering the physical example of motion in a uniform gravitational field, with Newton's Laws of motion presented at a later stage. The primary reason for this is that the former can be solved algebraically without the use of differential equations and is an appropriate place to introduce the concept of derivative. The latter requires a more complete understanding of the concept.

From the point of view of the physics this is highly inappropriate. The uniform gravitational field is a very special and peculiar case. Putting it first gives it a primacy which is both undeserved and misleading.[1] Many other examples could be given, including the delay of the presentation of the harmonic oscillator, the placing of electrostatics, the study of normal modes of oscillation, the treatment of quantum mechanics, etc.

On the other hand, physics content plays a significant role in the calculus sequence. Many standard calculus textbooks make heavy use of physics problems as examples and to help motivate students concerning the real-world relevance of the material presented.

Both the spiral character and the specific ordering of topics in the physics curriculum is primarily controlled by the mathematical sophistication the student is assumed to have.

Only a relatively small number of students actually major in physics. However, in most universities the same introductory physics course taken by majors is also taken by engineers, chemists, and mathematics majors. Almost all physical scientists trained in American universities take a physics course of the standard type. The structure and content of introductory physics with calculus therefore has a profound implication for the training of all our (hard) scientists.

Curricular Stability

Calculus-based physics for scientists and engineers is taught in a variety of formats and conditions, but it is almost always found in the format of a three or four semester (occasionally three quarters) introductory sequence with calculus as a corequisite (or occasionally with one semester as a prerequisite). For majors, this introduction is followed by the spiral described above.

The physics curriculum was stabilized in a semiformal sense in a series of conferences and articles [3], [19] in the late '50's and early '60's. At that time, a mini-revolution took place in the style of physics teaching, shifting emphasis towards concept and understanding and away from development of technical skills.

The content of this course is very stable. Despite the existence of dozens of introductory physics texts and the yearly publication of many more, their approaches differ only in detail. A recent survey of ten of the most popular first-year physics texts by Gordon Aubrecht [1] strongly confirms this.

The widely-held view among physicists that physics is the most cumulative and mathematical of the sciences ...leads to a strong tendency to teach physics semi-historically.

This stable content is a result of the widely-held view among physicists that physics is the most cumulative and mathematical of the sciences. Even scientific revolutions such as quantum mechanics and special relativity are viewed by most professional physicists as extensions rather than replacements of previous theories. In the research forefront, techniques from older models often coexist with current dogma. This leads to a strong tendency to teach physics semi-historically.

The present introductory curriculum gives the student an overview of the basic techniques of those parts of historical physics which still survive from the period 1600 (Galileo) to about 1940 (nuclear physics of fission and fusion). The content from after 1916 (the Bohr model of the atom) is usually slim, since going further requires treading on the slippery ground of quantum mechanics. Quantum mechanics is a less intuitive subject than the others, and one which relies heavily on the mathematics of differential equations and matrices. It is usually suppressed in the introductory course.

A Revolution Is Brewing

This characterization of the physics curriculum may lead one to project long-term stability in the way physics is taught at the introductory level. Nonetheless, there is evidence on the horizon of a major revolution in the teaching of physics that is driven by revolutions in three major areas of relevance: physics research, educational psychology, and available technology. Jack Wilson, Executive Director of the American Association of Physics Teachers (AAPT) refers [21] to these three drivers of change as "the three C's:"

- Contemporary
- Cognitive
- Computer.

Predicting the future is a chancy business under any circumstances, but predicting major changes in any human activity is a long shot. As I discuss the impact of each of these revolutions, the reader should bear in mind that these views are necessarily speculative and idiosyncratic.

Contemporary Physics

As a result of the perceived cumulative structure of physics, the current introductory course has a strong overlap with the courses taught at the turn of the century. (The changes agreed upon thirty years ago dealt primarily with style rather than content.) For example, the material covered in 21 of the 23 chapters in Robert Millikan's turn of the century book [16] is contained in current texts.[2] According to Aubrecht's survey, only about 20% of current introductory texts have chapters on the physics of the past 50 years and those only spend about 5% of their chapters on that material.

John Rigden, AIP Director of Physics Programs and former editor of the *American Journal of Physics* stated the issue compellingly in one of his editorials [18]:

> Great-grandparents and grandparents and parents took ... the same physics course as contemporary students are now taking. ... No new information appears in a new edition of a physics textbook.

This wouldn't be a problem if physics were a static field. It is not. In the past thirty years we have seen an explosion of new understanding and power in a variety of subfields of physics, ranging from the discovery of the substructure of protons and neutrons, to high temperature superconductivity, to the discovery of the three-dimensional structure in the clustering of galaxies. These discoveries cover all possible scales of length, time, and mass.

The excitement and vitality of contemporary physics is not conveyed in the current introductory course, nor are the student's skills developed in a manner appropriate for the new physics.

There have even been major breakthroughs in fields long thought to be well understood. Current developments in Newtonian mechanics are evolving into a theory of chaotic behavior which may lead to a change in our way of viewing physics as deep as that produced by the discovery of quantum mechanics.

The excitement and vitality of contemporary physics is not conveyed in the current introductory course, nor are the student's skills developed in a manner appropriate for the new physics. Recently, however, the leaders of both the research and teaching communities are beginning to take a broad interest in including contemporary physics in the introductory courses. This interest is displayed in two recent conferences held in Europe and the United States [2], [14].

Cognitive Science

Psychological studies of what students know and how they learn are producing fundamental changes in the theories of learning and education. The few studies that have been done on physics students indicate that there are profound difficulties in the way we teach physics [5], [15], [17]. The traditional assumption that the student is a *tabula rasa* on which new descriptions of the universe may be written appears to be false.

Students bring to their first physics course well-formed yet often incorrect preconceptions about the physical world.

Students bring to their first physics course well-formed yet often *incorrect* preconceptions about the physical world. The present structure of introductory physics does not deal with this well. A number of pre- and post-course tests taken at a variety of universities show that a course in physics does little or nothing to change the average student's Aristotelian view of the universe [10], [11]. As we learn how students learn and change their views from "naive" to "expert," we can design our courses so as to facilitate this transition.

Computer Technology

The immense growth in the power of high-tech tools in the past thirty years has had a profound impact on the way professional physics is done. The most powerful and influential of these tools is the computer. More computational power is packed into a desk-top computer the size of a breadbox than was available in the largest mainframes thirty years ago.[3]

Approximately 75% of contemporary research physicists use computers. At the time our current "stable" curriculum was designed, the number was more like 5%. (These numbers are based on informal surveys and consultation of the research literature.) The result is that the computational skills required by the professional must all be learned at the graduate level.

Certainly many students learn to program as undergraduates, and indeed, some are already superb programmers by the time they enter college. But programming as taught in high schools, in computer science departments, and learned on one's own to write games with is not the same as learning to do physics with the computer.

The physicist who wants to do physics with a computer needs a wide variety of skills, not all of them numerical. Estimation skills are essential, not only for the experimentalist who must have a good idea of the rate at which data must be taken, but for the numerical analyst who must have a reasonable idea of what an appropriate discretization is. An important skill is to be able to tell when to do an exact analytic calculation, when to approximate, and when to use the computer. This skill requires substantial experience.

Computers are necessary tools at the undergraduate level as part of a student's training as a professional. But in addition, the availability of the computer opens immense opportunities for the physics teacher. If the student has the computer available from an early stage, a larger class of problems and subjects may be introduced, avoiding the previously imposed mathematical constraints.

Alternative Curricula

At the University of Maryland, members of the Maryland University Project in Physics and Educational Technology (M.U.P.P.E.T.) have been investigating the impact of the computer for the introductory physics curriculum since 1984. Students are taught to use the computer in their first semester, and its presence has permitted a number of interesting modifications of the standard curriculum:

1. Newton's law can be introduced in discrete form as the first topic in mechanics. This allows the student to think about the physics before becoming involved in special cases (uniform accelerations) or mathematical complexities (calculus, vectors).
2. A wider variety of projectile motions can be studied than is possible when only analytic techniques are used. These include motion with air resistance, in a fluid, with electromagnetic forces, etc. This gives a better balance of analytic and numerically solvable problems and relates better to the real world than a curriculum with only idealized solvable models.

3. Non-linear dynamics can be studied. This permits us to teach fundamental concepts of numerical physics, to introduce some contemporary topics, and, by showing examples of chaotic systems, to make better connections between mechanics and statistical physics/thermodynamics than was previously possible.
4. Even first-year college students can develop enough computer power to do creative independent work on open-ended projects whose answers may not be known. This can provide an exposure to how science is really done.
5. Flexible, powerful, pre-prepared interactive programs can permit the student to investigate the solution of a wide variety of problems in a fraction of the time it would have taken with pencil and paper. This can permit even mediocre students to develop an intuitive feeling for problems previously accessible only to the best students.

Other uses of the computer have been made by other groups, including the development of Socratic tutorial programs [4], the modularization and organization of the physics curriculum into a Keller-plan no-lecture format (Project PHYSNET, Michigan State University), and the extensive use of microcomputer-based laboratories [20], [6]. Although these applications have a substantial effect on the physics curriculum, they do not produce obvious and significant modifications of the way the physics and mathematics curricula interact. I will therefore not discuss them here and leave the interested reader to seek out the references.

In addition to the Maryland project, new interest in the impact of computers on the teaching of physics is indicated by the large number of universities introducing upper-class courses in computational physics and a number of new textbooks taking that point of view.

"No new information appears in a new edition of a physics textbook."

A recent text by R.M. Eisberg and L.S. Lerner [7] provides a fairly standard introduction that includes a number of numerical examples and problems.[4] A widely used advanced text, *Computational Physics* by S. Koonin [13] contains numerous programs and numerical examples. The first truly computer-based introductory text is about to appear, written by Harvey Gould and Jan Tobochnik [9].

The AAPT leadership has called for a review of the current curriculum [12] and has been funded by NSF to organize conferences and workshops to discuss the shape of a new curriculum. While it may be a few years before a generally acceptable formula is found, and before enough colleges and universities get enough computer resources for the change to be broadly accepted, the smell of revolution is definitely in the air.

Implications for Mathematics

If we presume that the postulated revolution in physics teaching does in fact take place, and if we also presume that the service calculus courses are to continue to play a strong and relevant role in introductory physics, then the emphasis of the course should change somewhat. Although all of the subjects I propose adding to the traditional course are taught somewhere in the mathematics curriculum at many major universities, we must operate under the understanding that the primary service sequence for scientists is the three or four semester sequence in calculus. Given the heavy schedule of most scientists in the first and second year of their college, it is not possible to require that a significant number of additional mathematics courses be taken in the underclass years.

We must operate under the understanding that the primary service sequence for scientists is the three or four semester sequence in calculus.

Therefore, to make the introductory course more relevant, I propose that as the mathematical needs of scientists become broadened, especially by an expanded interaction with the computer, the content of the introductory mathematics service sequence be broadened to include topics which do not formally fall under the topic "calculus." However, for simplicity, I will continue to refer to this sequence as "the calculus course."

Specifically, I would like to see the traditional calculus course extended to include more emphasis on:

- practical numerical methods;
- qualitative behaviors;
- approximation theory;
- study of discrete systems;
- pathological functions.

Many individual mathematics teachers idiosyncratically include some discussion of one or more of these subjects in their courses; but most traditional calculus textbooks ignore them entirely.

Numerical Methods

Practical numerical methods form the heart of computer approaches to real-world problems, yet these are consistently ignored in the traditional introductory sequence. The problems of numerical integration and differentiation are suppressed in favor of extensive discussion of how to differentiate and integrate large numbers of special cases, despite the fact that the real-world problems most scientists face will almost certainly have to be treated numerically a large fraction of the time.

Even when practical rules, such as Simpson's rule of integration, are mentioned, almost never is there any discussion of where they are appropriate, how they can be improved, or the fact that there may be better alternatives. Practical solution methods for differential and integral equations are rarely discussed in the standard calculus sequence. Methods such as power series solutions, useful in proving general results but not very helpful in solving a new equation, are strongly emphasized. A better balance needs to be struck. The question of stability and improvability of methods should also be mentioned.

As discussions of numerical methods replace some of the discussions of analytic ones, a good understanding of the qualitative behavior of functions becomes even more important than before. Some of this is already present in traditional discussions of the theory of differential equations and the phase plane; but in the absence of discussions of numerical methods, their relevance is obscured.

By approximation theory I mean the process of how to understand the relevance of approximation schemes to specific situations, and of how to develop new schemes. If students are not even aware of the possibilities, for example, of the optimization of power series, they will be unable to seek out references on their own. Some broad introduction to the variety of practical and powerful methods available should be given.

Discrete Equations

Since students are usually introduced to discrete equations in the context of approximations to continuous equations in a numerical analysis, physics, or engineering course, they develop the mistaken impression that a discrete equation is a poor relation of the continuous one, and that any differences between them is a "failure" of the discrete equation. Discrete equations are themselves the relevant mathematical models in a number of circumstances, such as iterated function problems. They have their own unique set of characteristics.

Some of these discrete equations lead naturally to the discussions of functions which were considered "pathological" by mathematicians and scientists for many years, and which were thought to be irrelevant to physical phenomena. This has turned out not to be the case. For example, a discussion of the Cantor set and related fractals would be very valuable. Usually these are left to more advanced courses, such as topology, and there, their real-world applications tend to be ignored.

Functions which were considered "pathological" by mathematicians and scientists ... were thought to be irrelevant to physical phenomena. This has turned out not to be the case.

Anyone who proposes the addition of new material to a course currently packed to bursting with a superfluity of material has some obligation to identify what should be removed to make room. I would be happier to see less discussion of out-moded analytical techniques which have limited applicability. I also believe that some room could be made by not analyzing the very large number of analytical examples usually treated. Specific cuts will of course have to be decided by the community through extensive discussion.

Despite these proposed changes, I suspect that a large majority of physics teachers feel, as I do, that the calculus sequence provides a strong base for the fundamental mathematical skills of all physicists; that the primary emphases should be retained, including the emphasis on the study of smooth, non-pathological functions, on analytic techniques and theorems, and above all, on the rigor and structure of mathematical thought.

Notes

1. One of the few texts which reverses the traditional order is the classic Feynman *Lectures in Physics* [8] based on the teaching of Nobel Laureate Richard P. Feynman.
2. Although the content is similar, Millikan's emphasis is somewhat different from modern texts.
3. A contemporary student would probably feel more comfortable with the explanation "a breadbox is about the size of a desktop computer" than with the one given here. When was the last time you saw a breadbox?
4. This text may have come a bit before its time: it included a number of problems appropriate for the pocket programmable calculator. It is already out of print.

References

[1] Aubrecht, Gordon. "Should there be twentieth-century physics in twenty-first century textbooks?" University of Maryland preprint.

[2] Aubrecht, G. (ed.) *Quarks, Quasars, and Quandries.* Proc. of the Conf. on the Teaching of Modern Physics, Fermilab, April 1986 AAPT, College Park, MD, 1987.

[3] Bitter, F., *et al.* "Report of conference on the improvement of college physics courses." *Amer. J. of Physics* 28 (1960) 568-576.

[4] Bork, A. *Learning with Computers.* Bedford, MA: Digital Press, 1981.

[5] Champagne, A.B.; Klopfer, L.E.; Anderson, J.H. "Factors influencing the learning of classical mechanics." *Amer. J. of Physics* 48 (1980) 1074-1079.

[6] De Jong, M.; Layman, J.W. "Using the Apple II as a laboratory instrument." *The Physics Teacher,* (May 1984) 293-296.

[7] Eisberg, R.M.; Lerner, L.S. *Physics, Foundations and Applications.* New York, NY: McGraw-Hill, 1981.

[8] Feynman, R.P.; Leighton, R.B.; Sands, M. *The Feynman Lectures in Physics.* Reading, MA: Addison-Wesley, 1964.

[9] Gould, H.; Tobochnik, J. *An Introduction to Computer Simulation: Applications to Physical Systems.* New York, NY: John Wiley. (In press.)

[10] Halloun, I.A.; Hestenes, D. "The initial knowledge state of college physics students." *Amer. J. of Physics* 53 (1985) 1043-1055.

[11] Halloun, I.A.; Hestenes, D. "Common sense concepts about motion." *Amer. J. of Physics* 53 (1985) 1056-1065.

[12] Holcomb, D.F.; Resnick, R.; Rigden, J.S. "New approaches to introductory physics." *Physics Today,* (May 1987) 87.

[13] Koonin, S.E. *Computational Physics.* Menlo Park, CA: Benjamin/Cummings, 1986.

[14] Marx, G. (ed.) *Chaos in Education.* Balaton, Hungary, June 1987. (To appear.)

[15] McDermott, L.C. "Research on conceptual understanding in mechanics." *Physics Today* 37 (1984) 24-32.

[16] Millikan, Robert A. *Mechanics, Molecular Physics, and Heat.* Boston, MA: Ginn and Company, 1902.

[17] Reif, F.; Heller, J.I. "Knowledge structure and problem solving in physics." *Educational Psychologist* 17 (1981) 102-127.

[18] Rigden, John S. "The current challenge: Introductory physics." *Amer. J. of Physics* 54 (1986) 1067.

[19] Verbrugge, F. "Conference on introductory physics courses." *Amer. J. of Physics* 25 (1957) 127-128; "Improving the quality and effectiveness of introductory physics courses," *ibid,* 417-424.

[20] Wilson, Jack M. "The impact of computers on the physics laboratory." Int. Sum. Workshop: Research on Physics Education, La Londe les Maures, France, June 26-July 13, 1983. Proceedings, p. 445.

[21] Wilson, Jack M. "Microcomputers as learning tools." *Inter-American Conference on Physics Education,* Oaxtepec, Mexico, July 20-24, 1987. (Proceedings to appear.)

EDWARD REDISH is Professor of Physics and Astronomy at the University of Maryland, College Park. Former chairman of his department, he is creator and principal investigator of M.U.P.P.E.T.—the Maryland University Project in Physics and Educational Technology. He is a member of the Nuclear Science Advisory Committee and Chairman of the Advisory Committee for the Indiana University Cyclotron.

Calculus in the Undergraduate Business Curriculum

Gordon D. Prichett

BABSON COLLEGE

If there is widespread common concern that a core course in the standard undergraduate business school curriculum is irrelevant and unneeded, it is a concern over the calculus. Although there has been, for twenty-five years, a stalwart cadre of supporters of calculus in the business core, there has been little common agreement on either the level or the topics which are critical to the success of such a core course. To gain a clear insight into the appropriate role of calculus in the business curriculum in the 1990s, it is important to sketch the evolution of the present role played by calculus in the life of a business student.

The Past

In 1963 Richards and Carso [3] reported that only 4 of 71 AACSB (American Assembly of Collegiate Schools of Business) schools responding to a questionnaire required differential calculus. Tull and Hussain [4] in a

1966 report on AACSB schools show 38 out of 84 requiring differential calculus. Something seems to have occurred in the early sixties that suddenly convinced schools of business that calculus was important to their curriculum.

We should recall the tempestuous climate of educational concern of the early sixties, the emergence of many new mathematics incentives, and the widespread criticism of any curriculum that did not have "strong" mathematical underpinnings. It may be worth noting that the Harvard Business School introduced a calculus requirement for entering students in the early sixties which survived only until the late sixties!

Two questions are critical to our present concerns:

1. Why was calculus suddenly inserted into the business curriculum?
2. Who designed the appropriate calculus course for this specialized educational track?

In the business school reports of the 1950s, a few schools like Carnegie Mellon and the Sloan School of MIT emerged as models of futuristic curriculum emphasis. The background ambience of engineering and economics departments within both these schools, coupled with strong positions taken up by specific apostles, made one semester of calculus appear to be minimal mathematical background for management students.

The push toward calculus was further supported by works like Samuelson's *Foundations of Economic Analysis.* Some, like Kemeney, Snell and Thompson, decided that business applications made it preferable to teach finite vs. continuous versions of the same concepts and tools, but they were fighting two battles—the general battle for more emphasis on mathematics, and a battle for a variant approach which few of their fellow mathematicians endorsed.

Many business calculus courses ... are taught by mathematicians ... unfamiliar with the general business curriculum. ... This marriage of an orphaned course to a distant curriculum leads one to today's concerns.

The schools that took calculus most seriously expected students to do a full year, and they were content to let existing service faculty in mathematics departments teach it. Many business calculus courses began and still reside in mathematics departments and are taught by mathematicians, a great number of whom are unfamiliar with the general business curriculum.

Since business schools had not utilized calculus in the past and were not familiar with many of the applications of calculus beyond those in economics, course design was left to mathematicians and textbook authors. This led to a calculus course designed in a wilderness separating pure mathematics from a very specialized use of mathematics. This marriage of an orphaned course to a distant curriculum leads one to today's concerns.

The Present

There is not a standard introductory calculus course in today's business curriculum. Although AACSB requirements do not explicitly state that calculus is a necessary prerequisite for an accreditable school, any school not offering calculus in their core would most likely come under criticism for the exclusion. It should also be noted, however, that of the approximately 1200 business programs in the United States, only 237 are accredited by the AACSB. Many nonaccredited schools offer no calculus requirements, many claiming (justifiably) that their students are far too weak in algebra to undertake even a superficial calculus course. Business schools do not attract a great many quantitatively strong students, so weak arithmetic and algebraic skills plague teachers of the business calculus course.

Teaching business calculus today is, to most, a thankless task.

The present calculus course offered by those schools requiring calculus is frequently referred to as a watered-down version of calculus, and looked at with distain by many pure mathematicians. One must realize that much of what is taught in the standard calculus for mathematics students has evolved from studying the pathologies of special functions or special situations which enjoy an important role in the life of a mathematician, but which almost never occur in the life of a practitioner in management.

A typical business calculus course would be selected from the following menu, in which parentheses surround topics that are usually optional and often omitted:

1. Review of algebra and sets
2. Formulas, equations, inequalities, graphs
3. Linear equations and functions (Applications: Break-even analysis; Linear demand functions)
4. Systems of linear equations and inequalities (Applications: Supply and demand analysis; Introduction to linear programming)

5. Exponential and logarithmic functions
6. Mathematics of finance
7. Introduction to differential calculus
   a. Limits and continuity
   b. The difference quotient and definition of the derivative
   c. The simple power rule $d(x^n)/dx = nx^{n-1}$
   d. The derivative of $[f(x)]^n$
   e. The product and quotient rules
   f. The derivatives of exponential and logarithmic functions
   g. Maxima and minima of functions
   h. Maxima and minima applications
   i. (Sketching graphs of polynomials)
   j. (Relative rates of change)
   k. (The chain rule and implicit differentiation)
   l. (Calculus of two independent variables)
8. Integral calculus
   a. Antiderivatives: The indefinite integral
   b. Integration by substitution
   c. Integrals of exponentials
   d. The integral of $(mx + b)^{-1}$
   e. Area and the definite integral
   f. Interpretive applications of area
   g. (Improper integrals)
   h. (Numerical integration)
   i. (Integration by parts)
   j. (Differential equations)

A quick look at the preceding outline should shock any teachers sensitive to the fact that they have only *one* semester to bring a class of students with a wide variety of backgrounds and quantitative abilities to a common level of proficiency in most of the above topics. Teaching business calculus today is, to most, a thankless task. Add to this the pressure today to perform well on teaching evaluation surveys, the philosophy used by many schools that calculus is a primary sieve to weed out poor students, and the general discrediting of the integrity of the course by most mathematicians, and one wonders with whom the course is staffed in any college or university. Clearly the design and role of the calculus course in the business curriculum needs review and restructuring.

The Future

To understand why so much must be taught in one semester, one must have a better understanding of the scope of the entire business curriculum. As with engineering students, most business students have almost no room for electives in their program. As more and more emphasis is being placed on broadening the liberal arts base of the business curriculum, more courses are attached to the front end of the curriculum, but few courses yield space elsewhere in the curriculum.

At present most business students are required to take three quantitative methods courses: Business Calculus or some equivalent mathematics course; Statistics (this course is essentially descriptive and inferential, requiring little probability theory); and Information Systems (a course involving elementary programming and computer applications). Electives include Management Science (applications of operations research), Finite Mathematics (a course that might better be taught in the spirit of an applied discrete mathematics course), and further courses in Management Information Systems. Little physical or life science is required of or taken by business students.

Calculus is not a true curricular prerequisite to studying in an undergraduate business program.

From this it is clear that calculus is not a critical prerequisite for any of the technical courses in the standard business curriculum. In fact the only courses that require calculus, excluding case by case instances, are some introductory economics courses. In most circumstances the applications of calculus in these courses are what many term as "toy problems" and these applications could easily be presented without relying on the calculus. Hence calculus is *not* a true curricular prerequisite to studying in an undergraduate business program. (Recall the decision of Harvard Business School to drop the calculus requirement.)

A New Course

For years calculus has been used by colleges and universities as a credentialing hurdle, as a prerequisite to assure sufficient "mathematical sophistication." This remains the primary justification for maintaining a full semester calculus course in the business curriculum. It is time to design a new introductory quantitative methods course to measure this vital prerequisite. The treatment of calculus should be embedded in a course which will assure that all business students have sufficient mathematical sophistication, receive content of real educational value, and encounter the mathematics and quantitative problems of management education.

The design of such a course should be inspired by the goals of teaching clear reasoning, critical and analytical thinking, and good organizational and writing skills. The course must assure that students who complete it have sufficient mastery of fundamental arithmetic and algebraic skills, and can apply the use of calculators and computers to obtain and process quantitative information. In addition, this course must address specific mathematical techniques frequently applied to solve problems encountered in management education. Above all, the underlying intent of such a course is to impart educational challenge and value to first-year students of management, with utility and relevance as a catalyzing rather than driving force of the course.

Such a course may be properly called *Foundations of Quantitative Methods in Management;* a suggested course outline might be:

1. Linear equations and functions (Applications: Break even analysis; Linear demand functions)
2. Introduction to linear programming techniques
   a. Problem formulations and applications
   b. Graphical solutions
   c. Computer solutions using a computer package
3. Exponential and logarithmic functions
4. Mathematics of finance
   a. Use of the calculator
5. Introduction to differential calculus
   a. The difference quotient and definition of the derivative
   b. The simple power rule, $d(x^n)/dx = nx^{n-1}$
   c. The derivative of $[f(x)]^n$
   d. The product and quotient rules
   e. The derivatives of exponential and logarithmic functions (Applications: Response functions; Marginal analysis in business and economics)
   f. Computer computations using a package such as MACSYMA
   g. Maxima and minima of functions (Applications: Inventory control models)
6. Integral calculus
   a. Antiderivatives: The indefinite integral
   b. The integral $\int x^n dx, n \neq 1$
   c. Integral of e^x
   d. Area and the definite integral
   e. Numerical integration
   f. Computer integration using a program such as MACSYMA
   g. Interpretive applications of the integral (Applications: Marginal analysis)
   h. Difference and differential equations (Applications: Rumor spreading models)

Exclusion and Inclusions

Much material that has been prevalent in the past has been excluded from the above outline.

1. Limits: They should be taught in the context of the definition of the derivative, not in general
2. Continuity (define only)
3. Curve sketching
4. Chain rule and implicit differentiation; treat case by case
5. Calculus of two independent variables
6. Improper integrals: Discuss in the statistics course if probability is introduced as an integral
7. Techniques of integration

In place of these omitted topics, special emphasis is given to certain topics not highlighted in the past:

1. Linear programming: Problem formulation, with graphic and computer solutions
2. The use of programs such as MACSYMA to compute derivatives, integrals, and solutions to complicated equations
3. Numerical integration
4. Interpretations of the derivative and integral
5. Concept of optimization
6. Difference and differential equations with applications

An Agenda for the 1990's

Since the object of the above outline for a quantitative methods course in the business curriculum is the same as the object of recent suggestions by Graham [6] for changes in mathematical preparation for science, viz., "to help create new mathematics curricula that are interesting, sophisticated, innovative, and well integrated with the applications of mathematics," I will follow the example of Steen in a recent article on mathematics and science (7) and suggest that the following agenda is worth starting on:

1. *Firm prerequisites should be established for students entering the quantitative methods course for business.* A core course should not be used as a sieve. Students should be required to make up deficiencies, or not be admitted to the curriculum.
2. *Only teachers who are familiar with the business curriculum and appreciate the role of quantitative methods in business should teach business mathematics.* Mathematics departments must work more closely with the curriculum for which they are a service department.
3. *The design and maintenance of a good business mathematics program should be justly rewarded both*

professionally and financially. Ways must be found to reward quality performance and innovation within a mathematics department in pedagogical areas not classically considered mathematically pure.

4. *The chief objective of a core course in quantitative methods for managers should be the teaching of critical and analytical skills.* Clear reasoning, a sense of confidence in one's ability to solve mathematical problems, and well-developed organizational and expository skills are critical to success in management education in the 1990's.
5. *Any core course in quantitative methods for managers must incorporate the use of computers and calculators.* To exclude computers and calculators from solving quantitative problems in the 1990s would be equivalent to attempting solutions to trigonometric problems fifty years ago without tables.
6. *The quantitative methods course should be linked to other courses in the business school curriculum.* Understandable and believable reasons why skills learned from mathematics transfer directly to skills needed in management must be included in every quantitative core course.
7. *Teaching and evaluating writing and organizational skills should be an integral part of teaching any quantitative methods course.* The expository skills of most college students are dismal; it is time to correct the problem, not point the finger. We must embrace writing across the curriculum as a primary initiative of the 1990's.

References

[1] F.C. Pierson, *et al.* *The Education of American Businessmen.* McGraw-Hill, 1959.

[2] R.A. Gordon and J.E. Howell. *Higher Education for Business.* Columbia University Press, 1959.

[3] M.D. Richards and R. Carso, Jr. *Mathematics in Collegiate Business Schools.* Monograph C-10, Southwestern Publishing Company, January 1963.

[4] D.S. Tull and K.M. Hussain. "Quantitative methods in the school of business, a study of AACSB member institutions," *AACSB Bulletin,* January 1966.

[5] E.K. Bowen. "Mathematics in the undergraduate business curriculum," *Accounting Review,* October 1967, 792.

[6] W.R. Graham. *Not. Am. Math. Soc.* 34 (1987) 245.

[7] L.A. Steen. "Mathematics education: A predictor of scientific competitiveness," *Science* 237 (July 17, 1987) 251-252, 302.

[8] E.K. Bowen, G.D. Prichett, and J.C. Saber. *Mathematics: With Applications in Management and Economics,* Sixth Edition. Richard D. Irwin, 1987.

GORDON D. PRICHETT is Vice President for Academic Affairs and Dean of the Faculty of Babson College in Babson Park (Wellesley), Massachusetts. Previously, he was Chairman of the Department of Mathematics at Hamilton College, Chairman of the Department of Quantitative Methods at Babson College, and Secretary-Treasurer of the Northeast Section of the Mathematical Association of America.

Calculus for the Biological Sciences

Simon A. Levin

CORNELL UNIVERSITY

Any consideration of the teaching of calculus to biologists must recognize that calculus is just one of a variety of mathematical tools that increasingly are proving important to the biologist, and that therefore the teaching of calculus must be viewed within the context of a broader mathematics education. The relevant mathematical tools will vary depending on the biological specialty. For example, for the ecologist or geneticist, statistics and experimental design are integral parts of the curriculum; but for the biochemist, such courses are considered a luxury.

However, needs change, and often over very short time scales. For example, in the last decade, computing has become an essential part of any scientist's training. Probability and combinatorics, which within biology were once primarily the province of population geneticists and ecologists, have grown in importance for molecular biologists interested in sequence analysis. Even topological methods are finding applications within the core areas of biology. Such potential for change in the areas of mathematics that may prove useful argues for the value of training biologists broadly in mathematical methods, rather than restricting attention to narrow and classical areas.

There is a tradeoff, however. The typical biology curriculum, especially within areas such as molecular biology and biochemistry, is packed tight with required courses in biology and the physical sciences. There is little slack, and little room for flexibility. A year of mathematics is recognized as being essential, but it is unlikely that most biology programs can afford to add additional mathematics as a general requirement. Thus, new topics can be introduced only if one is willing to eliminate others, or to sacrifice depth for breadth.

Needs change, and often over very short time scales.

My argument is that some such compromise is essential, and that the way to achieve it is to emphasize conceptual understanding at the expense of analytical computational ability. This is a painful exchange, because understanding in mathematics cannot be achieved without repetitive drill and problem solving. Nonetheless, it must be recognized that the biologist's needs are different in kind than the physicist's or engineer's. The biologist needs to understand the role of models in biology and medicine; to know what mathematics is, what it can do, and what methods are available; and to know where to go to obtain deeper capabilities.

Mathematical Models

Models in biology serve as pedagogical tools, as aids to understanding, and as rough approximations that guide treatment or experimentation. On the other hand, although there are exceptions to the general rule, mathematical models rarely serve as devices for exact prediction in the way that models can in physics. Thus the most important mathematical problems facing the biologist are conceptual, involving the proper formulation of a model, rather than analytical.

Some examples should strengthen this argument. A physician, in prescribing a drug regimen, should understand the importance of the underlying kinetics, the meaning of the half-life of the drug within the body, and the notion that the system eventually will reach an approximate steady-state, although the plasma level of the drug may fluctuate substantially between dosages. This will alert him or her to the importance of testing plasma levels at a fixed point in the dosage cycle, and to the necessity of establishing quantitative relationships between dosage levels and steady-state plasma levels.

There is sufficient variation among individuals, however, that generic models cannot be completely reliable, even when standardized by weight or age. Steady-state levels must be determined empirically. The model, however, has performed an invaluable role in suggesting what needs to be measured. More generally, identifying what biological quantities need to be measured—the refractory time of a neuron following firing, the binding constant of a ligand in relation to a particular substrate, the heritability of a genetic character, the diffusion rate of a particular compound across a membrane, the maximal rate of increase of a bacterial population—is one of the most common uses of models.

Because of the difficulty of controlling individual and environmental parameters, the biological model more typically is used to describe an idealized situation. For example, the Hardy-Weinberg conditions, which specify the genotypic frequencies to be achieved in a population in the absence of any selective differences among individuals, describes an idealization that may represent an excellent approximation under some conditions, and a very poor one under others.

The most important mathematical problems facing the biologist are conceptual, involving the proper formulation of a model, rather than analytical.

Similarly, the Lotka-Volterra model, simplistically describing the interaction between predator and prey species, represents the biological analogue of the frictionless pendulum. The structural instability of the system makes it totally inadequate in describing any real biological interaction; but as a pedagogical tool, the model is sufficient to illustrate how a particular mechanism can lead to inherently oscillatory behavior. Often, the role of the model is simply to abstract and isolate a piece of a larger system, in the recognition that it is difficult or impossible to achieve such isolation in any real system, no matter how controlled.

The preliminary conclusions are that the teaching of calculus must be integrated with the teaching of other mathematical topics, such as analytic geometry, dynamical systems, linear algebra, and computational methods. Furthermore, considerable attention should be paid to discussion of the proper role of mathematical models, to the conceptual insights that can be achieved through the qualitative theory of dynamical systems, and to an appreciation of how analytical approaches interface with computational ones.

The availability of high speed computers has rendered archaic the practice of equipping the student with a slide rule; and similarly, it has reduced the necessity of training students to be facile in the clever manipulation of integrals. Indeed, there still is value in introducing the student to the slide rule and to log tables, in order to provide a perspective on logarithms and powers; similarly, there is considerable value to introducing the student to the techniques of integration, and to provide some drill. However, such techniques cannot be viewed as ends unto themselves, but must take their place in an integrated and balanced curriculum whose primary goal is to teach the relevance of mathematics in biological and medical applications.

Calculus for Biologists

Setting aside the importance of teaching discrete mathematics, which is outside the charge of this particular report, I turn attention next to the key ingredients of a two-semester sequence in calculus for biologists. In so doing, I do not distinguish numerical methods as a separate topic, since such methods are taught most effectively when integrated with mathematical applications. Assuming some background in analytic geometry and complex numbers, I propose that the essential ingredients are:

1. Introduction to limits, rates, and the differential calculus
2. Differentiation rules and formulas
3. Curve plotting and the theory of maxima and minima; Taylor's theorem
4. Introduction to first-order differential equations
5. Introduction to the integral calculus
6. Integration rules and formulas
7. Integration and the solution of differential equations
   a. Solution of linear equations
   b. Methods for nonlinear equations
8. Areas, averages
9. Introduction to vectors and matrices, with eigenvalues
10. Partial differentiation: Taylor's theorem for several variables, and the theory of maxima and minima
11. Systems of differential equations
    a. Solution of linear systems
    b. Qualitative theory for nonlinear systems
12. Differential equations of higher order
13. Partial differential equations, especially of parabolic type
14. Other topics, as time permits, including infinite series, volumes, and areas

Naturally, there is no single way to teach this material, and some aspects of the above outline—for example, the preference for teaching integration as motivated by the solution of differential equations—are matters of personal taste. Other substantive choices also have been made in the above syllabus, going beyond those identified in the previous section.

For example, the inclusion of parabolic partial differential equations as an essential topic is motivated by the widespread applicability of concepts of diffusion throughout biology. On the other hand, other than the use of Fourier analytical methods for time series analysis, infinite series play little role in the training of a biologist. The Taylor expansion is important as a concept; however, it need not be viewed within the context of infinite series, since in applications it almost always would be truncated after the first few terms.

Biological Examples

Most of the topics listed are standard for a basic calculus course, but their presentation to biologists can be improved considerably by the development of biological examples. One should build on models that are fairly simple in biological content, and that offer the potential for addressing a variety of mathematical issues within the same biological framework.

Often, the role of the model is simply to abstract and isolate a piece of a larger system, in the recognition that it is difficult or impossible to achieve such isolation in any real system, no matter how controlled.

For example, various methods for graphing the relationship between the amount of a ligand bound to a substrate and the dosage level are used by pharmacologists for discovering the mechanisms underlying binding, and for distinguishing among competing binding models. So-called allosteric models postulate that the receptor molecule undergoes a conformational change when a single drug molecule is bound to it, and that this change affects binding and dissociation rates for other sites on the receptor molecule. Depending on whether binding thereby is facilitated or inhibited, this is known as positive or negative cooperativity. It introduces a particular nonlinearity into the model, representing the fact that binding rates change with concentration. By examining binding curves directly, and various transforms such as double reciprocal plots, one can use qualitative information regarding curvature and extrema to distinguish among the various hypotheses.

Such examples are likely to be closer to the heart of the biologist than are the usual examples from engineering, although the latter should not be neglected completely. It is an unfortunate fact that most beginning biology students know very little biology, and may not be as strongly motivated by biological examples as one would hope. Examples therefore must be kept as simple as possible, so that the focus is on learning the mathematics rather than on learning the biology.

Furthermore, it is important that the student be clear in knowing whether the particular mathematics learned is of general relevance and applicability, or whether it is particular to the example used to motivate it. The importance of maintaining this distinction is an argument for presenting ideas fairly abstractly, using biological examples for occasional motivation.

Optimization

The theory of maxima and minima is important to biologists not only as an aid to graphing, but also in relation to fundamental biological and management concepts: principles of optimal design, maximal rates of increase of populations, optimal treatment regimes in medicine or optimal harvesting regimes in fisheries, and evolutionary adaptation. Most of these are best treated within the later sections of the course, embedded within a dynamical systems description. The change of populations through evolutionary processes is the single most important organizing principle in biology. The fundamental philosophical problem is to understand the relationships and distinctions between the process of adaptation and the possible attainment of optima.

It is an unfortunate fact that most beginning biology students know very little biology.

The simplest models of gene frequency change assume that there is a fixed selective value associated with each genotypic combination. For the dynamical systems appropriate to these assumptions, it is possible to demonstrate that the average fitness of all individuals in the population increases monotonically with time, tending asymptotically to a local maximum. Thus, such models can be used to illustrate a variety of concepts: maxima and minima, local versus global extrema, the asymptotic behavior of ordinary differential equations, multiple steady states, linearization, and stability.

More complicated models, incorporating the evolutionary tradeoffs and constraints that are familiar to the biologist, provide opportunities for illustrating the use of Lagrange multipliers. Other such examples come from fisheries or epidemic management, where there is an explicit economic or effort constraint. Similar models appear in behavioral biology, where time or energy must be allocated to tasks such as foraging for food, grooming behavior, or other activities, and in physiological ecology, where the principles underlying a plant's allocation of resources to growth and reproduction, or to roots and shoots, is of fundamental interest.

Dynamical Systems

Dynamical systems have a pervasive influence in the understanding of biology, and it is hard to think of an area where they are not an essential part of basic instruction. In biochemistry and pharmacology, the dynamics of enzyme-substrate and drug-receptor associations are among the most basic of concepts. Chemotherapy models, even with simple first-order kinetics, give rise to systems of differential equations representing the flow of the chemicals through a network of compartments. Formally identical models are applicable to the flow of materials and energy through ecosystems. Thus even the linear theory is of direct relevance to a wide variety of problems, spanning the biological spectrum.

Indeed, the most basic model of drug dynamics is of the form $dx/dt = I - kx$, where I is the dosage rate and k is the per unit rate of elimination. The physician is familiar with the fact that this model predicts a steady-state level I/k, that there is a characteristic half-life determined by k, and that this half-life determines the time to rid the body of the drug if I is suddenly set to zero, or the time to reach a new equilibrium if I is abruptly changed from one value to a new one.

Thus this model has immediate applicability and relevance, and furthermore serves as a starting point for the investigation of such extensions as allowing I to vary periodically (as would be the case for any dosage regimen other than a continuous intravenous), or allowing for an explicit delay representing the time it takes for the drug to make its way into the plasma. Another direction for extension is to consider the multi-compartment systems that are more appropriate to drug dynamics, and thereby to account for the distributed delays associated with the gradual entry of the drug into the plasma.

The consideration of nonlinear dynamical systems substantially broadens the range of applications, and provides a framework for the discussion of a variety of important concepts: steady states, linearization, transient dynamics, asymptotic stability, bifurcation, periodic solutions, and chaos. As advanced and esoteric as some of these topics may seem, they all are finding application in biological investigations. Understanding these concepts undoubtedly is of more value to the biologist than is the ability to make specific analytical computations.

The change of populations through evolutionary processes is the single most important organizing principle in biology.

Qualitative shifts in the physiological behavior of an individual, from one basin of attraction to another, may correspond to a shift from a healthy to an unhealthy state. In some cases, this may correspond to a breakdown of a regulatory mechanism, and the transition from a homeostatically maintained stable equilibrium to a fluctuating or even chaotic dynamic. In other cases, the healthy state may exhibit carefully regulated periodicity, as in cardiovascular dynamics, or in the daily circadian rhythms of the body. Qualitative changes in behavior may be associated with internal changes, such as the breakdown of normal regulatory mechanisms, or with external changes, such as the intake of a toxin.

Attention to dynamic phenomena is equally important in other areas of biology. For the neurobiologist, the sustained repetitive and sometimes chaotic firing in networks of neurons represents a behavior that can be understood only within the context of the mutual excitatory and inhibitory interactions within the network, and the transition from normal to more erratic behaviors can be understood in terms of bifurcations tracking changes in critical internal and external parameters. The genetic control of development similarly is mediated through networks of regulatory pathways.

For the ecologist or epidemiologist, periodic phenomena in the dynamics of populations or disease long have been central subjects for investigation, and the various hypotheses regarding the mechanisms underlying them have been well studied within the context of dynamical systems theory. Indeed, even simpler notions associated with transient behavior have been of critical importance.

An Example from Epidemiology

Again, it is worthwhile to illustrate with an example. One of the simplest and classical models of the transmission dynamics of infectious diseases assumes that the host population is of constant size, and broken into three categories: susceptible (S), infectious (I), and recovered (immune) (R). Individuals move among these classes according to certain rules, including the possibility of losing immunity and returning to the susceptible class. Because the time scale of a particular epidemic is assumed to be short compared to the change in population density, it is assumed that births balance deaths. The usual resultant model takes the form

$$dS/dt = b - kSI - bS$$
$$dI/dt = kSI - bI - vI$$
$$dR/dt = vI - bR$$

Here b is the birth (=death) rate, v is the recovery rate of infectious individuals, and kSI, the incidence function, represents the assumption that the rate that new infections occur is proportional to both the number of susceptible individuals and the number of infectives. Naturally, any of the assumptions can be modified, and so the above model again serves as a starting point for investigation, and as a way to introduce concepts.

Because population size is constant (it needn't be in more general models, such as those that incorporate disease-induced mortality), it is convenient to treat S, I, and R as fractions, so that $S + I + R = 1$; this assumption is implicit in the particular formulation above, and allows reduction of the apparent 3-dimensional system to two dimensions.

The simplest concept that emerges from the system above is that of the threshold transition rate. It is easily seen, because S cannot exceed unity, that no outbreak can occur unless $k > b + v$; that is, if $k < b + v$, the number of infectives will decrease monotonically, whatever the initial values. This simple concept is of great importance to disease management, because it gives a measure of the amount by which the transmission of a disease must be reduced in order to prevent outbreaks.

By similar reasoning, one can demonstrate that no outbreak can occur unless the proportion of susceptibles exceeds a threshold value $(b + v)/k$. Thus, vaccination strategies aimed at reducing the susceptible population below that threshold can succeed in the control of the disease. Finally, when the threshold condition is exceeded and outbreaks are possible, the above model predicts that the population will settle down to a situation in which a stable fraction $(b + v)/k$ of the individuals are susceptible.

Thus, as with some of the other examples mentioned, this example has sufficient reality to motivate the importance of the underlying concepts. More complicated versions, for example those that incorporate a latent period, or different susceptibility classes, or seasonality, can be used to introduce more complicated dynamics; and equally importantly, to demonstrate how models are used in the investigation of biological phenomena, and in the management of applied biological problems.

Diffusion Processes

Finally, the above outline makes a case for the early consideration of partial derivatives, and for some discussion of partial differential equations. In defense of the latter recommendation, I make reference to the ubiquitous nature of diffusion phenomena in biology. Diffusion models have been used successfully to describe the flow of heat and materials within animals and plants, the movement of solutes in ecosystems, the spread of particulates from smokestacks, and the passive spread of animals and plants. Early applications of these models addressed the spread of species introduced into new habitats, the rates of spread of advantageous genes, and the chemotactic movement of bacteria.

Fundamental work in neurobiology has used diffusion models to describe the transmission of neural impulses, and basic models in genetics have used diffusion approximations for stochastic processes to determine the gene frequencies to be expected under the primary influence of random genetic drift. Diffusion-based models have been fundamental to the understanding of geographical gradients and patterns in the distribution of terrestrial and oceanographic species, and in the frequencies of different genetic types. Finally, some of the most original and stimulating models of how development takes place have been structured on a discussion of pattern formation in systems involving chemicals that react with one another while diffusing through a medium. Thus, there is considerable motivation for including at least an introduction to these concepts in the basic calculus course for biologists.

Demonstrating Relevance

Although I made specific recommendations concerning topics to be included in a one-year calculus course for biologists, my fundamental recommendations concern the objectives of such a course. The primary goals of teaching biologists mathematics should be to teach them how models can be used in biology and medicine, and to introduce them to a broad range of basic concepts even at the expense of training that makes students facile in performing calculations.

Too often, the biologist takes calculus as a freshman, and has forgotten all that he or she learned long before graduation from college, primarily because the issue of relevance has never been addressed. Our goal in teaching mathematics should be to demonstrate relevance, by our choice of topics and examples, by our method of presentation, and by our emphasis on concepts.

The primary goals of teaching biologists mathematics should be to teach them how models can be used ...even at the expense of training that makes students facile in performing calculations.

As biologists so taught realize the importance of mathematical concepts, such concepts will become better integrated into biological investigations, and ultimately into the teaching of biology. This has occurred in some instances, for example in the teaching of enzyme kinetics, population genetics, and population ecology; but there is considerable potential for improvement.

Our goal should be to provide biologists with a better appreciation and understanding of mathematics and models, and thereby to move the subject in the direction of more quantitative rigor. In my opinion, the failure to do this in the past has been because we have assumed that the needs of biologists are identical to those of physicists and engineers, and that biologists are somehow mathematically more backward and less quantitative.

We need to recognize the different nature of the subject, and to tailor our courses so that they are of maximal use to the biologist, and can convince the biologist of the need to think mathematically and rigorously. Only then will we have met the challenge facing us.

SIMON A. LEVIN is Charles A. Alexander Professor of Biological Sciences at Cornell University in Ithaca, New York. He is President of the Society for Mathematical Biology, and is Director of the Center for Environmental Research and of the Ecosystems Research Center at Cornell, and a member of the Center for Applied Mathematics. He is a member of the Commission on Life Sciences and of the Board on Basic Biology of the National Research Council.

The Matter of Assessment

Donald W. Bushaw
WASHINGTON STATE UNIVERSITY

The content of a calculus course, and even the spirit in which it is taught, do not exist in isolation. They have external associations with prerequisites, courses for which calculus itself is prerequisite, applications, the *Zeitgeist,* career plans of students, numbers and quality of students, and all sorts of other pressures and constraints; about these the aspiring reformer of calculus can do little. But things *can* be done about the choice of instructional materials, technological aids, and modes of instruction.

The content and spirit of any course, calculus for example, are also commonly associated with many kinds of evaluations: of students who want to enter the course, of students in the course, of students who have completed the course, of texts, of teachers, of the content and spirit themselves.

Any ... reform should anticipate the need to keep the evaluation aspects of the new calculus in harmony with its intended content and spirit, and include ways of meeting that need.

A reform of calculus that ignores evaluation—that attempts to put the new wine of a "lean and lively calculus" into an old bottle made of evaluation techniques developed, perhaps inadequately, for old courses—is very likely to perish. Any such reform should anticipate the need to keep the evaluation aspects of the new calculus in harmony with its intended content and spirit, and include ways of meeting that need.

For this reason, it is important that anyone considering calculus reform be aware of some information about academic evaluation. The subject is large, and the allotted space is small, so this treatment is perforce sketchy. The most useful part of the essay may be the list of references, which provide openings into pertinent parts of the literature.

A Few Definitions

Talk of evaluation and assessment is very much in the higher education air these days, and rightly so. The terminology, however, is muddy. In this essay the term "assessment" will be taken as generic, and will be meant to apply to any systematic procedure for judging the quality of an educational process or of the outcomes of that process.

It will be useful for our purposes to identify three major forms of assessment:

- *In-course Testing:* In-course or final assessment of student progress;
- *Teaching Evaluation:* Assessment of the quality of the educational process, usually by direct observation in courses;
- *Outcomes Assessment:* Assessment of student abilities, attitudes, accomplishments, etc., presumably acquired through some specific educational experience or experiences.

These three categories of assessment are fuzzy sets. For example, a final examination, surely a form of in-course testing, can also be considered a form of outcomes assessment. It can also be regarded as a type of teaching evaluation: when a whole class's performance on a final examination is disappointing, we tend to ask ourselves what we did wrong; and examination results in multi-section courses are sometimes used quite deliberately as measures of the effectiveness of the instructors involved.

In-course Testing

The most common type of assessment is in-course testing, which uses many devices: graded homework, quizzes, examinations, oral presentations, team projects, term papers, etc. It provides a base for grading; it provides students with information about "where they stand;" it can provide the instructor with valuable clues about individual or epidemic student difficulties; and, when properly done, it can provide experiences of much intrinsic educational value. Good advice on this subject is contained in books like those by Eble [6], McKeachie [14], and Lowman [13]; more technical information appears in such books as that by Gronlund [11].

Teaching Evaluation

The purpose of teaching evaluation is to obtain information about the quality of instruction in a particular class, or by a particular instructor. The two purposes for teaching evaluation most often mentioned are the identification of areas where improvement is in order and the provision of part of the base for personnel decisions.

Teaching evaluation is *not* limited to the use of questionnaires to be filled out by students. "Student evaluation" can be, and usually is, an important element in teaching evaluation, but the latter can also include systematic evaluation by colleagues, self-evaluation (or, better, self-reporting), and assessment of student progress—not to mention subjective judgments by administrators, or the venerable supposition, still unfortunately current in some quarters, that instruction is satisfactory unless there is clear contrary evidence, e.g., an exceptionally large number of student complaints.

Justice simply cannot be done to some "central ideas" ... by using only elementary mathematical symbolism or multiple-choice responses. Writing is often the natural vehicle.

Teaching evaluation is discussed seriously in a large number of articles and several recent books, including those by Centra [4] and Seldin [16] and that edited by French-Lazovik [9]; see also the recent article by McKeachie [15]. A report on teaching evaluation addressed specifically to instructors of postsecondary mathematics is in preparation by a committee of the Mathematical Association of America chaired by the author of the present paper.

Outcomes Assessment

The broadest form of assessment is outcomes assessment, which is usually done for purposes of student certification, program evaluation and improvement, or accountability to some monitoring agency or to the public.

Placement testing, as a form of assessment of outcomes of previous educational experience, really belongs here, and in some schools is the most prominent form of outcomes assessment. Many of the things that will be said about outcomes assessment or assessment in general therefore apply also to placement testing, which nevertheless will not be emphasized in this paper.

Outcomes assessment can be performed in many ways through a variety of generally available instruments: many placement examinations, GRE general tests and subject tests, the Fundamentals of Engineering Examination (formerly EIT), the LSAT, the MSAT, the COMP system of ACT, and the Academic Profile just now being launched by ETS. There are also locally produced written examinations as well as senior theses, oral examinations, surveys of graduates, analysis of job placement data, and various special ways of assessing writing proficiency.

Until recently outcomes assessment has not been conspicuous in higher education, except in areas relating to placement, licensing, testing for admission to graduate and professional schools, and the like—primarily for the benefit of the individual student.

But in very recent years, partly in response to widespread dissatisfaction with—or at least uncertainty about—the quality of higher education, there has been an upsurge of interest in "assessment" as a vehicle for program improvement or institutional accountability. This is often expressed in state mandates or in criteria newly adopted by accrediting agencies. "A year or two ago, only a handful of states had formal initiatives labeled 'assessment.' Now, two-thirds do" (Boyer [3]). Useful information about assessment in this sense is contained in the volumes edited by Ewell [8] and Adelman [1].

Basic Principles

Although the three major types of assessment are usually discussed separately, the literature presents a near-consensus on certain principles that apply to all of them, principles based in part on research and in part on practical experience and good common sense. Here are some of them:

1. *Before any program of assessment is developed, that which is to be assessed should be clearly identified.*

In-course testing should really be designed to measure attainment of class objectives; teaching evaluation should not be practiced without some reasonable understanding (shared by assessors and assessed) on what characterizes effective teaching; and outcomes assessment should presuppose some standards related to student competency.

An important corollary to this principle is that emphases in in-course tests should agree with emphases in the course syllabus. Tests in a possible new calculus course that is more sharply focused on central ideas and on the role of calculus as the language of science should themselves be focused on the same themes. Students can learn quickly to give short shrift to "central ideas" and "roles" if it is known that "central ideas" and "roles" do not come up at exam time. Likewise, instructors in such a course should be evaluated on their ability to teach those things that are considered most important in the course, and in a manner compatible with the intended spirit of the course. Outcomes assessment should be conducted under the analogous requirement.

2. *Ideally, an assessment program should be multidimensional.*

For in-course testing, for example, one test is hardly ever enough. Even if tests were infallible, as they never are, the common practice of giving several tests in a course would be methodologically sound as well as humane. Studies show that when there are several tests before a final examination, performance on the final itself is improved. But there are various kinds of written tests, and also such potential assessment devices as assigned homework, separate essays, oral exams or interviews, evaluations of student notebooks or journals, and old-fashioned observation of classroom participation.

The common view that mathematics should perhaps be exempted from the current call for "writing across the curriculum" is questionable. Courses such as the "lean and lively calculus," where there is to be increased emphasis on ideas and decreased emphasis on mere memorization of procedures, will be natural settings for testing systems that rely not merely on problems with numerical answers or multiple-choice questions but also on writing.

If writing about mathematics is to be an important element of assessment, then it should be an important element of the rest of the course. Justice simply cannot be done to some "central ideas" of the calculus by using only elementary mathematical symbolism or multiple-choice responses. Writing is often the natural vehicle, and its use in a calculus class will be both a contribution to the improvement of writing skills generally and an opportunity to use writing as an effective device for learning. See for example Griffin [10], Herrington [12], or Emig [7].

In any case, a mix of reasonable testing methods can be expected to give more valid evaluations, and a richer educational experience, than any one of them alone.

With obvious modifications, the same remarks apply to teaching evaluation and outcomes assessment, where exclusive reliance on one assessment instrument or practice is almost always inadequate and is strongly discouraged by specialists in these fields.

3. *Assessment instruments should be designed with care and if possible checked for validity and reliability. When commercially produced instruments are available, they are often better—and cheaper, all things considered—than local productions.*

There is probably no need to dwell on the first of these observations; any experienced teacher knows how easy it is for something to go wrong in the production of a test, and how much damage can be done by a defective test.

To my knowledge, there now exist no generally available examinations that come anywhere near matching the vision of calculus emerging from the Tulane conference. This does not mean that there will not be any; indeed, the production of such an examination might be a good way of redefining that vision—of implicitly defining its objectives—for instructors and writers.

Let there be no facile talk about the evils of "teaching to the examination." If an examination is well crafted, ...then "teaching to" it can be a very good thing.

And please, let there be no facile talk about the evils of "teaching to the examination." If an examination is well crafted, and is complemented by other means of assessment in accordance with Principle 2, then "teaching to" it can be a very good thing. In general, performance criteria—and a good test should be a faithful expression of performance criteria—should provide guidance for the performers as well as for those who are responsible for evaluations of performance.

4. *Finally, ways should be sought to get something out of assessment besides ratings or scores of some kind.*

It is a commonplace of the literature on teaching evaluation that teaching evaluation should be coupled with a system of faculty development opportunities, so when the instructor is found to need improvement in some respect, there is a resource through which that improvement can be accomplished effectively. As suggested earlier, outcomes assessment is often part of a system of program review and planning and may lead to program improvements attained by diversion of resources or by some other means. Even in-class testing, except perhaps at the very end of a course, should help both student and instructor identify weaknesses and should provide a base for more effective teaching and learning afterwards.

All three of the major types of assessment that have been identified, and many of the specific methods of implementing them that have been mentioned, with corresponding methods of follow-up, might arise in the development of a new or substantially redefined course such as the "lean and lively calculus." Carefully selected assessment methods should be expected to contribute in important ways to the educational experience and also to the review and continuing improvement of the course. Thus anyone who is expected to benefit from advice on what forms the new calculus might take and on how it should be taught can also be expected to benefit from good advice on the assessment activities that go with it.

References

[1] Adelman, Clifford (ed.). Assessment in higher education: issues and contexts. Washington, DC: Office of Educational Research and Improvement, Department of Education, 1986.

[2] Astin, Alexander W. "Why not try some new ways of measuring quality?" *Educational Record* 63 (Spring 1982) 10-15.

[3] Boyer, Carol M., et al. "Assessment and outcomes measurement—a view from the states." *AAHE Bulletin* (March, 1987) 8-12.

[4] Centra, John A. *Determining Faculty Effectiveness.* San Francisco: Jossey-Bass, 1982.

[5] Donald, Janet G., and Sullivan, Arthur M. *Using Research To Improve Teaching.* San Francisco: Jossey-Bass, 1985.

[6] Eble, Kenneth E. *The Craft of Teaching: A Guide To Mastering the Professor's Art.* San Francisco: Jossey-Bass, 1976.

[7] Emig, Janet. "Writing as a mode of learning." *College Composition and Communication* 28 (May 1977) 122-27.

[8] Ewell, Peter T. (ed.) *Assessing Educational Outcomes.* San Francisco: Jossey-Bass, 1985.

[9] French-Lazovik, Grace. (ed.) *Practices That Improve Teaching Evaluation.* San Francisco: Jossey-Bass, 1982.

[10] Griffin, C. Williams. (ed.) *Teaching Writing In All Disciplines.* San Francisco: Jossey-Bass, 1982. See especially Barbara King, "Using writing in the mathematics class: Theory and practice," 39-44.

[11] Gronlund, Norman E. *Measurement and Evaluation In Teaching*, 4th ed. New York: Macmillan, 1981.

[12] Herrington, Anne J. "Writing to learn: Writing across the disciplines." *College English* 43:4 (April 1981) 379-87.

[13] Lowman, Joseph. *Mastering the Techniques of Teaching.* San Francisco: Jossey-Bass, 1984.

[14] McKeachie, Wilbert J. *Teaching Tips: A Guidebook for the Beginning College Teacher*, 7th ed. Lexington, MA: Heath, 1978.

[15] McKeachie, Wilbert J. "Instructional evaluation: Current issues and possible improvements." *Journal of Higher Education* 58:3 (May/June 1987) 344-350.

[16] Seldin, Peter. *Changing Practices In Faculty Evaluation.* San Francisco: Jossey-Bass, 1984.

DONALD BUSHAW is Vice Provost for Instruction and Professor of Mathematics at Washington State University in Pullman, Washington. He has served as chair of the Committee on the Undergraduate Program in Mathematics (CUPM) and as a member of regional accreditation teams. Bushaw chairs a committee on Evaluation of Teaching of the Mathematical Association of America, and is a member or chair of three other MAA committees.

Calculus Reform and Women Undergraduates

Rhonda J. Hughes

BRYN MAWR COLLEGE

One of the goals of the current initiative to reform the calculus curriculum in colleges and universities is to make science-based careers more accessible to women. Nevertheless, any program for reform should reflect evidence that women appear to do as well in calculus, as it is *now* taught, as do men. There is much to be learned from the relative success of women in calculus and in college mathematics, and this paper will examine evidence for that success, offer some explanations for it, and explore pedagogical possibilities for preserving that success in the midst of major reform of calculus as we now know it.

For most women, enrollment in calculus represents survival of their interest in mathematics despite overwhelmingly negative cues from our culture—women in calculus are already going against society's expectations for them. Statistics indicate that for most young women, interest in mathematics wanes in the adolescent years, and that the trend of avoiding mathematics continues throughout high school, accompanied by lower SAT scores in mathematics and reduced expectations of success in mathematical and scientific endeavors [1,

3]. There is some recovery in college (where calculus is usually taken), and then the attrition recurs in the graduate years [2]. Any program for calculus reform must bear in mind this complicated picture, and take into account the strengths and needs of women students.

The present interest in calculus reform was generated by a variety of concerns, including the widespread use of computers and the instructional possibilities they present, the realization that other topics might better serve the needs of undergraduates, and the failure of many calculus courses to capture the interest of students. Nevertheless, it is also a reflection of a general soul-searching among members of the mathematics community.

There is considerable evidence, both anecdotal and statistical, that women are doing well in calculus.

A study of Undergraduate Programs in the Mathematical and Computer Sciences (1985-86) showed that the number of computer science degrees in 1984-85 has more than tripled since 1979-80, while the number in mathematics and statistics was up only slightly from the level in 1980, and below the level in 1975 [4]. In a recent report on graduate mathematics enrollments by the Conference Board of the Mathematical Sciences (CBMS), it was noted that in 1986, 45.8% of graduate students enrolled in mathematics (at the top thirty-nine institutions) were foreign, as compared with 19.6% in 1977 [5]. The report states that "it is clear that we are failing to attract many of the very best potential [American] mathematicians." The number of Ph.D.'s awarded to U.S. citizens has fallen steadily from 774 in 1972-73 to 386 in 1985-86, while the relative percentage dropped from 78% to 51% [6]. In addition, the 1984 David Report warned that the community faces serious and urgent problems of revitalization and renewal, in the face of inadequate federal funding [7]. It is only natural that this pressing need for self-examination should spawn a movement to rethink and improve the manner in which we teach calculus, presently the focal point of mathematics education at the college level.

For those of us who have been concerned for years about the failure of mathematics to attract young women and minorities, this climate of reform and the accompanying flurry of activity is somewhat ironic. In the CBMS report, the committee cites a letter from one American-born graduate student stating that the foreign graduate students form close-knit groups, and he has few students to talk with [5]. This form of isolation is particularly damaging in the sciences, but it is by no means a new experience for women and minority graduate students. Indeed, many women I have known have experienced the same isolation as the above-mentioned American male. Thus the enthusiasm of the entire mathematical community for this issue is most welcome. As a consequence, calculus may undergo rejuvenation in a manner that will attract more young people, regardless of sex or race, to careers in mathematics, science, and science-based fields.

Women Succeed in Calculus

There is considerable evidence, both anecdotal and statistical, that women are doing well in calculus (the criteria I use to determine that a student is "doing well" are grades and continuation in mathematics or science courses). Indeed, in 1986 46% of baccalaureate mathematics majors were women [2]. If calculus were an obstacle for women, one would expect the percentage of women majors to be somewhat lower. Moreover, the attrition that occurs at the graduate level, reflected in the fact that (in 1986) 35% of mathematics master's degrees, and only 17% of Ph.D.'s in mathematics were awarded to women [2], can hardly be attributed to difficulties in calculus. A casual survey of calculus grades for women and men in several calculus courses taught at a variety of institutions over a ten-year period yielded the following data:

Grade	Women	Men	% Women	% Men
A	50	79	30.67	19.85
B	55	106	33.74	26.63
C	34	112	20.86	28.14
D	18	57	11.04	14.32
E	6	44	3.68	11.06

The data are from twenty-three first-year calculus sections at Harvard University, the University of Ohio, Tufts University, Temple University, and Villanova University; the instructors were both women and men. The data supports in a variety of ways anecdotal impressions that women do well in calculus: more than half of the women received A's or B's, and the failure rate for women is extremely low. In fact, the women appear to do better than the men, although the larger pool of men in the sample may account for the latter's higher failure rate and lower grades. A thorough investigation along these lines would be most informative.

There are several possible explanations for the phenomena observed in this modest experiment. The most obvious one has already been mentioned: those women who get to calculus are already somewhat motivated, having survived educational and cultural influences that may have been less than encouraging of their mathematical and scientific interests. The low failure rate for women convincingly supports this hypothesis. Another explanation is offered by the fact that (in 1983) almost 60% of college freshwomen who intended to major in mathematics reported "A" averages in high school, while the corresponding figure for men is about 45% [1]; thus, based on high school performance, the women in college mathematics courses are as strong as (if not stronger than) the men.

Even the brightest women often exhibit a marked lack of self-confidence, and are disproportionately discouraged by setbacks. ...For this reason, encouragement and support in calculus ...may fuel women students for the road ahead.

A discouraging postscript is that despite their motivation, ability and successful performance, even the brightest women often exhibit a marked lack of self-confidence, and are disproportionately discouraged by setbacks; the lack of encouragement at earlier stages seems to take its toll. For this reason, encouragement and support in calculus are vital elements in counteracting the damage that may have already been done, and may fuel women students for the road ahead.

Teaching Methods

The Methods Workshop at the 1986 Tulane Calculus Conference offers several goals for Instruction in Calculus that involve modifications in both the manner in which calculus is taught, and in the course content. One of the difficulties cited by this Workshop in teaching calculus effectively is that calculus is a "stepchild" in many departments, with large enrollments and limited attention from tenured faculty [8]. These courses, given low priority by departments or institutions, are frequently taught by TA's, instructors, or non-tenured faculty. The language difficulties of some foreign TA's may exacerbate this problem [5].

In my own case I benefitted significantly from what might have been the failure of a department to give highest priority to the teaching of calculus. Indeed, both of my calculus teachers were women, both untenured, and one of them gave me considerable encouragement and individual attention (the classes were small—about twenty, as I recall). Alas, for the remainder of my mathematical education (nine years worth), I did not have another woman teacher.

Junior faculty often have greater rapport with students than do older faculty, and this can benefit women, as well as men. As only about 6% of tenured faculty in the mathematical sciences at four-year colleges and universities are women [1], as opposed to 15% of all full-time faculty [4], a move towards having calculus taught by higher-ranking faculty significantly reduces the chance of a woman teaching calculus. In fact, provided TA's can speak English, and have some teaching experience, I do not feel that their use as calculus instructors is bad *per se* [9].

The reformed calculus would be a streamlined course, more conceptual, more relevant to real-world problems, and taught in a more open-ended, probing fashion than current versions [8]. These aims are difficult to fault; women students are certainly equal to the challenge of the "new calculus," and should benefit from these changes as much as men students.

A Case Study

However, as far as women are concerned, changes in content may be less crucial than changes in methodology. As women progress along the path towards more advanced mathematics, we know that they fall by the wayside. Wide-scale change could undo some of the subtle positive influence of a successful experience in first-year calculus. Therefore, given the conflicting goals of smaller classes with more individual attention, and the presence of tenured faculty, I feel there is more to be gained by both women and men students from the personal attention smaller classes afford. (An interesting point, made by one of my colleagues, is that when classes are *too* small, there may be only one or two women, resulting in feelings of isolation by the women students. Perhaps the optimum class size for women is somewhat larger than one might expect.)

When classes are too small, there may be only one or two women. ...Perhaps the optimum class size for women is somewhat larger than one might expect.

A women's college provides a convenient setting for considering the effect of various programs on women students. Many of the suggestions made by the Tulane

Methods Workshop are already in place at my institution, with varying degrees of success. In spite of our small size, we still face many of the same problems inherent in a larger institution. (I spent ten of my most formative years at a state university, and have experienced some of those difficulties first-hand.)

For us, a large calculus class has at most sixty students. We teach both large and small calculus sections, a small section having a maximum of thirty students. There is at present only one course to meet the needs of aspiring mathematicians and scientists, as well as those of students fulfilling a mathematics requirement. Despite the administration of placement exams, a wide variation in preparation of the students still occurs in the first-year course. The Methods Workshop states that "it is essential to have an effective diagnostic and placement program," and that the success of the proposed calculus curriculum hinges on this [8]. We have found this goal elusive, for a variety of reasons: (i) overprepared students, (ii) underprepared students, and (iii) ABT's (All But Trigonometry).

Students in the first category have already had calculus in high school, in some cases the equivalent of one year, but they have not received advanced placement credit. Well-meaning advisors who are not mathematicians often project their own uneasiness about mathematics onto the students, encouraging them to take courses for which they are overprepared. If the student is really able, she should be strongly urged to try the next course, for the presence of such students makes those who have never had calculus (a small minority of our students) quite uneasy, with good reason. (I do not remember this being a problem at a state institution.)

Underprepared students are usually identified by the placement exam and enrolled in precalculus, but some are only marginally underprepared, or may merely lack confidence. We usually send these students to calculus, but keep an eye on them for the first few weeks. The third category is a subset of the underprepared student, and there are usually several such students each year.

A precalculus course is usually inappropriate, and quite boring for most of these students; we do not want them to sit through a course waiting for the last three or four weeks of trigonometry. In order to deal with this problem, we have devised a series of non-credit minicourses, taught by a graduate or advanced undergraduate student, that treat trigonometry, exponentials and logarithms, and other topics in separate two-week modules. If these are carefully coordinated with the calculus sequence, a student has the opportunity to brush up on material she feels uncertain about, or become acquainted with the basics of trigonometry before it is introduced in calculus. (The logistics here are difficult, for trigonometric functions appear fairly early in the syllabus). In institutions with teaching assistants, the minicourses should be easy to organize. Of course, the student doing the teaching should have some experience, as well as solid evidence of being an effective teacher. Even though we have a small student body, we are fortunate in usually having such students available.

Another advantage of manageable class size is that papers may be regularly graded, providing the valuable feedback recognized by the Tulane Workshop as crucial [8]. We employ large numbers of mathematics and science students to grade papers for us. For a class of sixty students, it is not uncommon for homework to be collected twice a week and be thoroughly graded. Since homework is generally graded by the student graders, I usually give quizzes (I don't count them) so the students have some idea of how I grade, and what types of questions I think are important, before the actual exams.

The difference between "feedback" and "support" is like the difference between "eating" and "dining."

A reasonably successful means of helping students in calculus, as well as actively involving mathematics majors in reviewing courses they have already taken, is our Math Clinic. This operation is run solely by undergraduate students, on a voluntary basis. They are available a couple of nights a week to help students with questions about their mathematics courses. Although the faculty is occasionally tempted to intervene, we have not done so for many years. Directorship of the Clinic passes from one student to another, and they do an admirable job of keeping the operation afloat.

Conclusion

These details are offered in the hope that they give some picture of how calculus is presently managed at a small women's college. Roughly 30% of our students major in mathematics and science, and most of them pass through our calculus course. Nevertheless, far more effective than any of these structural aspects of our course is the patience, encouragement, and support I see my colleagues offering their students. While this is easier to do at a small college, I have vivid memories of dedicated professors trying to do the same thing with classes of one or two hundred students.

While the MAA report frequently mentions the importance of "feedback," there is virtually no mention of offering support to students who need it, perhaps because the latter is far more difficult to "package." However, the difference between "feedback" and "support" is like the difference between "eating" and "dining." If mathematics as a profession is to recover from its current malaise, and the vast untapped resource of women and minorities is to be realized to the fullest extent, we must allow our students to dine on the fruits of mathematics. A "lean and lively" calculus, thoughtfully implemented with the needs of students in mind, could well contribute to this goal.

Acknowledgement. The author gratefully acknowledges the assistance of Lori Kenschaft, Executive Director of the Association for Women in Mathematics, for supplying the data in references [1] and [2], and to my colleague Kyewon Park for carefully and thoughtfully reading this paper, providing data on calculus grades, and offering numerous helpful suggestions. The author also thanks Professors Bettye Anne Case, Joan Hutchinson, Anthony Hughes, Anne Leggett, Paul Melvin, Judith Roitman, Alice Schafer, and Ms. Jaye Talvacchia for informative conversations. Above all, she thanks Patricia Montague, her first calculus teacher.

References

[1] *Women and Minorities in Science and Engineering,* National Science Foundation, January 1986.

[2] *Educational Information Branch.* US Department of Education; data for 1985-86 academic year.

3] "Girls and math: Is biology really destiny?" Education Life, *New York Times,* August 2, 1987.

[4] *Undergraduate Programs in the Mathematical and Computer Sciences, The 1985-86 Survey,* MAA Notes Number 7. Washington, D.C.: Mathematical Association of America, 1987.

[5] "Report of the Committee on American Graduate Enrollments." Conference Board on Mathematical Sciences, December, 1986.

[6] "Report on the 1986 survey of new doctorates," *Notices of the American Mathematical Society* 33 (1986) 919-923.

[7] Edward E. David, Jr., "Toward renewing a threatened resource: Findings and recommendations of the ad hoc committee on resources for the mathematical sciences." *Notices of the American Mathematical Society* 31 (1984) 141-145.

[8] Douglas, Ronald G. (ed.) *Toward a Lean and Lively Calculus.* MAA Notes Number 6. Washington, D.C.: Mathematical Association of America, 1986.

[9] *Teaching Assistants and Part-time Instructors: A Challenge.* MAA Notes. Washington, D.C.: Mathematical Association of America, 1987.

RHONDA HUGHES is Associate Professor and Chair of Mathematics at Bryn Mawr College, Bryn Mawr, Pennsylvania, and President of the Association for Women in Mathematics. She has been a member of the Institute for Advanced Study at Princeton, New Jersey, a Fellow of the Bunting Institute of Radcliffe College, and has lectured and written extensively on functional analysis and perturbation theory. She is currently a member of the MAA Committee on the Participation of Women.

Calculus Success for All Students

Shirley M. Malcom and Uri Treisman

AMERICAN ASSOCIATION FOR THE ADVANCEMENT OF SCIENCE
UNIVERSITY OF CALIFORNIA AT BERKELEY

Consider these facts about student achievement in mathematics:

- In 1985 according to the National Research Council, 31,201 doctorates were awarded by U.S. universities; 689 or 2.2% of these were awarded in mathematics. Of the 689 Ph.D.'s, 15.4% went to women and 34.7% to non-U.S. citizens with temporary visas.

- Not one American Indian or Puerto Rican received a doctorate in mathematics in 1985; 7 Blacks and 5 Mexican Americans received degrees for a total of 12 doctorates (1.7%) awarded to members of minority groups under-represented in science and engineering.

- Between 1982 and 1983, 2,839 master's degrees were awarded in mathematics, 953 or 33.6% to women and 23.2% to non-resident aliens. Minorities—68 Blacks, 48 Hispanics, and 6 American Indians—received 4.3% of the degrees.

- In 1983, 11,470 bachelor's degrees were awarded in the mathematical sciences—629 to Blacks, 27 to American Indians, and 253 to Hispanics. Under-represented minorities received 7.9% of bachelor's degrees in mathematics.

- In 1985, 23 American Indians, 173 Blacks, 41 Mexican Americans, and 31 Puerto Ricans received doctorates in all natural sciences and engineering fields.

The 268 minority Ph.D.'s (2%) in natural sciences and engineering had successfully negotiated the calculus barrier, as did those persons who received master's and bachelor's degrees in these fields. As the minority proportion of our school age population increases to over a third, our concern must be with the hundreds of thousands of capable minority students who are felled by the calculus hurdle.

Those of us who are concerned about increasing the diversity of persons involved in science and engineering in this country and in broadening the pool of talent available to an economy based increasingly in science and technology must also be concerned about the direction of efforts by the mathematical community to reform the calculus.

Hundreds of thousands of capable minority students [are] felled by the calculus hurdle.

As the base for science and engineering, mathematics holds a unique position among the science and engineering fields; as the base for commerce and trade, mathematics holds a unique position in business and finance; and as the critical filter to an increasing number of careers for women and minorities, mathematics also holds a unique position in the movement for social and economic equity for these groups.

It is critical that the larger social context of education, the economy, demographics, and national need be factored into what might otherwise be considered just an attempt by the discipline to reform itself in response to a changing intellectual context.

Social Context for Calculus Reform

Mathematicians have been concerned for some time about problems that have appeared within the discipline:

- The increasing image of mathematics departments as primarily "service departments" in most institutions;
- The decreasing proportion of students who declare an intention to major in mathematics;
- The decreasing number of bachelor's, master's, and doctorate degrees in mathematics;
- The decreasing proportion of U.S. citizens among those receiving Ph.D.'s in mathematics;
- The increasing proportion of remedial course work being offered in college mathematics departments and the corresponding decrease in upper-level course work in mathematics.

These and other indicators have led to serious discussion both within and outside of the mathematics community. The link between these issues and equity concerns have been the subject of discussion at some of the highest policy levels. These included discussions by the Mathematical Sciences Advisory Committee of the National Science Foundation and by the 1987 American Association for the Advancement of Science (AAAS) Congressional Seminar, "Reclaiming Human Talent."

Guess Who's Coming to College?

If one starts to look at the changing population trends in this country, concern about social equity issues in relation to higher education becomes a national call to action. Lynn Arthur Steen cites many of these trends in his recent *Science* article [1] "Mathematics education: A predictor of scientific competitiveness," including:

- The declining number of 18-24 year olds, the proportion from which the college age population is largely drawn. The nearly 30% decline between now and the end of the century will take place just at the time large numbers of teachers retire and the baby boom echo produces a 30% increase in the size of the school age population. Who will teach these children?

As the critical filter to an increasing number of careers for women and minorities, mathematics also holds a unique position in the movement for social and economic equity for these groups.

- The increasing proportion of this school age population who are members of minority groups. According to Harold Hodgkinson [2], by around the year 2000 "Americans will be a nation in which one of every *three* of us will be non-white." If we add to these groups those others within the society who have tended to be less well served by the educational system (e.g., women and persons with disabilities), we are looking at over two-thirds of the school population.

In an issues paper included in *The Condition of Education* [3], Phillip Kaufman pointed out that a decline in the 18-24 year old age group does not automatically translate into a decline in college enrollment. He cited as evidence the fact that the projected decline for the 1980's did not take place. As a matter of fact, college enrollments increased in the early 1980's due largely to the increase in college going rates by 18-21 year olds (mostly among whites) and increased enrollment by women, particularly older women.

Women increased as a proportion of all college students, from 49.9% in 1978 to 52.9% in 1985 and accounted for 63.3% of the increase in college attendance between 1978 and 1985. While the proportion of part-time students has remained fairly constant over the 1978-1985 time period (around 35% of all college students), women's share of part-timers increased.

Minority college-going rates have remained stable during this time; in the case of Blacks, the rate has actually declined. American Indian, Black, and Hispanic students are more likely to be found in two-year college programs than are white students.

Our national need for scientifically and technically-trained talent requires that we succeed in addressing the issue of mathematics as a gateway or barrier.

Despite the counter-example that Kaufman cites for the nonlinear relationship between 18-24 year old cohort size and college enrollment, he concludes that the projected declines were merely postponed. He suggests that while national declines may be moderate, the effects will likely be different on different types of institutions. He further suggests that prestigious institutions with large applicant bases and low-priced community colleges are less likely to feel the enrollment declines, while all other institutions are likely to experience considerable losses.

The growing minority proportion of the college age population must be enabled to grow at least proportionately among the college attending population or the continued existence of many institutions (or the viability of departments within those institutions) will be seriously jeopardized.

Mathematics: Gateway or Barrier

As critical as the question of who is coming to college is the question of the skills that they will bring with them. At present, large proportions of American Indian, Black, and Hispanic students leave high school under-prepared to pursue quantitatively-based fields of study in higher education, having neither taken the appropriate courses nor obtained the requisite skills to enter a calculus sequence in college. Young women also lag behind young men in high school mathematics course-taking, but this is largely at the highest level of courses, such as trigonometry and calculus.

Early in the 1970's Lucy Sells coined the phrase "critical filter" to describe the overwhelming effect that mathematics preparation has on the career aspirations of women and minorities, especially in limiting their participation in careers in science and engineering. Although it was her interest in barriers to careers in science and engineering fields that led Sells to examine the high school mathematics preparation of women and minorities entering the University of California at Berkeley and the University of Maryland, she quickly discovered that inadequate mathematics preparation also limits participation in other fields as well, such as business, architecture, and the health professions.

The trend is clearly toward an increasing number of majors that require calculus as an entry ticket: if you cannot obtain that ticket, an increasing number of fields are closed to you. The implications of failing go far beyond grade point average or eligibility to play sports. And the implications for our national need for scientifically and technically-trained talent requires that we succeed in addressing the issue of mathematics as a gateway or barrier.

Students Failing Courses

Success in college-level calculus obviously depends on the past preparation of students as well as on the amount of productive effort that students spend on their course. The Professional Development Program (PDP) at the University of California at Berkeley has demonstrated that "good" backgrounds in mathematics and "smarts" are necessary but not sufficient conditions for success in calculus courses.

Minority students can be successful in calculus.

By looking at Asian students at Berkeley who were succeeding, PDP staff could see that Black and Hispanic students were employing unproductive behaviors in their calculus courses: studying alone without feedback from peers, studying for much less time than was needed to master the material; passing over material that they did not understand; and failing to ask for help.

With each passing day, such students found themselves deeper and deeper in the hole.

Contrast this with the performance seen after intervention when capable minority students replaced unproductive study behaviors with productive ones. These same minority students succeeded. Berkeley's PDP shows that minority students can be successful in calculus. Other examples, such as the extensive participation in AP Calculus programs at mostly Hispanic Garfield High School in East Los Angeles, reinforce the lessons from Berkeley. There are no inherent barriers to minority student success in calculus.

Achieving Proficiency

The key to the success of Berkeley's PDP program is the discovery that one can promote high levels of achievement among Black and Latin calculus students by adapting techniques being used by Chinese undergraduates. A significant number of Chinese students achieve proficiency in mathematics largely through their work in informal study groups. These groups provide students the opportunity to discuss with other students their mathematics homework and their performance on tests and quizzes.

These discussions accomplish a number of important objectives: they provide students with critical feedback about the accuracy and quality of their work, and they force students to explain to the satisfaction of others, how complex proofs were derived or how difficult problems were solved.

The process of forming clear explanations was the key to the success of these groups: one cannot explain a difficult concept to another unless that concept is well understood.

The PDP Mathematics Workshop attempts to create an instructional setting where students would be forced to talk with others as they did mathematics. The workshops enroll 15-20 undergraduate calculus students who are all members of the same calculus lecture section. Workshops meet for two hours per day, twice a week. Since workshop participants are in the same class, they are assigned the same homework, must prepare for the same tests, and must struggle with the same material. This common experience allows them to share a common foundation for the work that they are assigned in the workshops.

The focus for this work is a worksheet—a collection of difficult, challenging problems that test skills and concepts that students must master if they are to be successful in later work in calculus. When students arrive at a workshop session, they are given the day's worksheet. Students work alone at first and then are encouraged to share the results of their labors with four or five others.

Those who have successfully completed problems must explain to others how a solution or proof was derived. Listeners challenge what they hear and critique what has been presented; if the explanation is clear (and correct), others in the group will repeat what has been said until each person can replicate the steps to a successful solution or until a better, more elegant means of working with the problem has been discovered.

A workshop leader—who is usually a graduate student in mathematics or physics (or some other quantitative field)—is responsible for both the creation of these worksheet problems and for monitoring the work of students in the workshops. The leader walks about the meeting room and observes the groups closely, paying particular attention to the content of conversations students are having.

These conversations provide important insights into the students' thought processes: they provide a window through which one can observe how well or how poorly important concepts have been understood. When students appear to be "on target" they are left alone with their deliberations; when they appear to flounder, the workshop leader can intervene. Once on target, students return to their labors.

The process of forming clear explanations was the key ... one cannot explain a difficult concept to another unless that concept is well understood.

The workshops provide students with the opportunity to practice skills that they are expected to exhibit on tests and quizzes; moreover, their practice occurs in the presence of a skilled mathematician who is able to provide them with instant feedback on their efforts. Thus, bad habits can be quickly dealt with and good habits can be immediately strengthened and reinforced.

Expectations of Competence

An additional, critical feature of the workshop program's success is that there is no hint of a "remedial" focus in any of the work that students do. Students are encouraged to see themselves as competent and as capable of achieving high standards of academic excellence. Significantly, the performance of workshop students since the program's inception more than justifies this assumption of competence.

Black students at Berkeley (and in most predominantly white universities) are at greater risk of academic failure and are more likely to drop out of college than any other comparable group of undergraduates. It is particularly significant, therefore, that since 1978 approximately 55% of all Black students enrolled in the workshops have earned grades of B- or better in first year calculus, while only 21% of non-workshop Blacks earned comparable grades. The mean calculus grade point average of Black workshop students was 2.6 (N=231); the comparable mean for Blacks not in the workshop was 1.9 (N=284).

During this period, only 8 workshop students in 221 failed calculus. Comparable failure rates for Blacks not in the program were substantially higher (105 of 284 failed the course). Significantly, the program has had dramatic impact on students with a weak foundation in mathematics: the mean grade in calculus among poorly prepared Black workshop students (students with SAT Math scores between 200 and 460) was four-tenths of a grade point higher than that of non-workshop students who entered the university with strong preparation in mathematics (SAT Math scores between 550 and 800).

There are no inherent barriers to minority student success in calculus.

Participation in the workshops was also associated with high retention and graduation rates. Approximately 65% of all Black workshop students (47/72) who entered the university in 1978 and in 1979 had graduated or were still enrolled in classes as of spring semester of 1985. The comparable rate for non-workshop Black students who entered the university in those years was 47% (132/281). Of particular significance was the fact that 44% of the workshop graduates earned degrees in a mathematics-based field such as engineering, environmental design (architecture), or one of the natural sciences. Comparable findings have been reported for Hispanic workshop students who appear to have persisted in the university and earned grades in mathematics at rates that are almost identical to those reported for Black workshop students.

Courses Failing Students

Other critical issues in the success of students in college-level calculus include the nature of the course and the nature and quality of the instruction. There has been a tendency to blame lack of success almost exclusively on the student. This has been especially true for students who are non-traditional. While lack of background or lack of effort may account for many problems, we must at least consider the possibility of the courses failing the student.

If we look at programs at the precollege and college level that promote success in science and engineering by minority and women students (including mastery of the calculus) we see a number of characteristics emerge which suggest lines of inquiry that should be explored in any re-examination of the intellectual underpinnings of calculus. Unless the restructuring starts with a commitment to results that serve *all* students who are intellectually capable of mastering calculus, the effort may only have been an interesting intellectual exercise doomed to failure.

Strategies That Succeed

Research and experience with special programs tell us that there are no inherent barriers to success in mathematics, science, or engineering by women, minority, or disabled students if they are provided with appropriate instruction and support systems:

- High expectations by students and their teachers that the student will succeed. There is no "presumption of failure."
- Presence of a support structure and safety net. Bridge programs ready the students for the rigors of college-level work, especially in mathematics. Diagnostic testing identifies problems which are addressed in summer programs. Problems in class performance are not allowed to snowball but are handled through immediate feedback, referral, and tutoring.
- Capable and appropriate instruction that links mathematics to science and engineering. Where the relationship between a science experiment or a design problem and the mathematics is made clear, students seem to perform better and are more highly motivated. Too often the trend has gone in the wrong direction—not only a separation of mathematics from the hands-on activities by faculty in mathematics, but also a substitution of mathematics for hands-on experience and practical understanding by faculty in the sciences.
- Purpose of instruction is to enable students to succeed, not to weed them out to reduce the numbers in highly-competitive programs.
- Cooperative learning, peer tutoring, and other forms of instruction are utilized, including the use of technology where appropriate.

- A willingness to take students from where they are to where they need to be.

The key to success is to make sure that the learning environment can work for them; that the real goal of instruction is learning; that the best possible teaching is available at this entry level; that content is tied as much as possible to real-world problems as they are likely to be encountered, utilizing the tools that are likely to be available; and that the institutions and departments are committed to providing the resources necessary to support learning.

One can address the problem of poor precollege mathematics preparation by raising entry standards, by working with precollege-level instructors to improve their content-base and instructional abilities, or by accepting the need to provide remedial instruction on a transitional basis until reforms of the entire system take effect. Many institutions have chosen the first option, preventing contact with the under-prepared. Unless changes are instituted now, this option may soon be unavailable.

Recommendations for Action

1. Keep in mind the results of research on what it takes to help women, minorities, and disabled students succeed in mathematics.
2. Start restructuring by considering the needs of those students who form the overwhelming majority of the future talent pool for science, mathematics, and engineering, even though they have been inadequately utilized in the past.
3. Include in the discussions people who have experienced success in teaching calculus to significant numbers of students from these groups.
4. Recognize the need to build a system of precollege instruction that supports the reforms of calculus (including content, skills, and pedagogy).
5. Promote dialogue on these issues among mathematicians, mathematics educators, and mathematics teachers.
6. Promote interaction of mathematicians with faculty from engineering, physical, biological, and social sciences as well as non-science fields that require calculus, as a source of real-world problems to be included in instruction as well as allies in reconnecting mathematics to the other content areas.
7. Work to change the image of mathematics from that of a difficult and abstract subject.
8. Work to change the image of mathematicians and other people who use mathematics extensively to include the widest variety of people possible. Be particularly conscious of the need to show examples of people working in groups or as teams to solve mathematics-based problems.
9. Study the structure and promote replication of effective models such as the Professional Development Program of the University of California at Berkeley, and successful efforts at women's colleges and minority institutions.
10. Implement new models which address the serious recruitment and retention issues for mathematics majors, especially among women. When intervention programs have been instituted to increase participation of members of under-represented groups in science and engineering, they have been found to be equally effective and valid for all students.

References

[1] Steen, Lynn Arthur. "Mathematics education: A predictor of scientific competitiveness." *Science* 237 (17 July 1987) 251-252, 302.

[2] Hodgkinson, Harold L. *All One System: Demographics of Education.* Institute for Educational Leadership, Washington, DC, 1985.

[3] Kaufman, Phillip. *The Condition of Education.* Center for Statistics, Washington, D.C., 1985.

SHIRLEY M. MALCOM is Head of the Office of Opportunities in Science of the American Association for the Advancement in Science (AAAS). A biologist by training, Malcom has served as a Program Manager for the National Science Foundation and as Chair of the National Science Foundation Committee on Equal Opportunities in Science and Technology.

PHILIP URI TREISMAN is Associate Director of the Professional Development Project at the University of California at Berkeley. for which work he recently received the $50,000 Dana Foundation Award for pioneering work in higher education. He received a Ph.D. degree in mathematics education from the University of California.

The Role of Placement Testing

John G. Harvey

UNIVERSITY OF WISCONSIN

Had I, about 30 years ago, been placed in a state of suspended animation when I completed my introductory calculus courses and had I been revived this year, I would be as much at home with the present entry-level college mathematics curriculum as I am presently.

It is true that as a student I took only two 3-credit introductory calculus courses that were preceded by a 3-credit analytic geometry course and that today I teach a sequence of three 5-credit introductory courses for engineering, physical sciences, and mathematics majors. However, the instructional techniques I used and the primary expected outcomes—a repertoire of analytical skills and techniques—are about the same.

In the intervening 30 years analytic geometry, vector calculus, multivariable calculus, and elementary differential equations have joined the standard core curriculum and account for most of the increase in the time devoted to introductory calculus. But, even as a graduate student, I discovered that these topics were already covered in introductory calculus courses at some universities. Thirty years has changed surprisingly little in introductory calculus.

Thirty years has changed surprisingly little in introductory calculus.

In contrast, the introductory curriculum in many mathematically related areas has not been static. After a stint in suspended animation I would not even recognize introductory college courses in physics, chemistry, or electrical engineering. In physics these courses now include the elements of quantum mechanics and elementary particle theory; in chemistry they include new knowledge about the structure of atoms and the interaction of molecules. And electrical engineering has been transformed into computer engineering where there is less an emphasis on power and motors and more on computer technologies.

It may be unfair to make these comparisons or to conclude that introductory mathematics courses need changing. After all, what was true 30 years ago in mathematics is still true. Although new knowledge in physics and chemistry has altered our conception of those worlds and computer technologies have completely changed the way electrical engineering is done, mathematics describes the world as well today as it did then. Perhaps the introductory college mathematics curriculum may not need changing—only fine tuning, as Sherman Stein's colleagues suggested ([17], p. 167). However, most participants in the 1987 Tulane Conference to develop curriculum and teaching methods for calculus at the college level concluded otherwise ([17]), and I believe that many college faculty agree with them.

In addition, it seems to me that this is a propitious time to initiate changes intended to improve both our entry-level courses and instruction in those courses. The cost of computers seems to decline steadily; presently, I can do many complex tasks on a computer costing less than $4,000 that could only be done on a computer costing $400,000 in 1977 and $4,000,000 in 1967. In addition, calculators are becoming portable computers, are increasingly able, and are very inexpensive. (At present the Casio *fx-7000G*, the Sharp *EL5200*, and the Hewlett-Packard *HP-28C* can be purchased for $55.00, $69.99, and $175.00, respectively.) These tools make it possible for us to forsake tables, point plotting, and routine computation in favor of teaching concepts, problem solving, and more detailed and more realistic applications.

Confronting Diversity

If I had been in state of suspended animation, there are a few things that I would not recognize. Among them are mathematics placement testing, remedial courses, disproportionately large enrollments in precalculus courses, and the (limited) use of calculators and computers. Two of these changes—remedial courses and precalculus enrollments—are ones that most collegiate mathematicians view with disfavor. The use of calculators and computers is a change that many of us would like to encourage. Mathematics placement testing may help us to accomplish these goals.

Colleges and universities now admit a much more diverse group of students than they did, say, in the 1950's.

Colleges and universities now admit a much more diverse group of students than they did, say, in the

1950's. A majority of our students do continue to resemble the college-intending high school senior of the 50's; such students are white, from middle-income families, have completed a college preparatory curriculum that includes at least three years of high school English, a year of algebra, a year of plane geometry, and possibly, a second year of algebra and trigonometry, and enter college in the fall of the year they graduate from high school. But we also have a large group of students who come from minority groups; by the year 2000 they will comprise at least 30% of public high school enrollment ([1], p. 39). In addition, we now have a heterogeneous group of non-traditional students. Twenty-nine percent of all students enrolled in four-year institutions ([16], p. 80) are part-time students. Other characteristics of our non-traditional students are shown in Table 1, adapted from Boyer ([1], p. 50).

Characteristics:	Part-Time	Full-Time
Age: 25 or older	67%	13%
Dropped out since entering college	58%	16%
Employed full-time	59%	4%
GPA of at least B	61%	55%

Non-Traditional Students at Four Year Institutions

TABLE 1

As the entering college population became more diverse it became difficult (and often impossible) to place students accurately in mathematics courses by using existing data such as high school transcripts or grade point averages. Data about high school mathematics courses also proved to be unreliable for students who had recently graduated from high school since these courses varied widely in content. As a consequence, it was not possible to conclude that they had provided adequate preparation for college-level courses. Moreover, non-traditional students often remembered little about the mathematics courses they had previously taken.

Faced with rising course dropout and failure rates, many colleges and universities introduced mathematics placement tests. To help colleges and universities develop their placement programs, the Mathematical Association of America (MAA) established a Placement Testing (PT) Program in 1977 administered by the Committee on Placement Examinations (COPE). The growth of MAA's PT Program is probably a good indicator of the overall growth of mathematics placement testing in the United States; Table 2 indicates the number and kind of schools who subscribed to the PT Program since 1980.

Year	Total	Two Year Colleges	Four Year Colleges	Universities
1980	129	–	–	–
1981	226	–	–	–
1982	203	–	–	–
1983	295	36	125	134
1984	327	50	107	146
1985	321	59	131	97
1986	301	62	110	96
1987	379	72	143	119

MAA Placement Testing Program, 1980-1987

TABLE 2

Effectiveness of Placement Testing

Placement testing programs do seem to be effective. I base this conclusion on several kinds of information.

- Many institutions have well established placement testing programs that seem to work well. For example, the University of Wisconsin at Madison has a placement program that is more than 20 years old. We use the scores from three tests as the sole source of the data we use to place students in our entry-level mathematics courses: intermediate (i.e., high school) algebra, college algebra, trigonometry, engineering calculus, and business calculus.

- Second, a recent survey of subscribers to the PT Program indicated that 93% of the responding institutions used at least one of the PTP tests for placement during the previous year ([3], 1987). Other kinds of information were also used:
  1. The number and kind of previous mathematics courses (62%),
  2. Grades in previous mathematics courses (58%),
  3. SAT quantitative score (39%),
  4. ACT mathematics score (37%),
  5. SAT qualitative score (17%),
  6. High school rank in class (17%),
  7. High school grade point average (14%),
  8. ACT verbal score (8%).

  If these data are representative of all colleges and universities that have a placement testing program, then it is clear that placement test scores are the most heavily used among the factors most usually considered in making placement decisions.

- Several studies document the effectiveness of placement programs. In a study conducted by the American Mathematical Association of Two-Year Colleges, it was concluded that the PT Program tests BA/1B and SK were "useful placement instruments" [14]. In a study conducted at the United States Coast Guard Academy [13] it was shown that the PTP test CR was most highly correlated with grades cadets received in the introductory calculus and analytic geometry course.

Finally, in a study conducted at Tallahassee Community College [2] it was shown that seven of eight students who disregarded the placement advice, but only nine of 37 students who followed the advice, failed to complete intermediate or basic algebra with a grade of C or better.

From evidence such as this, I conclude that present placement testing programs are successful in helping colleges and universities to place students in entry-level mathematics courses. However, these placement programs are based upon *present* high school and college curricula, the instructional techniques used in teaching those curricula, the technologies presently utilized, and present college and university entrance requirements. Thus, any noticeable change in any one of these areas may make it necessary—at least, very important—to change the placement testing programs.

Changing Mathematics Curricula

Any change in the introductory calculus courses will necessitate a change in both college precalculus courses and in the high school courses we normally consider as prerequisites for calculus: algebra, geometry, and precalculus. This seems clear to me when I examine the proposals for change in the report of the Tulane Conference [11], especially when I consider the history of the "new math" revolution that sprang from the report of the College Board's Commission on Mathematics [7].

Two sets of recommendations made by Tulane Conference participants deal with changes in the content of and instruction in introductory calculus. Even though these two sets were authored by disjoint groups of conference participants, there is remarkable agreement between them. On the surface the changes suggested by these two groups may seem minor and would not appear to require a noticeable change in the curricula we teach or the ways we teach it. However, I believe this is not true even if the recommendations made about the use of technologies are disregarded.

Both groups suggest that a revitalized introductory calculus is one (a) *in which* students see a broader range of problems and problem situations, become more precise in written and oral presentations, and better develop their analytical and reasoning abilities, and (b) *from which* students gain a better understanding of concepts, develop a better appreciation of mathematics and its uses, and learn better to use mathematics resource materials ([11], pp. vii-ix, xvi).

We are not teaching many ... higher-order skills: the problem-solving abilities we teach to and expect from students are, generally, ones requiring routine application of procedures and techniques.

These goals involve both the teaching and the learning of higher-order thinking skills and an improvement in students' problem-solving abilities. At present we are not teaching many of these higher-order skills: the problem-solving abilities we teach to and expect from students are, generally, ones requiring routine application of procedures and techniques.

If calculus is revitalized along the line suggested by the Tulane Conference, then students will need to enter those courses with a better (i.e., higher order) understanding of and ability to apply the algebraic, geometric, and analytical concepts they have already learned. Thus, it will be necessary for placement tests to change so as to determine if students have met these new prerequisites for calculus since at present most placement tests examine primarily low-level skills.

For example, my analysis (see Table 3) of the test items on the MAA's PT Algebra and Calculus-Readiness Tests revealed that *all* of the items tested low-level skills: recall of factual knowledge, mathematical manipulation, and routine problem-solving.

Level of Thinking	Algebra	Calculus-Readiness
Recall factual knowledge	0	8
Perform math. manipulations	22	10
Solve routine problems	10	6

Classification of PT Program Test Items

TABLE 3

A revitalized calculus will also require changes in the courses prerequisite to it. The changes needed appear to be those already endorsed by the National Council of Teachers of Mathematics [15] and outlined by The College Board [4], [5]), and so it seems possible that high school college preparatory courses will change. However, we will also need to change entry-level college courses so that they, too, reflect the changed prerequisites for calculus.

Changing Technologies

At the Tulane Conference the Content Workshop recommended the immediate use of calculators in calculus and envisioned that a symbolic manipulation calculus course would eventually become the norm. It is clear that if calculators are going to be used in entry-level college mathematics courses, then students should be able to use them when taking placement tests and that the placement tests should be designed for calculator use.

If calculators are going to be used in entry-level college mathematics courses, then students should be able to use them when taking placement tests.

With the financial support of Texas Instruments, the MAA Calculator-Based Placement Test Program (CBPTP) Project is presently developing a calculator-based version of each of the six present PT Program tests. The project has progressed far enough to yield some results:

- When a calculator is used during testing, some items on present placement tests *must* be replaced;
- It is possible to design items that test mathematics content—not just calculator skills;
- Use of a calculator can permit expansion of the testing of both higher-order thinking and problem solving skills.

The three present CBPTP Project test panels are basing their development of "calculator-active" items on the assumption that students will have a scientific, non-programmable calculator like the Texas Instruments *TI-30.* When it can be assumed that students in entry-level college mathematics courses actively use graphics calculators like the Casio *7000G* or the Sharp *EL5200*, or graphics and symbolic manipulation calculators like the Hewlett-Packard *HP-28C*, then a next generation of placement tests will be needed that accurately reflect both the way that mathematics is being taught and learned and the prerequisites of the courses in which students are placed.

Test Administration

As soon as present technologies are better incorporated into education, they will affect not only the content but also the administration of placement tests. For one thing, it will be easy to produce parallel or equivalent versions of placement tests. To anticipate this ability, for the past three years COPE has been working on the distractor analyses for each item on each test in order to describe each multiple-choice response with a formula. This makes it possible to develop computer programs that can, within parameters specified by the distractor analyses, randomly generate parallel items and parallel forms of each of the PT Program tests.

It is already possible to administer placement tests using computers. Indeed, the College Board has developed the Computerized Placement Tests [6] that uses a limited, fixed pool of test items; the tests are fairly expensive to use. However, it seems unlikely to me that large colleges and universities will use computer administration of their placement tests because they need to test large numbers of students simultaneously and testing sessions are held only infrequently.

Prognostic Testing

In their most recent report, the Carnegie Foundation for the Advancement of Teaching related that the first problem they encountered was the discontinuity between schools and colleges ([1], p. 2). According to the Carnegie report, school and college educators work in isolation from each other, students find the transition from high school to college haphazard and confusing, and there is a mismatch between faculty expectations and the academic preparation of entering students. All of these claims are probably true and are already influencing college entrance and placement testing.

Prognostic testing is one way of smoothing the transition from high school to college.

Prognostic testing is one way of smoothing the transition from high school to college. In a prognostic testing program, college-intending high school juniors are given a placement test; based upon the score on that test, the junior student is given a "prognosis" of what mathematics course he or she would enroll in if he or she takes no additional mathematics courses during the senior year and if he or she performs as well on the college placement test that is used to place students into entry-level mathematics courses.

Prognostic mathematics testing originated at Ohio State University (OSU) in 1977 when students in one Columbus area high school were tested. In 1985 the Ohio Early Mathematics Placement Testing (EMPT) Program tested 65,217 students in 80% of Ohio's public, private, and parochial high schools. Until 1986 the EMPT Program only advised students whether they would be placed in a remedial course. From 1979 to 1985, the percentage of the OSU freshman class with remedial placement dropped from 43% to 25% [9]. The EMPT Program, directed by Bert Waits at OSU, is a program of the Ohio Board of Regents; the results of EMPT testing are used by all state-supported colleges and universities in Ohio.

The results from the Ohio EMPT Program are so encouraging that several states, colleges, and universities have started or are planning to start similar programs. Similar programs exist in Illinois, Louisiana, and Oregon. The University of Wisconsin at Madison is planning to initiate a program in 1987. Several states, colleges, and universities have contacted COPE for information on prognostic testing.

Changing Entrance Requirements

One response to large remedial enrollments and the disparity between faculty expectations and student academic preparation is to raise entrance requirements. In 1982, admission requirements were being changed or reviewed by the public higher education systems in 27 states [18]. This change is probably overdue; at present, only 67% of all colleges and universities require any mathematics course for admission ([1], p. 88). However, most students presently entering our institutions already exceed any mathematics entrance requirement we might reasonably impose. Consider these two cases:

Only 67% of all colleges and universities require any mathematics course for admission.

• The University of Wisconsin at Madison is a restricted admissions university; at present it requires entering students to have completed two years of high school mathematics—one year each of algebra and geometry. In order to graduate, Wisconsin students must complete an additional mathematics course if they did not have two years of high school algebra and a year of geometry when they were admitted. Table 4 shows for the years 1985 and 1986 the number of high school courses our entering freshmen reported they had taken and the entry-level courses in which they were placed.

College Placement		Years of H. S. Mathematics < 2.5	< 3.5	< 4.5	≥ 4.5
Inter. Algebra	1985	98	281	452	69
	1986	85	316	397	30
College Algebra	1985	2	209	1468	580
	1986	0	318	1561	473
Trigonometry	1985	0	11	185	151
	1986	0	14	225	129
Calculus	1985	1	17	672	932
	1986	0	14	625	750

Course Placement of Wisconsin Freshmen

TABLE 4

As is easily seen from these data, most University of Wisconsin freshmen report that they had taken at least two and one-half years of mathematics in high school. Yet in each of 1985 and 1986 a majority of the students who had taken at least three and one-half years of high school mathematics were placed into intermediate or college algebra.

• Ohio State University is an open admissions university. However, data from OSU are similar to those from Wisconsin. In 1986, 206 OSU freshmen had taken less than two years of college preparatory mathematics, 478 had taken at least two but less than three years, 1615 had taken at least three but less than four years, and 4280 students had taken four or more years of college preparatory mathematics.

Increasing the mathematics entrance requirement may not necessarily eliminate the need for remedial mathematics courses.

The placement procedures at OSU assign students to one of five placement levels indicating readiness for calculus, for precalculus mathematics, or for remedial mathematics. More than 60% of the students with less than four years of college preparatory mathematics, about 54% of the students with three years of college preparatory mathematics, and about 86% of the students with less than three years of college preparatory mathematics had remedial mathematics placement scores [10].

These two cases indicate that increasing the mathematics entrance requirement may not necessarily eliminate the need for remedial mathematics courses or for an effective placement testing program. However, increasing the entrance requirement may reduce the number of students who end up in remedial mathematics courses.

Conclusion

1. Present placement programs successfully help place students in entry-level mathematics courses.
2. A revitalized calculus will make necessary new placement tests that assess higher-order thinking skills and problem solving abilities.
3. A revitalized calculus will make it necessary to revitalize both high school and college mathematics courses that are prerequisite to the calculus.
4. Placement tests will need to be revised in order to provide for the use of new technologies in learning and teaching mathematics.
5. Prognostic placement testing of high school juniors has shown some encouraging results and should be expanded.
6. Raising college entrance requirements will not change the need for an effective placement testing program.

References

[1] Boyer, E.L. *College: The Undergraduate Experience in America.* New York: Harper and Row, 1987.

[2] Case, B.A. "Mathematics placement at Florida State University." *PT Newsletter* 7:2 (Spring 1984) 1, 6-8.

[3] Cederberg, J. and Harvey, J.G. "Questionnaire results." *PT Newsletter* 9:1 (Spring 1987) 1-2.

[4] College Entrance Examination Board. *Academic Preparation for College: What Students Need to Know and Be Able To Do.* New York, 1983.

[5] College Entrance Board. *Academic Preparation in Mathematics: Teaching for Transition from High School to College.* New York, 1985.

[6] College Entrance Examination Board. *Computerized Placement Tests.* New York, 1985.

[7] Commission on Mathematics. *Program for College Preparatory Mathematics.* New York: College Entrance Examination Board, 1959.

[8] Committee on Placement Examinations (COPE). COPE Meeting Notes (1984-1987) (unpublished).

[9] Demana, F. "News from the campuses: The Ohio State University." *EMPATH,* I:1 (Sept. 1986) 4.

[10] Demana, F. and Waits, B.K. "Is three years enough?" *Mathematics Teacher* (to appear).

[11] Douglas, R.G. (ed.) *Toward a Lean and Lively Calculus.* (MAA Notes Number 6) Washington, DC: Mathematical Association of America, 1987.

[12] Harvey, J.G. "Placement test issues in calculator-based mathematics examinations." Paper presented at The College Board and The Mathematical Association of America Symposium on Calculators in the Standardized Testing of Mathematics. New York, 1986.

[13] Manfred, E. "Using the calculus readiness test at the U.S. Coast Guard Academy." *PT Newsletter,* 7:2 (Spring 1984) 1-3.

[14] Mathematical Association of America. *PT Newsletter* (volumes 1-9). Washington, DC, 1978-1987.

[15] National Council of Teachers of Mathematics. *An Agenda for Action: Recommendations for School Mathematics of the 1980s.* Reston, VA, 1980.

[16] Ottinger, C.A. *1984-85 Fact Book on Higher Education.* New York: American Council on Education, 1984.

[17] Stein, S.K. "What's all the fuss about?" In R.G. Douglas (ed.), *Toward a Lean and Lively Calculus* (MAA Notes Number 6). Washington, DC: Mathematical Association of America, 1987.

[18] Thompson, S.D. *College Admissions: New Requirements By the State Universities.* Reston, VA: National Association of Secondary School Principals, 1982.

JOHN G. HARVEY is Professor of Mathematics and of Curriculum and Instruction at the University of Wisconsin at Madison. He is currently Director of the Calculator-Based Placement Test Project of the Mathematical Association of America, and a member of the MAA Committee on Placement Examinations, and of the joint MAA-AMATYC Task Force on Remediation in College Mathematics. He is a former editor for Mathematics Education of the *American Mathematical Monthly.*

Calculus from an Administrative Perspective

Richard S. Millman
WRIGHT STATE UNIVERSITY

Unfortunately, what administrators hear are complaints. To be sure there are complaints both enlightened and unenlightened about all subjects. My favorite comes from the 1820 letter of Farkas Bolyai to Janos Bolyai [3]:

> You should detest it ... it can deprive you of all your leisure, your health, your rest, and the whole happiness of your life. This abysmal darkness might perhaps devour a thousand towering Newtons, it will never be light on earth ....

This criticism is from a man who studied the independence of the parallel postulate and was warning his son not to pursue it. Fortunately, like many children, Janos didn't listen and eventually showed that the parallel postulate is independent of the others by finding a non-Euclidean geometry.

Scholars all agree that calculus is a fascinating subject which has had a profound effect on science in particular and human intellectual development in general. Even the skeptic, Bishop Berkeley [1], said that calculus "... is the general key by help whereof the modern mathematicians unlock the secrets of Geometry, and consequently of Nature."

Ever since the inception of calculus, there has been and will always be criticism of it from learned individuals and those of lesser intellectual skills. Certainly the calculus curriculum is not "devouring a thousand towering Newtons"—our students are not quite at that level (at least at public universities). On the other hand, many valid discussion points are derived from both the informed and uninformed questions which are addressed to mathematicians.

Calculus shows its many faces in different ways, depending upon the angle of the viewer. As an administrator, as a department chair, and as a professor of mathematics, I've seen the subject from different perspectives. While mathematicians are well aware of the latter two viewpoints, it is useful to understand the concerns that a dean has with calculus.

No matter what your vantage point, what is ultimately important, of course, is whether the students who finish the course understand and retain the material well enough to use this knowledge in their future work. There are, however, many issues which complicate this primary objective. In my present position as a Dean and from discussions with administrative colleagues, I have come to appreciate the myriad problems which surround calculus, none of which have to do with the chain rule, *et al.*

An administrative view of calculus should, I believe, focus on four areas in this order of priority:

- quality of curriculum;
- accountability of the course;
- attitude of the department;
- effectiveness/cost of the program.

Many of the items on this list would apply far more broadly than just to calculus.

Quality of Curriculum

First and foremost is the quality of material presented in the course. I don't mean that we should react to occasional lapses in a professor's response to a maximum/minimum question, but rather that we need to address the problem, when it arises, of long-term passage of misinformation. Fortunately, this is the aspect of calculus that is both the most easily monitored and, in the unusual event that there is a problem, the most easily corrected. Mathematics departments do this efficiently and well.

A second point concerning quality is the idea of "realism" in calculus. We have all heard at AMS meetings of Professor X who decides that differential forms are the "only real way" to present integration at the freshman and sophomore level. While my prejudice as a differential geometer is in favor of tensor notation and manipulations, the junior or senior levels would be far better for such vigorous pursuits (and I have my doubts even there!).

We need to refrain from teaching "big C" calculus—that is, calculus for engineers, physical scientists and mathematicians—when the audience is business students or prospective biologists. This is the notion of realism. Can we expect biology majors or business students to work hard on subjects that they will never use and for which they will not have a real appreciation?

Fortunately we mathematicians have, over the last twenty years, realized the error of our ways and split calculus into a plethora of different sequences to address these matters. The lesson is that we must listen to the advice of our customers.

Although the students may think they are the customers, it is actually the university who is the consumer of the service aspect of any department. It is the university faculty who decide whether calculus, physics, or chemistry will be required of various majors. University faculty members in all disciplines are the ones who need to be listened to carefully about what they would like in a mathematics course.

It is not enough for us as mathematicians to know that the course is a good one—we must be able to show it to others.

To be sure, outsiders may not know the mathematics that we do and cannot dictate to us what is most valuable. For example, there may be items they would like excluded which are actually prerequisites for other subjects that they prefer to include. The key point, however, is that they do have a good feeling for what is necessary. They may be myopic, but some myopia is not necessarily bad.

As we design curriculum, we should beware of the beast that loves to pack courses with "favorite topics." (I must confess to feeding the beast on occasion: I can't seem to get through the entire undergraduate calculus sequence without emphasizing the notion of torsion and curvature of plane curves. *Mea culpa.*)

During my naive youth, I tried to persuade a group of mathematicians that conic sections could be learned easily by anybody who wanted them in later life and were not needed in college calculus. By dropping the week devoted to ellipses, parabolas, etc., the calculus curriculum would loosen up a bit. That turned out to be someone else's favorite topic and I was summarily cut off at the knees. Not only do we pack in the number of hours, but because our students have trouble with a vast amount of the material covered, we are forcing many of them into a five-year degree program.

Changes in student preparation constitute another important item. No one can deny that the infusion of the "new math" fifteen years ago had a negative effect on the competence of mathematics students as they entered college. Whether this should have happened or not, whether it is the fault of the high school teacher, the math educator or mathematician is no longer relevant. In the short run, we must deal realistically with entering students as they present themselves.

We must understand what these individuals know, what they don't know, and what they are prepared to do. As administrators, it is important that we realize the various departments are well aware of the changing caliber and preparation of the students for studying their disciplines. In a nutshell, do these students have a chance to pass or are they forced to fail? Because the long run problem is so important to calculus, a dean would applaud efforts of an academic department (especially mathematics) to work with middle schools and high schools, as well as with the College of Education to obtain better prepared college students.

Accountability

It is not enough for us as mathematicians to know that the course is a good one—we must be able to show it to others. While we may resent this intrusion on our expertise there is no longer a choice, especially for public institutions. In addition to the administration of the university (even from the non-academic side, such as Vice Presidents for Student Affairs, University Legal Counsel, etc.) there are also legislators who are quite concerned with the value of an education. The standard questions that are asked are often naive—sometimes to the point of ignorance—but that is not the issue. These questions must be answered and deserve to be answered.

One way to respond to the outside pressures on low level mathematics courses (calculus, in particular) is to ask if the department is willing to consider mastery levels, fundamental learning levels, or exit examinations for any of the sequences. For calculus, this could mean a modular approach; that is, students will go to a certain level (say the chain rule) and must pass an exam on that level before proceeding to the next one. In addition, an increasingly popular notion is the "value addedness" for all courses of study. Can we really show that students who have finished the calculus sequence with a C know it in enough detail that they will be able to handle subsequent physics, chemistry, or business courses?

Our courses must be demonstrably of the highest quality.

I'm not suggesting that we all need to move toward these modes of education; I'm only asking whether a department is willing to consider it every once in a while. I hope, but am not convinced, that administrators recognize the danger of thinking of all innovative changes as good ones. Innovation really means that something is quite new, not necessarily better. On the other hand, we do need to be willing to consider alternate means of presenting calculus, even if we are ultimately to reject them. Our courses must be demonstrably of the highest quality. ("To be thus is nothing; But to be safely thus." *Macbeth* III, i 48).

When I'm asked, as dean, to defend a particular instructional decision, it is very useful to be able to point to measures of the quality of instruction. Does the mathematics department quantify in some fashion and adequately reward teaching at all levels or is there merely lip service paid? Is there a mechanism in the department to ensure that new instructors are teaching at an appropriately high level? Are pedagogical articles and thoughts rewarded? Of course, the reward for an article in a mathematics education journal must depend on its quality just as that for a research article. Just doing *something* shouldn't be enough—even deans aren't that gross. However, it is important that faculty realize that good pedagogy in all of its forms is considered meritorious.

Attitudes

The next major point is that of the attitude of the calculus instructors. Students really appreciate an excited, dedicated teacher in the classroom. If the attitude is "Calculus is a chore and I'm here just to get my time in so that I can look at my advanced courses and do my research," the students will complain bitterly, with good reason.

While we rarely learn something new about the subject of calculus when teaching it, we do discover quite a bit about how students think and learn calculus. This can be fascinating and is a partial reward for good teaching. We do a disservice to both our students and our subject when we regard calculus as a chore and then compound it by communicating that attitude to our students.

Unfortunately, there are many mathematicians who, while they enjoy the discipline tremendously, do not convey the excitement they feel to their students. One can present an exceptionally clear and well thought out lecture on trigonometric substitution without being enthusiastic. Students will learn from such a lecture. However, it is so much better to have presented the ideas with a certain panache, so that the students can get an idea of *why* someone might become a mathematician (as well as learn the specific calculus skills). Certainly the instructors in all disciplines who are the most popular with students are those who are the most enthusiastic about their teaching. Enthusiasm is contagious. While it won't substitute for content, we need more excitement in the classroom.

Unfortunately, these are litigious times in higher education. Thus we must not only treat all students uniformly, but must do so across all sections and show that the treatment is indeed homogeneous. This responsibility usually falls on the department chair, rather than any individual instructor, but it is one that needs to be monitored extremely carefully because of the heavy usage of adjuncts, part-time people and new faculty. [2] Most departments have a common syllabus which is a fine first start at a uniform treatment. Many have common final exams and a few insist that all exams in a course be the same across all sections. While the latter is difficult to administer, the former is not and is an excellent way to demonstrate that students are treated equitably.

I've learned to loathe any conversation which starts with "I was always bad in math, but ...." The usual comments describe how hard mathematics is and how it's not surprising, therefore, that little Johnny or Janie is having trouble with it. A useful response is to quote the international study of high school students [4] which deals not only with accomplishments, but attitudes. It is clear that mothers' attitudes in Japan ("Work harder and it will come to you") are quite different than the mothers' attitudes in the United States ("It's just too hard for you"). Pointing out the careful research that has been done through this study helps to enlighten people from outside of the physical sciences. They even begin to understand what a "spiral curriculum" is and what its manifold drawbacks are. (One doesn't have to do differential geometry to have a manifold drawback.)

Effectiveness vs. Cost

A final point to be made concerns cost analysis of the effectiveness of the calculus curriculum. This is not meant to be an equation which describes the cost per left-handed student credit hour, but rather emphasizes that there are real financial implications which come not only from class sizes but from grading structures and repeat/incomplete policies. Most departments do not have enough resources to offer the number of sections that we would like and so need to teach to full classrooms. Students who must drop or repeat a class or take an incomplete are occupying chairs that others need. They use up an instructor's time and effort and are detrimental to class morale in the long run.

Some drop out is certainly unavoidable. On the other hand, it severely impacts the faculty work load, both in terms of efficiency and in forcing the students into curricula which are becoming five-year programs instead of four. The Accreditation Board of Engineering and Technology (ABET) is extremely concerned that the engineering curriculum is becoming standardized at five years rather than four, even though we all think of it as a four-year course of study.

The double-hump camel curve for grades happens in mathematics, chemistry and physics, but as dean I only hear complaints about the mathematics aspect. Why is this? Does it happen at your university? If so, why? Answers to such questions would be useful not only to mathematicians, but to anyone who has experienced unenlightened inquiry from legislators and others.

Students who must drop or repeat a class ... are occupying chairs that others need. They use up an instructor's time and effort ....

Changing class size is one way to economize. What is the difference between teaching classes of 30, 40, 60 or 200? I have taught calculus in sections of 40 and in sections of 80 and quite frankly I see no difference in effort except when grading papers. I do recognize that visits outside the classroom to one's office can take up an enormous amount of time and many of my colleagues have complained about such things. But when teaching large classes, I didn't notice any more office visits than usual. (Does this mean that students' visits during office hours are not a linear function of students?)

I am not advocating large lecture sections (in fact, classroom structures at some universities would prevent this), but rather that people look carefully at these ideas. It is important not to use conventional wisdom, but rather to be able to point to various comparative studies that have been done in the past. While some have been done in general, there are none I know of that are specific to college mathematics.

Conclusion

It is increasingly clear that we need to resolve issues through a more organized study of them than has been done in the past. As resources get tighter, and as legislators look more and more carefully at what we are doing, we will no longer have the luxury of saying "It's clear to anybody who knows mathematics that this procedure is right." This translates into an obligation to devote some of our scholarly energy towards mathematics education at the collegiate level.

I am delighted that the National Science Foundation is prepared to infuse money into such efforts. The challenge to us as mathematicians is to decide what specific projects and comparative studies need to be done; to set up unbiased studies to explore the issues; and to reach some conclusions. The challenge to us as college administrators is to find the resources to implement conclusions of these studies in a fair and equitable manner. The task cannot be left to our colleagues in the College of Education or to government bureaucrats.

A knowledge of the MAA's committee structure is very useful in this regard. I'm delighted that there are subcommittees that deal with service to engineers (chair, Donald Bushaw), preparation of college teachers (chair, Guido Weiss), role of part-time faculty and graduate teaching assistants (chair, Bettye Anne Case), and many others. These committees provide valuable facts to quiet some of the criticism that we hear. This allows mathematicians to cite specific sources when responding to complaints. We will thus appear to be listening carefully, caring and responding in good faith with a tremendous spirit of cooperation. I can't overemphasize this point.

As an administrator I say there's a real world out there. As a mathematician I say that boundary conditions exist. Whichever way one puts it, the implications are that we must work within a certain context, listen to people, and present well thought-out responses to others' concerns, whether those concerns be frivolous, unenlightened or substantive.

Can you hear the shape of calculus from its complaints?

In 1966 Mark Kac won a Chauvenet prize for his beautifully written paper, "Can You Hear The Shape Of A Drum?" This article asked whether you could tell what the shape of a domain in Euclidean space was just by knowing the eigenvalues of its Laplacian. Although that problem is unresolved in 3-space, the analogous one in our context is clearly false.

Can you hear the shape of calculus from its complaints? No, absolutely not. What you can hear, however, are the concerns of people who have spent time and money in an effort to get a good education. They have a right to ask us questions about our calculus curriculum, even if those questions are naive. It is crucial that we not only provide our students with a high quality calculus experience, but that we answer their questions in a responsive, thoughtful manner.

One of the early (1734) critics of calculus, Bishop Berkeley [1], asked: "And what are these fluxions? The velocities of evanescent increments. And what are these same evanescent increments? They are neither finite quantities, nor quantities infinitely small, nor yet nothing. May we not call them the ghosts of departed quantities? ...." To not answer questions carefully is to run the risk that we become ghosts of depleted qualities.

References

[1] Berkeley, Bishop, "A discussion addressed to an infidel mathematician," contained in David Eugene Smith *Sourcebook in Mathematics*.

[2] Case, Bettye Anne, *et al*, "Teaching Assistants and Part-time Instructors: A Challenge." Mathematical Association of America, 1987.

[3] Struik, D., *A Concise History of Mathematics*. Dover Publications, 1948.

[4] Travers, Kenneth, *et al. The Underachieving Curriculum: Assessing U.S. School Mathematics From An International Perspective*. Champaign, IL: Stipes Publishing Company, 1987.

RICHARD S. MILLMAN is Dean of the College of Science and Mathematics at Wright State University in Dayton, Ohio. Previously he served as Program Director for Geometric Analysis in the Division of Mathematical Sciences at the National Science Foundation. Millman has been a member of the Council of the American Mathematical Society and Associate Editor of the *American Mathematical Monthly*. He is currently Chairman of the MAA Committee on Consultants, and a member of the JPBM Committee on Preparation of College Mathematics Teachers.

Innovation in Calculus Textbooks

Jeremiah J. Lyons

W. H. FREEMAN AND CO.

How can publishers participate in and support innovation within the calculus course? I want to examine the current state of affairs within textbook publishing, giving special attention to the calculus. By demonstrating the assumptions and the current standards which we bring to bear on our editorial actions, we will be better able to assess the options open to us—instructors and publishers—as we consider ways of providing textbooks for emerging, innovative courses.

Clearly, innovative textbooks will not be published in a vacuum, in the hope that courses might be coaxed into existence by their appearance. With rare exceptions, textbooks are *adopted* by instructors for existing courses; instructors do not *adapt* course outlines to accommodate idiosyncratic textbooks.

What is both refreshing and challenging about these times in publishing is trying to assess the direction of change. Publishing is a profession in which information is constantly being gathered and evaluated. Editors work within a strange time system: decisions about publishing must be made today, and the outcome of our choices or decisions will not become visible for several years. Authors, too, are partners in this process of inference making. Correct decisions will result in the publication of the new generation of authoritative mathematics textbooks. The increased national concern for the state of mathematics education at all levels, and the creation of boards and councils, are very welcome signs of the seriousness of these discussions.

Pressure for Conformity

Throughout my twenty years in publishing, the organization and texture of calculus texts have changed very little. And yet, every sales representative and mathematics editor will tell you no mathematician is ever pleased with the calculus textbook currently in use. Every instructor has a pet topic or two which is not done correctly, or a particular notation convention which immediately signals the worthiness of a given book. For years we have heard dissatisfaction about the standard books, and yet there has been no radical change in the books we are publishing. And, it goes without saying, there has been no serious departure from the conventional text in adoption patterns. This stasis is demonstrated vividly by the fact that many of the better-selling texts are in third, fourth, and fifth editions.

Quantitative—not qualitative—expansion has been the striking change in the past twenty years: there are many more calculus texts, and each text is approximately 1100 pages in length, or greater. When there were fewer books, instructors had some familiarity with the texts and committees had the time to give careful consideration to the few new texts or new editions which were published each year. A text had an identity, a set of distinguishing features that were known to instructors. And these characteristics came from the background and the interests of the author. Calculus texts were relatively easy to rank by a few simple measures: theoretical or applied; rigorous or less rigorous (there were no "short" or easy calculus texts); books intended for students in physics and engineering; honors calculus, and so forth.

Beginning in the early 1970s, enrollments grew and more publishers entered mathematics. The number of books published each year increased. With more and more calculus texts being published, instructors were forced to become more critical of the new entries. After all, not even editors spend evenings and weekends reading through introductory textbooks. There was and is tremendous selective pressure (to borrow a trope from Peter Renz) to converge toward the mean, as established by the best-selling books. The successful books, after all, passed the scrutiny of the curriculum gatekeepers. Therefore, these texts must be filling the perceived textbook needs of the course. It follows that for a new textbook to gain a profitable measure of adoptions, the new text must come close to the best sellers, and offer just a bit more of one feature or another.

"The major revolution in calculus texts in the last decade has been the introduction of a second color."

One thing leads to another. If publishers are dissuading their authors from innovation, because innovative books do not sell, what aspects of a textbook can be improved upon? Sherman Stein [1, p. 169] characterized this dilemma from a calculus author's perspective:

> It seems that a calculus author has the freedom to make only two decisions: Where to put analytic geometry and whether the title should be "Calculus with Analytic Geometry" or "Calculus and Analytic Geometry." Thus the major revolution in calculus texts in the last decade has been the introduction of a second color.

Competition without Innovation

There are a few other arenas in which authors and publishers sought ways to distinguish their books while leaving the core calculus content and outline intact. These are the very features which, in the end, conspire to inhibit innovation and change. Let's examine the features which are found described in the promotional material announcing a new calculus textbook or revision.

The number of exercises has been increased significantly, and they now number six to eight thousand. The exercises are placed following each major section within a chapter, and a review set is found at the end of each chapter. Does anyone actually count the number of exercises, including the lettered A, B, C..., subparts of questions? Yes, they do. Is there an optimum number of exercises for a text? No, in this dimension of textbook making, the guiding rule seems to be that one cannot have too many exercises.

We can give you more book, for more money. Not different books: more book.

The same can be said for the number of worked-out examples. The range of applications has been widened considerably, to include examples drawn from biology and economics, for instance. These additional examples are not replacing existing ones from engineering or physics. Usually, there are simply more examples (and step-by-step solutions) added to each succeeding edition.

Publishers are asking for acceptance for our texts from a calculus textbook marketplace which is becoming increasingly crowded, and in which modifying the traditional content of the calculus is out of bounds for us. And yet the impetus to publish a new text which is a commercial success in this lucrative market continues. What other aspects of a book can be revised and improved upon, if the mathematical content must be standardized in relation to existing books and course outlines? As I mentioned above, we can give you more book, for more money. Not different books: more book.

There have been dramatic improvements in the number and quality of the graphics in calculus texts. Almost all are done in two colors now, and it won't be too far into the future before even more colors will be used. The use of two colors and high quality airbrush techniques have resulted in figures of exceptionally good quality. This trend towards more and better illustrations will continue. Computer-generated curves are becoming routine, and color graphs of functions will soon be incorporated into our textbooks.

The other area in which publishers can modify the existing textbook model, without cutting into content, is by incorporating pedagogical devices. More heads, use of margins for key terms, learning objectives and summaries, lists of applied examples—all these elements are an essential part of the textbook presentation.

I have described some dimensions of the textbook-making process in order to demonstrate several points. In an increasingly crowded and standardized publishing marketplace, publishers seek to outdo each other and capture your attention and your adoption in ways which ensure the preservation of content, in an acceptable order, and at the same time create some special identity for each book. It is no accident that each of the features cited here contributes to a longer, more costly book. Herein is our dilemma.

Ronald Douglas [1, p. 13] stated the dilemma of publishers in these words: "Everyone is dissatisfied with the current crop of calculus textbooks. Yet, if an author writes and manages to get published a textbook which is a little different, most colleges and universities will refuse to use it. How can we break out of this dilemma?"

Economic Constraints

The current models for calculus publishing have been in place for twenty years. Each year brings new books and more revisions. In order to break out of this quantitative growth pattern, two crucial factors must be considered. First, publishers must be confident that new directions in calculus teaching and course organization, though many of these courses will be experimental in design, will be in place for awhile. Second, no publisher is going to sponsor the publication of an innovative textbook in calculus that demands an investment comparable to that of a mainstream text. Instructors seeking textbooks for their alternative calculus courses must be willing to accept a few tradeoffs. The reasons for this should be obvious.

The investment in a mainstream, three semester calculus textbook published in two colors will easily surpass $350,000 before the first copy is off the press. In addition, the unit cost for each copy (paper, printing, and binding) will be $6.00, average. Given these costs for entry into a market, no wonder there is enormous pressure to position a textbook close to the center of the market.

There are several means by which publishers can respond to the coming evolutionary changes in calculus. Our strategy for undertaking innovative publishing projects can be stated very directly: publishers will sponsor innovative projects intended for modest segments of an introductory market, but only at an investment level commensurate with the expected rate of return on our investment.

Publishers and authors now have expedient and economical modes of production by computer available.

Publishers and authors now have expedient and economical modes of production by computer available. For the first time we are seeing microcomputer-controlled laser printing of text pages which are of satisfactory appearance. It is because of this recent development that the option of publishing innovative alternatives to the big textbooks is a reality. The computer is having a forceful impact not only on the content of mathematics courses, but also on the text materials made available for the teaching of these same courses.

Stephen Maurer was direct in his assessment of publishers and our willingness to respond to change in the curriculum: "It's no use telling publishers to change their ways. They are hemmed in by market forces. We must show them that a new type of book will attract a market before we can expect them to help." [1, p. 81] Publishers are always responsive to numbers, so that is very sound advice.

Making Innovation Affordable

If we agree that substantial investment is not possible for small segments of the calculus market, what customary features are instructors willing to give up in order for a publisher to keep the costs of publishing down? Let's look at the potential for changing textbook requirements, along with alternative and more affordable modes of book production.

First, we should not be thinking of new textbooks of the same bulk as existing ones. Eleven hundred pages is too much, even of a good thing. In addition to being very selective about what topics to include or not (precalculus topics, for instance), an obvious area to realize reductions in page length is to sharply reduce the space given to exercises. Does anyone need eight thousand exercises? Most departments have exercises on file and these can easily be put into an exercise bank on a microcomputer. Exercises can be generated as needed. This would reduce the publishing costs directly related to length.

Next, are we willing to dispense with the pleasing but ultimately excessive use of color in our new generation textbooks? The preparation of fine line graphics is very expensive. And the additional cost of separations and two- and four-color printing drives the investment in books up, in several ways. Not only are two-color books more expensive for artwork and film, but also the cost of two-color printing is prohibitive unless we do print runs of 7500, 10,000, or more copies. In order to justify printing in those minimum quantities, we are rapidly departing from the realm of "innovative publishing." Can we live with one-color printing, and use the flexibility of type styles and sizes currently available in many scientific word processing software packages instead of a second color? Imaginative use of different type styles and sizes can provide the same visual distinctions and schemes as two-color type.

The next obvious area to look for ways to economize is in artwork. The average investment in a complex calculus illustrations program is $75,000, for a first edition. Substantial savings can be realized if authors can provide illustrations which are of reproducible quality, thus alleviating the need for costly rendering. I have seen excellent computer-generated figures in many manuscripts. It seems wasteful for publishers to be redrawing from perfectly acceptable camera-ready art. In fact, experienced textbook authors will appreciate the precision and control they have over the preparation of figures which they generate by computer.

Eleven hundred pages is too much, even of a good thing.

If we can agree that our innovative texts will not be as big as their more orthodox relatives (a virtue in any event), if we can be satisfied with one-color printing, and if we rely wherever possible on computer-generated illustrations, then we are beginning to describe an affordable publishing venture.

Gratuitous Ancillaries

What we have not considered are the additional materials which are part of the standard textbook "package." You have no doubt noticed the size of the cartons which arrive, sometimes unbidden, from publishers. It is no longer sufficient to publish only a textbook. So-called ancillaries provide another means for each new or existing textbook to distinguish itself. The development, production, and distribution of ancillary materials for an introductory mathematics textbook can represent additional costs approaching $100,000.

The ancillary or supplementary package consists of a complete and a partial solutions manual; a student study guide; a computerized testing program; a computer-based tutorial program; and overhead transparencies. Most of these items are given free to potential adopters of a text, although the student study guides and, in some instances, the student solutions guides are sold to students. The free supplements are published at the publisher's expense, usually provided for in the marketing budget. This is another stimulating set of reasons to ensure the standard outline and content of the book which stands at the center of the book package: the text has a family to support.

If we can agree to dispense with some ancillary items, we will be relieved of one constraint in our budget. No one admits to really using the supplements, anyway. I sometimes have the feeling that publishers undertake the production and distribution of supplements and other pedagogical aids, in some course areas, on the basis of matching other publishers and their packages.

Presumably, new calculus textbooks will have quite distinctive features in terms of mathematical content, organization, and approach to the subject. Therefore, in functional but unadorned book form, instructors will be able to see what's been done, and either accept or reject on those bases. In this more specialized and more modest segment of the calculus market, textbook decisions will not depend on such incidentals as the availability of supplements.

Viable Markets

Reducing the physical requirements of textbooks and their attendant supplements will remove some of the financial disincentives facing publishers, as we consider new publications. The wonders and portability of desktop publishing provide us with alternative composition and sources for accurate line illustrations, tables, and graphs. These economies, too, will encourage publishers to enter emerging course areas.

The question of markets still confronts us, though. While the investment facing publishers might be more modest, and therefore more attractive than undertaking large textbooks, our vision is national in scope. In established publishing companies it costs a certain amount of overhead just to turn on the publishing machine. Contributions to operating costs and other expense requirements preclude our undertaking the textbook equivalent of vanity publishing. No publisher is willing to publish for a single course being offered in one college. That kind of limited enterprise does not require the editorial, production, and marketing resources of even a modest-sized college publisher.

The stability and predictability of standard courses are two of their assets, in the eyes of publishers. If we publish a differential equations text from an engineering college in the northeast, we can be confident the course looks the same in other colleges. One publisher, in partnership with an author and several manuscript reviewers, can enter a national, coherent market.

The more cooperation and sharing of ideas that exists among instructors at different colleges, the more attractive such innovative courses will be for publishers. While we all decry the monolithic nature of introductory textbooks, publication for fragments of small, isolated experimental markets is *not* a viable alternative. If cooperation among mathematicians helps publishers identify several, rather than many, fragmentary proposed future directions in calculus instruction, we will be responsive.

At the outset of this paper I spoke of publishing as a data-collecting and data-evaluation business. Publishers are relieved, frankly, to see that concerted efforts are at work and national commissions are in place to support changes in the calculus courses. Too often our relations with a discipline are limited to one-on-one ventures with authors, and we have little sense or spirit of cooperation with a discipline.

Some publishers are entrepreneurial risk-takers. New, interesting departures from existing textbook models do find their way into print, eventually. But that is a very slow and unpredictable process. Having access to the proposed "high-level" information base will be of tremendous help to publishers (and authors) in evaluating the extent of a potential market for a calculus project.

With access to such data, and a willingness to use efficient and economical methods of production, innovative textbooks will be available. The publication of innovative alternatives to traditional calculus textbooks will certainly help the diffusion process for alternative approaches to the course. With the right data in hand, and a moderate investment requirement, publishers will be your enthusiastic partners in fostering change in calculus instruction, and publishing.

Reference

[1] Douglas, Ronald G. (ed.) *Toward A Lean and Lively Calculus.* MAA Notes, No. 6. Washington, D.C.: Mathematical Association of America, 1986.

JEREMIAH J. LYONS is Senior Editor at W. H. Freeman and Company and Scientific American Books. Previously he held editorial positions at Addison-Wesley and PWS Publishers. He was a member of the Committee on Corporate Members of the Mathematical Association of America, and is a former chairman of the Faculty Relations Committee of the Association of American Publishers.

Perspective from High Schools

Katherine P. Layton

BEVERLY HILLS HIGH SCHOOL

The population of high school students who are affected by calculus can be broken into three groups: mathematics students who take AP calculus, those who are enrolled in a non-AP calculus course of some kind, and those in the precalculus classes such as second year algebra, mathematical analysis, or trigonometry who will take calculus at college.

If a school has an AP calculus program, the students enrolled will be the very best mathematics students. These students make up the main pool for future mathematics majors. Many are already turned onto the excitement and beauty of mathematics. This joy of mathematics must be carefully cultivated. These students must be nurtured both at the high school and college levels. Special consideration should be given to them when they enter college.

At the college level, mathematics departments need to seek out these students and give them special counseling to get them properly placed in mathematics in their freshman year. Remember, these students have come from small mathematics classes where they have been nourished and have had much opportunity to interact with their teacher and with each other. Usually their high school teachers are among the strongest teachers in the mathematics department.

If a school has an AP calculus program, the students enrolled will be the very best mathematics students. These students make up the main pool for future mathematics majors.

Students enrolled in a non-AP calculus course also need special care and careful counseling which leads to correct college placement in mathematics. They have experienced some calculus and may believe they know it all; some may become complacent, cut class, and end up being unsuccessful. These students need to be reminded that attending class is important and that homework must be done; they too need the involvement and concern of their professors.

I would like to see non-AP calculus courses at the high school eliminated and students enrolled in other mathematics courses such as probability, statistics, discrete mathematics, or elementary functions including an introduction to limits.

Preparation for Calculus

If the calculus course is to change, then so must the precalculus courses. (By precalculus courses, I mean algebra, geometry, advanced algebra, trigonometry, and mathematical analysis.) Currently, much of the content in these courses is geared to preparing students for calculus. A large portion of the content is devoted to gaining skill in numerical and algebraic operations. Students learn much of their mathematics by memorization and may not have had much experience with being expected to understand concepts. Instructional time is spent on factoring, simplifying rational expressions, graphing, simplifying radicals, solving equations, and writing two column proofs.

How much skill is really needed in these areas in light of the new calculators and computer software? Students still need some proficiency in numerical and algebraic skills and they need some expertise in symbol pushing. In order to make the best use of a calculator, students need computational facility with paper and pencil, and the ability to do mental arithmetic.

Yet to use a software package effectively, students need strong conceptual understandings of the subject matter and then computers can take care of the mechanical details. The question arising here involves one of instructional psychology: how much expertise does a student need in performing mechanical procedures in order to understand concepts and to be an effective problem solver?

Understanding Concepts

If the desired products are students who can think, who have an understanding of concepts, who have developed logical maturity, and who have the ability to abstract, infer, and translate between mathematics and real world problems, then this type of training must begin early in their mathematical experience.

More time in high school needs to be spent on understanding concepts, developing logical reasoning, guiding students' thoughts, and helping them develop thinking skills. Students need numerical and graphical experiences to help them develop intuitive background. Students need guidance and experience with the use of both calculators and computers. These tools should be an integral part of the overall mathematics program. Issues of reasonableness of answers, appropriate use of technology, and round-off errors must be considered.

In changing the precalculus program, one needs to be careful not to overload the courses. Selected content must be eliminated if new material is added or if teachers are to teach more for concepts and understanding.

I believe students need considerable facility with basic algebraic skills, geometric facts, and numbers facts but the time and drill spent in these areas can be reduced, as can the current content. Both Usiskin and Fey have made suggestions for content in the precalculus area ([2], [3]).

A New Precalculus Course

The current calculus course is too full of techniques which students often memorize without understanding the fundamental concepts. Changes in calculus will require and depend upon changes in the precalculus program.

New courses must be constructed that stress development of concepts as opposed to purely mechanical understanding. Students and teachers will spend more time setting up problems, analyzing and interpreting results, and creating and solving realistic problems. Students will have more experiences with estimation, algorithms, iterative methods, recursion ideas, experimental mathematics, and data analysis (how to get data, what do these data mean, and how to transform, compare, and contrast the data).

Much more experience with functions will be included. Students will consider those given by a formula, generated by a computer, and arising from data. There will be more emphasis on analysis of graphs. Experience with a deductive system will be included.

While in groups students will have the opportunity to read about mathematical ideas new to them and work out problems using these ideas. They need to construct examples illustrating concepts and find counterexamples. There should be opportunities to solve non-standard problems and problems which are multistep; these problems must push students beyond the blind use of formulas.

Goals of Calculus

In the publication, "Toward A Lean and Lively Calculus" [1], there is a statement of goals giving competencies with which students should leave first-year calculus. These include the ability to give a coherent mathematical argument and the ability to be able not only to give answers but also to justify them. In addition, calculus should teach students how to apply mathematics in different contexts, to abstract and generalize, to analyze quantitatively and qualitatively. Students should learn to read mathematics on their own. In calculus they must also learn mechanical skills, both by hand and by machine.

As for things to know, students must understand the fundamental concepts of calculus: change and stasis, behavior at an instant and behavior in the average, and approximation and error. Students must also know the vocabulary of calculus used to describe these concepts, and they should feel comfortable with that vocabulary when it is used in other disciplines.

Students must understand the fundamental concepts of calculus: change and stasis, behavior at an instant and behavior in the average, and approximation and error.

At the Sloan Conference where these competencies were thought through, the content workshop developed a suggested calculus syllabi for the first two semesters of calculus [1]. This is a course I would like to see implemented. However, if a new calculus course is implemented without changes in precalculus courses, then the calculus program must place additional emphasis on functions, approximation methods, recursion ideas, data analysis, interpretation of graphs, use of calculators, developing number sense to recognize incorrect answers, checking answers for reasonableness, and dealing with the question of round-off error. Helping students to build a conceptual understanding will be extremely important.

Students

Many students are used to getting through mathematics classes by memorizing recipes for doing problems. These techniques seem easier to them than having to reason through a problem using mathematical concepts, i.e., having to think their way through a problem. They find mathematics difficult and look for an easy way out; memorization appears to work.

Students often are more computer literate than many of their teachers. Many will have computers at home and will have used some of the mathematics software packages. New high-powered calculators will be in the hands of many students. Even now students are questioning the value of learning certain arithmetic operations since they know their calculators can do them more quickly and accurately than they can. Calculators capable of displaying graphs were in the classrooms last fall and, of course, some students questioned the time spent on curve sketching. All this calculator and computer power must be used to make time for a more useful, exciting, and relevant mathematics curriculum for the students.

Parents

Parents also will need to be educated. Many parents were students during the "new math" era and may have been burned. Others will say "that's not what *we* learned." Parents have expectations about the skills their children should know. The mathematical community must convince them that it is all right for students to use technology wisely and with discretion.

Testing

There are two areas to consider here: teacher-made tests and the various kinds of standardized tests.

Testing influences both student and teacher behavior in the classroom. If students know they will only be tested on techniques, they will listen (or day-dream) during concept building experiences but will not worry about them. When they study for tests, some will only memorize procedures and techniques. Currently teacher-made tests test mostly techniques or skills. These are easier to write, easier to grade, and are the types most teachers experienced when they were students.

Students should be tested on the understanding of concepts, should be able to explain ideas, and should be expected to write about mathematics. The difficulty of questions should be graded from quite easy to challenging. Since many teachers use or model their tests after tests that come with textbooks, publishers must also be educated. Calculator use should be encouraged in testing: tests must include questions that make good use of the calculator.

Tests that test mostly techniques or skills ... are easier to write, easier to grade, and are the types most teachers experienced when they were students.

Teachers need to try various forms of testing: student portfolios containing work completed, open-ended exams, take-home exams, open-book exams, group exams, oral exams, and standard questions in nonstandard settings, i.e., giving a graphical version of the derivative of a function and asking for the graph of the function.

Standardized tests, as the saying goes, tend to drive the curriculum. For the curriculum to change, these tests must change too: AP Calculus, Scholastic Aptitude Test, College Entrance Examination Board Mathematics Achievement Test, National Assessment of Educational Progress, American College Test, and state and local competency tests. When will effective calculator questions be included on these important national tests?

Materials

The proposed new courses suggest heavy use of calculators and at least classroom demonstration of computer experiences. The latter necessitates the availability of well-thought through and simple-to-use software. Teachers want to know quickly how to use a piece of software and to know if it is dependable. If something goes wrong, instructions for what to do must be clear and efficient. Guidelines for use of the software with a given topic are also needed. Calculator exercises and suggestions for use in discussing a concept and exploring mathematical ideas are needed.

When will effective calculator questions be included on ...important national tests?

I would like to see both computers and calculators become an integral part of the curriculum. This, of course, implies that appropriate assignments are available which allow students to use these tools to explore mathematical ideas. These materials need to be very user-friendly so even the inexperienced student can use them with ease. We are concerned here not with teaching how to program the computer but with using the computer to teach and explore mathematics.

Textbooks integrating the technology and stressing the concepts approach are necessary. In many states the schools must furnish each student with a book free of charge. In planning a new calculus program and suggesting changes in the precalculus program, one must remember that high schools cannot change textbooks very often. Most schools are on a four to eight-year cycle.

Equipment

Each student will need a calculator. I would like to see a scientific one with graphing capability used in the precalculus classes. Calculus students would need one with a "solve" key and an "integrate" key. Computers need to be available for both classroom demonstration and for student use. In addition, a large screen monitor is necessary so a class of 30-40 students can see classroom demonstrations of materials on the computer. For the most effective use of a classroom tool, a computer and large screen monitor should be in each mathematics classroom.

I cannot leave the discussion of materials and equipment without stressing the shortage of funds. Mathematics departments in many schools cannot afford the necessary technology, software, and textbooks. Funding from industry, business, and state or federal governments is necessary.

Teachers

Teachers are the key factor in a successful mathematics program. There are many competent high school mathematics teachers doing a fine job with 150 young people 180 days a year. Teachers need to be convinced that a change is necessary and then must be given the time, tools, resources, and training to do the job. I have been told by many mathematics supervisors that high school teachers are the hardest to change and are least responsive to in-service education. They are reluctant to change what they believe has worked.

Many current teachers were educated in the late '50's and '60's, and if my experience is typical, most have a pure mathematics background. This early training does influence what content they feel is important. These teachers have been teaching for eighteen to twenty-eight years. They are committed, experienced, capable, but some are very tired!

Teacher Training

Teacher training will be a key to success for a new calculus program and precalculus curriculum. In-service, summer programs, or college courses available during the school year are several ways this training could be packaged. California has had considerable success training teachers during the summer and then using these experienced teachers to conduct workshops, speak at conferences, and give in-service programs. Much training will be needed, not only for the calculus teachers, but also for the teachers of the precalculus classes of algebra, geometry, advanced algebra, and trigonometry.

A good training program requires several features. Teachers must be given released time or be compensated for attending; materials and registration fees should not come out of their pockets. Courses should be presented by outstanding teachers familiar with the high school environment who are modelling what should happen in the high school classroom.

Teachers need continual support such as monthly meetings to share experiences and to discuss what works, what does not work, and what went wrong. They should have the opportunity to observe each other. Care needs to be given as to when and where this training is provided. A late afternoon or evening workshop is not always the best experience since teachers are tired after teaching five classes during the day and often must prepare for the next day.

Teacher training will be a key to success for a new calculus program and precalculus curriculum.

A vital issue is the teachers' feelings of confidence and comfort. This issue must be addressed. If classes become more open-ended and more exploratory in nature, teachers will feel less in control and will not always be an authority on the topic. There are teachers who fear something different or new. They need to be willing to learn with their students and must be given the necessary tools and confidence to do so.

Calculators and computers must become integral parts of calculus and precalculus courses; this is not true in many schools at this time. Often it is not because the equipment is not available, but because teachers are untrained and thus uncomfortable using the computer or calculator. Another critical issue is one of time: teachers rarely have time to find ways to implement use of this technology in class.

There are teachers who fear something different or new.

It is easier to teach mechanical material. It is much harder to get students to "think" and understand concepts. It is also easier to write and grade exams testing basic skills. One aspect of "easier" is the issue of time: time in class, time to plan lessons, and time to write tests. It takes less time in all these areas to deal with skills and mechanics instead of concepts. Teachers will need guidelines and models for writing new types of test questions, and for conducting different types of classroom experiences which help in concept building.

In planning the new curriculum, one must also be aware of the quality of people entering the teaching profession, since many have weak mathematical backgrounds that need to be strengthened.

Teaching

We want students who can think, reason, apply concepts, express themselves with clarity, and use technology effectively. To accomplish these goals, students need to be given opportunities to talk, write, and think about mathematics.

Students should be expected to read and write using the mathematics vocabulary of the course. Other expectations should include complete and coherent answers which are well thought out, well developed, and well written on both tests and homework, and well expressed in class discussions. Students should spend time at the board explaining their work, and there should be opportunities for group work.

As mathematics teachers we need to take responsibility for all aspects of students' mathematical development. We must not ignore algebraic errors when students are writing up solutions to mathematical problems or explaining their reasoning in solving a problem. We should not accept sloppy work; we must encourage students and work with them to improve their work.

Problems need to be assigned that are thought provoking, not just skill oriented, and that require detailed answers. These must be graded carefully. In all courses, teachers should use a variety of techniques such as concrete materials, chalk and talk, group work, computer demonstrations, films/videos, games, and discussion to help students learn and understand mathematics. In this environment students will put thoughts into words, help each other refine answers, and explore ideas.

References

[1] Douglas, Ronald B. (ed.) *Toward A Lean and Lively Calculus.* The Mathematical Association of America, Washington, DC, 1986.

[2] Fey, James T. (ed.) *Computing and Mathematics: The Impact On Secondary School Curricula.* National Council of Teachers of Mathematics, Reston, VA, 1984.

[3] Usiskin, Zalman. The University of Chicago School Mathematics Project for Average Students in Grades 7-12.

KATHERINE P. LAYTON teaches at Beverly Hills High School in Beverly Hills, California. She is a member of the Mathematical Sciences Education Board and of the Committee on the Mathematical Education of Teachers of the Mathematical Association of America. In 1983 she served on the National Science Board Commission on Precollege Education in Mathematics, Science, and Technology.

A Two-Year College Perspective

John Bradburn

ELGIN COMMUNITY COLLEGE

I start from the premise that it is intuitively obvious from the most casual observation that both the teaching and learning of calculus is in serious disarray. A further premise is that if anything of significance, however slight, is to be done in the attempt to rectify the situation, then funded pilot projects are a necessary step in the process.

"If you don't know where you are going, any road will get you there."

This paper comments on some of the questions which need to be addressed in the design of these projects, offers suggestions of ways that two-year college faculty can effectively help in designing and carrying out the projects, and makes recommendations on disseminating the results of the projects in a manner which would be useable by the greater mathematics community, especially by two-year college faculty.

As a first step, I recommend a complete perusal of the report [1] of the 1986 Tulane workshop on calculus by anyone interested in joining the trek toward a lean and lively calculus. Although this report covers many of the questions and ideas that must be considered, I shall repeat some of them in this paper as a means of further emphasizing them.

My instructional design professor used to say, "If you don't know where you are going, any road will get you there." Where is calculus going? Where does calculus fit in the development of mathematical ideas? Where does calculus fit in the development of students' mathematical maturity? What are the proper service functions of calculus? These questions are partially answered in the papers in [1], but I do not feel that complete enough answers are given there to move to the instructional design stage for calculus.

The Role of Proof

The central issue that must be addressed in any reform of calculus concerns the nature of proof. Where is the development of the idea of the nature of proof *deliberately* emphasized in the sequence of mathematics courses from beginning algebra through calculus? When I took plane geometry as a high school sophomore and solid geometry as a high school junior, everyone knew that you do proofs in those two courses. It now appears that someone looked at the titles of those courses and decided that since "geometry" but not "proof" is in the titles, then proof is not an appropriate topic in a geometry course.

One of the problems we deal with in attempting to teach calculus today is that proof is not a part of most students' mathematical backgrounds. Proof is no longer a large part of the geometry course and the algebra courses deal not with proof but with manipulative skills. The debate over the appropriate level of rigor in teaching calculus needs to be preceded by a discussion concerning whether calculus now represents students' first introduction to mathematical proofs. Certainly *some* proofs are still done in calculus, either using intuitive premises or very formally constructed premises.

I believe that proof is the very essence of mathematics.

I believe that proof is the very essence of mathematics. Many students end their formal mathematics training with calculus or shortly thereafter and the nature of proof should be a part of that training. As a two-year college teacher, I regularly teach all the courses in the sequence from basic algebra through calculus. In terms of the textbooks available, with the possible exception of analytic geometry, I do not see the idea of developing the nature of proof deliberately being given a high priority.

Since most two-year colleges teach that whole sequence, such institutions need answers to the question of where the nature of proof is deliberately and consciously taught in the sequence. Those answers may be contained in decisions on the content and style of the calculus or in statements about the prerequisite skills and knowledge of entering calculus students. However the answers are given, whether traditional or newly formulated, we do need them.

Establishing Priorities

Questions about the role of proof as well as other questions raised in [1] need to be answered before design of instructional content and style for the various projects can begin. Information needs to be gathered from various sources, especially about the service functions of calculus. When the questions are answered, the expected outcomes of calculus instruction need to be clearly prioritized—there are techniques for doing that—in order to make the difficult decisions about what topics to include or not include, and which ideas to give primary emphasis and which to give lesser emphasis.

Another reason for having a clearly prioritized list of expected outcomes is that it facilitates the process of evaluating the success of the project. The prioritized list is also a necessary part of the information to be disseminated in follow-up workshops in order for others to judge the effectiveness of the instruction and to determine if they wish to try the instructional package developed in that particular pilot project.

I have been teaching calculus for twenty years and I no longer have favorite topics to teach. I have fun days in teaching calculus, but I cannot count on any particular topic being fun to teach on any given day. Therefore, I would not argue for inclusion of any topic because I enjoy that topic, but would rather point to a prioritized listing of what calculus is supposed to accomplish.

Certain topics may be included for completeness, if completeness is high enough on the list of outcomes to outrank competing topics which meet other outcomes. Proof, as I have argued, should have high priority. Since most instructors rely heavily on calculus proofs based on intuitive premises, the course should also include some convincing examples of cases where intuition has led to incorrect conclusions.

Student Characteristics

Another question which needs to be answered before the actual design of the instruction can proceed concerns the prerequisite mathematical skills and knowledge on which the instruction is to be built. Although students in calculus have widely differing mathematical backgrounds, student's actual mathematical skills and knowledge, study skills, and learning skills fit into a fairly narrow interval.

For example, a couple of years ago I was really disappointed in my students' learning abilities and discovered that my colleagues in other fields agreed that this group of students was the worst ever. We assumed that the four year schools had dipped lower into the student pool to keep their enrollment figures up, leaving less able students for the two-year colleges. Yet our registrar assured me that on paper this was the best group of students we had ever had. Realism rather than wishful thinking in stating the prerequisite skills and knowledge leads to a more effective instructional design.

I may appear to be pointing an accusing finger in the preceding paragraph, but a two-year college mathematics teacher usually cannot afford such a luxury. We have enough students in our calculus classes who have taken, at our own schools, the sequence of courses leading up to calculus who do no better than the other students to keep us from pointing fingers. My earlier suggestion that two-year college mathematics teachers can use answers to the questions of where the nature of proof is deliberately taught in the sequence of courses also applies to most questions about the development of basic mathematical ideas.

The characteristic of two-year college students that is most often mentioned is that they tend to be older than other lower division undergraduates. Age is both a blessing and a hinderance: these students tend to be more serious about school, but have more responsibilities and demands on their time. Consequently, the majority of them are part-time.

Most two-year colleges are commuter schools. These two characteristics—part-time and commuter—are the variables that most affect instructional planning. Two-year college students need to plan schedules in advance (babysitting, jobs, transportation, etc.) and need a fairly fixed schedule and fixed time commitment for schoolwork for the entire semester. A large independent project assigned on short notice shows a lack of instructor's awareness and planning.

Style and Content

Peter Renz' paper, "Style versus Content: Forces Shaping the Evolution of Textbooks" [1, pp. 85-100], speaks specifically to the important issue of style of instruction. Other papers in the same volume speak to this issue by discussing sample questions and examples used in teaching calculus. The order of topics, for example, has a great effect on the instructional outcomes and is an important part of instructional design. I have seen useable textbooks issued in a new edition with the changes consisting primarily of a reordering of topics. Many times the new edition is not a workable text even though the style of presentation is the same.

The papers in [1] which deal with the utilization of computers and symbolic manipulation packages for teaching and learning calculus point out many possibilities for change in calculus instruction. However, care must be taken so that the computer does not take over the role of performing Mathemagic, that is, "snowing" the students, which the teacher can too easily do already.

Computers can show many things quickly, but students still need time to think. They need guidance in how to look at computer output just as they need guidance in reading mathematics texts. The prioritized list of outcomes, the choice of topics selected, and the availability of hardware and software are the major determining factors in designing instruction which uses computers in a meaningful way. The first two factors are far more important than the third, since hardware and software continually change.

Computers can show many things quickly, but students still need time to think.

One needs to exercise caution when deleting topics from a course. Most topics originally had valid reasons for being included, although those reasons may no longer be apparent from the treatment of the topic or even in the topic itself. One must be aware of the reasons the topic was included in the first place and make conscious decisions about those reasons when deleting a topic. I give two examples—one from trigonometry and one from calculus—to clarify this point.

In trigonometry one can be asked to give the value of sin (75°) or $\sin(5\pi/12)$. It is a very simple matter to get a value by punching a few buttons or, as in my day, by looking it up in a table. However, one purpose of that type of problem is to give students an opportunity to practice with various formulas:

(a) $\sin(5\pi/12) = \sin(\pi/6 + \pi/4)$

(b) $\sin(5\pi/12) = \sin(2\pi/3 - \pi/4)$

(c) $\sin(5\pi/12) = \sin(1/2)[5\pi/6]$.

In deleting this type of problem from trigonometry, one is in effect saying that adequate practice is provided elsewhere or that practicing with these formulas to become better acquainted with them is not worthwhile.

Techniques of integration is a frequently mentioned candidate for deletion from calculus. In terms of its importance to calculus, I agree. However, I have told my students for years that the real purpose of that chapter is to sharpen their skills in algebra and trigonometry. If that chapter is removed, we are saying implicitly either that other topics provide adequate sharpening or that students do not need to have their skills in algebra and trigonometry sharpened.

Teaching

If one is looking for experienced calculus teachers to help in the design and implementation of pilot projects, I would suggest a heavy dose of two-year college mathematics teachers. Two-year college faculty are first and foremost teachers. A typical regular teaching load is 15-17 hours per semester, often with an overload class or part-time contract on top of that.

A large number of two-year college mathematics instructors regularly teach calculus, are conscious of what it takes to be an effective teacher, and are willing to help improve the outcomes of teaching calculus. In designing and implementing different instructional strategies for calculus, the knowledge and experience these teachers bring to the task is very useful.

The feedback that experienced teachers can give on what is working and what is not working is invaluable in revising the material for the second and subsequent classes using the material. Sometimes students send mixed signals. At that point, decisions about the instructional process need to be based on teaching experience rather than on the mixed signals.

Two-year college faculty are first and foremost teachers.

Since teaching is such a large part of the professional lives of two-year college instructors, we want teaching to go well. We stand ready to help

- to improve calculus instruction;
- to select topics and teaching strategies for calculus;
- to provide feedback concerning new instructional packages;
- to disseminate information about successful pilot projects in a useable form.

The designers of calculus projects need to be very careful in the way that success is defined. Assuming that each pilot project has a prioritized list of expected outcomes, such a list for a particular project would be used in evaluating the success of that project. I strongly recommend that statements about the success rates of students *not* be included in the definition. The success rates of generally ill-prepared students who do not put in the necessary study time are not going to change dramatically in the short-term, so project success (or failure) should not be tied to so insensitive an indicator.

Dissemination

If calculus is to be reformed, then information about successful projects needs to be widely disseminated in a useable form. Not every school has to jump on the calculus bandwagon at the first opportunity. There is a small list of particular schools that will strongly influence any change that is proposed. If an overwhelming majority of the schools on the list accept the change, then the change will become general and widespread. I recommend that leaders of this reform movement write down their own minimal lists of schools and persistently work on those schools to be involved as much as possible in each step of the development and implementation process.

My hope is that many two-year colleges will be actively involved in the coming changes. Whether a two-year college is involved in the initial phases or not, it will change its calculus content when the schools to which most of its calculus students transfer have changed, since calculus is clearly in the transfer track.

Second, as part of promoting general acceptance of a change in calculus and also as a consequence of a general acceptance of that change, workshops need to be held to explain the change in calculus and to explain new ways to think about calculus and new ways to teach calculus. A final written report for each pilot project is useful in spreading the word about the projects. Talks and panel discussions about the projects at regional and national meetings are helpful in selling the ideas generated by the projects. However, I do not feel that those types of efforts are enough.

With the heavy teaching load that two-year college faculty carry, we tend not to be as well read as we would like. For us, relying primarily on printed material to spread the word is not enough. I recommend that each funded project contain, as part of the design of the project, a workshop component to be used in disseminating information about the project and about the continuing results of the project. The workshop component needs to be funded in such a way that the workshop can be given several times in various locations. I further recommend that the workshops be given at national, regional, affiliate, and section meetings of AMATYC and MAA. Cooperation of these organizations should be readily available.

I look forward to the improvement of calculus instruction over the coming years and to a return to enjoyment in my own teaching of calculus.

Reference

[1] Douglas, Ronald G. (ed.). *Toward A Lean and Lively Calculus.* MAA Notes, No. 6. Mathematical Association of America, Washington, D.C.

JOHN BRADBURN teaches mathematics at Elgin Community College in Elgin, Illinois. He has served as a member of the Illinois Board of Higher Education Committee on the Study of Undergraduate Education, as Governor of the Illinois Section of the Mathematical Association of America, and as Chairman of the MAA Committee on Two-Year Colleges. In 1982 Bradburn received the first national "Outstanding Faculty Member" award from the Association of Community College Trustees.

Calculus in a Large University Environment

Richard D. Anderson

LOUISIANA STATE UNIVERSITY

There is no single large university environment. The large university category includes the Harvards, the MITs, the Berkeleys, and the Michigans, all with highly select student bodies, as well as state universities of the South and Midwest with open admission policies. The roles of beginning calculus necessarily differ, from those of a course by-passed by many students in select universities to a course for which only a small percent of entering freshmen are eligible as in universities with open admission policies.

But in their lack of initiative for significant educational reform, the universities of the country are remarkably similar. Except for the addition of some new (optional) topics and the deletion of a few (generally harder) technical topics, the mainstream calculus books of today look remarkably like those of fifty years ago. And yet, in that time frame, the technical and scientific world has changed radically.

There has been one important development in calculus reform in universities over the past 15 or 20 years, namely the growth of non-mainstream courses, e.g., business or life science calculus. The 1975 and 1985 Undergraduate Surveys show a growth in the percentage of non-mainstream calculus from 22.5% of all university calculus enrollments in 1975 to 31% in 1985. The 1970 Survey did not list such "soft" or terminal calculus at all—presumably because it was not yet sufficiently well recognized for the committee then in charge to have listed it separately.

One can conjecture that, in universities, the growth in enrollments in calculus courses (or sequences) not leading to upper division mathematics played a role in reducing the pressure for reform in mainstream calculus since it was easier to believe that the engineering and physical science students really did need the emphases and drillwork of traditional calculus.

In their lack of initiative for significant educational reform, the universities of the country are remarkably similar.

Large universities share certain characteristics which affect their propensities for change in the calculus curriculum.

- They are big, with multiple sections taught by a wide variety of faculty.
- Calculus is the core of the mathematics service course sequences for students of engineering, physical sciences, computer sciences, business, life sciences, and mathematics itself.
- The user faculties on campus are sensitive to any changes which run counter to their ideas of what students should know. Since they don't have much occasion to consider details of service course curricula, they are inclined to be conservative about calculus—what was good for them is good for their students. Because they have diverse individual criteria as to what is important, any change will step on many toes. They are more willing to add new thrusts, e.g., linear algebra, to their students' programs than to make hard decisions about dropping old thrusts. They almost all find calculus useful as a "screen" in winnowing their students.

Role of Universities

It is important that the large university environment be a vital part of any efforts to change the nature of the teaching of calculus. The universities of our country are the focus of a very major part of basic research in science, in engineering, and certainly in mathematics.

They should also be an important origin of educational innovation. However, the reward system in universities is such that faculty time and effort spent on research totally dominate time and effort spent on educational reform. And thus educational reform has suffered. Very few faculty have either interest in educational reform or time for such work.

There is now much evidence that important elements of the research community are paying attention to basic educational needs. We must all work to see to it that the initiatives now begun are actively carried through at the university level as well as at the college level.

Science and Engineering

With the help of the engineering community, ever present in the university environment and highly cognizant in their own educational programs of the roles of technology, we in mathematics have an opportunity to ally ourselves to forces of change. It is odd indeed that, by all accounts, upper division education in engineering has changed radically in the past third of a century, whereas the basic science and mathematics courses for engineers in the lower division have changed very little.

From discussions with a number of people in various disciplines at various universities, it seems likely to me that the engineering community is readier to accept changes in calculus induced by the age of technology than is the physical science community. While many physical science faculty do employ computers in their research, the nature of research in the physical sciences as well as in mathematics requires individuals to get away by themselves in order to think.

Upper division education in engineering has changed radically in the past third of a century, whereas the basic ... mathematics courses for engineers in the lower division have changed very little.

Much of that process in mathematics, at least, involves conjecturing, drawing pictures, and trial-and-error methods and generally these require paper and pencil activities. What we seek in calculus for the next century is a balance between the old and the new: paper and pencil activities to assist in understanding and in problem solving but not in routine computations or algorithmic processes better done by computers and calculators.

The coming reformation of calculus in large universities is complicated by the fact that there are many local fingers in the pie. Engineers, computer scientists, physical scientists, life scientists, and business administration people as well as mathematicians all have their own special and differing needs for calculus level courses. Calculus is the dominant introductory mathematics course for students in the first two years in these various disciplines.

Technology has brought great changes in the way mathematics is used in the work place, away from paper and pencil procedures toward the use of calculators and computers. Symbolic manipulation and computer graphics are not only changing the way engineers and others use mathematics in the work place, but they clearly will force major changes in university level calculus—students' first step on the access route to the work place.

The rote algorithmic paper and pencil procedures which have been ... largely unchanged in calculus over the past 50 years, at least, are irrelevant in terms of even current work place use.

The rote algorithmic paper and pencil procedures which have been developed over the past century-and-a-half and have been largely unchanged in calculus over the past 50 years, at least, are irrelevant in terms of even current work place use. That does not say that the concepts and ideas of calculus are irrelevant, but only that we must design our educational practices in calculus to conform to students' future needs, both in the post-calculus learning environment and in the future work place.

Calculus in Universities

The 1985-86 Survey on Undergraduate Programs in the Mathematical and Computer Sciences [1] reveals much background statistical data on calculus in universities. Although the "university" category used there is from Department of Education lists not identical with AMS Group I, II, and III Institutions, percentage figures for enrollments, faculty and teaching phenomena are close to those applicable for any reasonable definition of "large universities." The data is for calculus taught in mathematics departments in the fall term of 1985 and include enrollments in the first, second, or third terms in the engineering and physical science calculus, and in the business and life sciences calculus.

In universities 40% of all undergraduate mathematics enrollments are in calculus, with a 27% to 13% split between engineering-physical science calculus to business and life-science calculus. 72% of all mathematics enrollments at the calculus level or above are in calculus itself; 81% of all such enrollments are in calculus or its natural successors, i.e., differential equations, advanced calculus, and advanced mathematics for engineers.

Thus calculus is overwhelmingly the dominant mathematics course taken by undergraduate students in universities. Total enrollments in calculus as well as total undergraduate mathematics enrollments in universities were essentially unchanged from 1980 to 1985.

In university calculus courses, approximately 38% of all students are taught in sections of under 40, 20% in sections of size 40 to 80, 12% in lectures of more than 80 without recitation sections, and 29% in lectures of more than 80 with recitation sections. (The remaining 1% are taught in self-paced or other format.) Only 5% of calculus sections in universities have any required computer use.

The average age of the full time university mathematics faculty is 44 with 65% over age 40 (up from 45% in 1975). 90% of all full time mathematics faculty in universities have doctorates, 63% of whom are tenured. Presumably almost all of the non-doctorate full time faculty would teach only courses below the calculus level.

In universities ... 72% of all mathematics enrollments at the calculus level or above are in calculus itself.

No specific data is available from the Survey on the percentage of calculus sections taught by teaching assistants, but anecdotal information suggests that whereas teaching assistants teach about one-fifth of all separate university sections in mathematics, the majority of these sections are at a level below calculus. In many universities, only select advanced graduate students are assigned calculus-level courses. However, there are a few (selective) major state universities with very little course load below the calculus level; at such places, teaching assistants teach much of the introductory calculus sections.

What Do Students Learn?

Student learning procedures consist primarily of working textbook problems—usually several or many of the same sort—following model procedures given in the textbook or by a teacher. Thus students learn various paper-and-pencil algorithms for producing answers to special types of problems. They customarily read the text only to find procedures for working such problems.

Almost no student below the A-level (and few A students) can cite, much less accurately state, and even less prove, any of the theorems related to elementary calculus. Student dependence on memorized procedures to produce answers follows a similar pattern of learning in pre-calculus mathematics. In calculus, however, it requires much wider and more readily recalled background information, primarily from algebra, trigonometry, and elementary functions.

Almost no students below the A-level can cite, much less accurately state, and even less prove, any of the theorems related to elementary calculus.

The intellectual achievement for most students in learning calculus is, nevertheless, considerable. They have had to learn about new concepts much more rapidly than in earlier courses, and they have had to develop command of a wider and more diverse "bag of tricks." They have learned to solve problems from a variety of geometrical and physical applications. Unfortunately, many of the tricks in the calculus bag appear irrelevant in an age where the computer and the calculator are rapidly replacing paper and pencil as the tools of the trade.

Roles of Calculus

In considering any significant change, much less a reformation of calculus, it is important for all to keep in mind the various roles traditionally played by calculus in our universities. We should not, unwittingly, reform calculus without taking into account the many useful by-products of the study of calculus. Here are some roles to keep in mind.

- Calculus embodies a unity and beauty as one of the great and useful intellectual achievements of mankind. However, many current courses pay only lip service to these aspects of calculus. In reforming calculus, we should overtly seek to acquaint all students with the unity, beauty, and power of calculus.
- Calculus represents modelling of a mathematical system with a richly diverse set of applications; physical, geometric, biological, and managerial. Calculus, together with the real number system on which it is based is, along with Euclidean geometry, an ultimate mathematical model.
- Calculus as currently taught has been the course in which most engineering and physical science students really learn algebra and trigonometry: algebraic topics are used and reviewed until they come together as necessary background for a new and more powerful subject. In the age of symbolic manipulation and computer graphics, we must seek to identify and strengthen those aspects of calculus as well as those aspects of algebra and trigonometry which will be important in the work place of the future.
- With its many graphical representations, calculus has been a rich source for the (further) development of students' geometric intuition. Along with geometric aspects of linear algebra, introductory multivariate calculus offers vital exposure to three and higher dimensional geometry.

Calculus as currently taught has been the course in which most engineering and physical science students really learn algebra and trigonometry.

- Calculus has been the mathematical and intellectual screen by means of which students in engineering, in the sciences, and in mathematics are judged as ready for more advanced work. It is "the universal prerequisite." Both the rapid assimilation of new concepts and the control of a broad framework of background information required in successful study of the calculus are student experiences involving characteristics which are manifestly important for further study.

Forces Against Reform

Inertia. One should never underestimate the inertia and resistance to change in a large system. The educational system is naturally conservative, with all universities structured along traditional departmental and college lines. Significant changes affecting introductory courses like calculus which are prerequisite to almost all courses in both mathematics and user disciplines must have at least grudging approval of most or all of the departments and colleges concerned.

For example, even interchanging the order of introduction of topics has great implications for normally concurrent physics courses. Thus any major changes in emphasis in calculus must be carefully planned and coordinated with user departments.

Tradition. Virtually all current mathematicians and users of mathematics have grown up with a calculus course largely unchanged in both content and presentation over at least the past half-century. There have been cycles of relative emphasis on proof and of varying orders of presentation of topics but, by and large, student learning has been almost totally involved with repetitive use of paper-and-pencil processes for producing answers to special types of problems. Based on earlier mathematics experiences, students, mathematics faculty, and user faculty expect standard skills and judge success or failure by a student's ability or inability to work special types of problems. In my experience, when questioning users or mathematicians about the content of calculus, there is strong instinctive reaction to judge as important those topics the faculty were taught and were successful in learning (or, for mathematics faculty, in teaching).

Qualifications and Attitudes. The bulk of students taking calculus in universities are taking it as a means of gaining access to other subjects. Their primary concern is to get through the course with the least diversion from their other interests.

Virtually all current mathematicians and users of mathematics have grown up with a calculus course largely unchanged in both content and presentation over at least the past half-century.

Furthermore, most have achieved what they have by working lots of problems by memorized paper-and-pencil algorithmic procedures following textbook or teacher examples. Their natural inclination is to want more of the same. They don't want to be shown more than one way to do a problem, they don't want to know *why* but only *how,* and few of them have been exposed to problems about which they need to think hard or long.

Many have not learned how to read the text except to follow the step-by-step illustrative examples. Many have difficulty even reading the word problems. A change to conceptual calculus with a downplaying of formal paper-and-pencil procedures will run into student resistance and must be phased in with companion changes at the school level.

Textbooks. Texts are written and published to be sold. They represent the publishers' views of what the teaching community wants. Thus current texts cover a wide variety of topics (for faculty choice) and have pages of standard, similar problems, following illustrative examples.

Texts are frequently adopted via committee recommendation, with committees looking for "teachable" books, those similar to ones liked in the past and those consistent with somewhat traditional values. New texts with radical changes only rarely get adopted, thus authors don't write them and publishers don't publish them.

There is uncertainty among us and among many of our colleagues about what topics are important in calculus for tomorrow and also uncertainty about the need for gradual continuing change after an initial "reformation." We do not yet understand how completely or rapidly symbolic manipulation and computer graphics will effect either societal use of calculus or the teaching of calculus nor, at an even more fundamental level, do we understand which aspects of calculus will turn out to be most important for users in the evolving and forever changing age of technology.

Reference

[1] Albers, Donald J.; Anderson, Richard D.; Loftsgaarden, Don O. (eds.) *Undergraduate Programs in the Mathematical and Computer Sciences: The 1985-1986 Survey.* MAA Notes No. 7. Washington, D.C.: Mathematical Association of America, 1987.

RICHARD D. ANDERSON is Professor of Mathematics, Emeritus, at Louisiana State University. He is Past President of the Mathematical Association of America, and served as Executive Director of the 1985-86 MAA-CBMS Survey of undergraduate programs in the mathematical and computer sciences. Currently he chairs a subcommittee of the Committee on the Undergraduate Program in Mathematics (CUPM) charged with making recommendations concerning the first two years of college mathematics.

A Calculus Curriculum for the Nineties

David Lovelock and Alan C. Newell
UNIVERSITY OF ARIZONA

There is little disagreement with the premise that the mathematical fluency of our educated population is near an all time low. More serious than the lack of general literacy is the fact that America is not producing a generation of students who will become the mathematical scientists and innovators of the future. Our best brains are being siphoned off into business, medicine, and other careers before the mathematical science community has even had an opportunity to persuade them that understanding nature, social behavior, economics, not to mention the vast changes in technology from a quantitative point of view, is a worthwhile pursuit.

There are many reasons for this. First, it might be argued that, except for rare periods, America has never encouraged its bright, young people into science. Immigrants have always done the job for us. Second, the peer pressure of material success suggests that tough, rugged Lee Iacocca is a better role model than the nerd who teachers mathematics on the television at six o'clock on a Monday morning. Young people are not exposed to role models of a sufficiently heroic type in the mathematical sciences.

America is not producing a generation of students who will become the mathematical scientists and innovators of the future.

Third, and perhaps the only area in which we might take action so as to have immediate impact, is the early experience of students with mathematical thinking and ideas. Presently, the preparation of our young people in English and mathematics, the two subjects which should surely dominate a high school education, is pathetic.

Changes in Attitude

Therefore the first suggestions which we offer address less the details of the curriculum in calculus and more the attitudes that we must adopt:

- There must be a concerted effort, working with local high schools and high school teachers, to see that students are better prepared in algebra, geometry and trigonometry and, most importantly, that they are aware of the self-discipline required to learn mathematics. Real learning requires work.
- We must stress that mathematical knowledge is cumulative and that scientific knowledge has a vertical structure. All tests and examinations should reflect this fundamental premise. Regular homework should be a priority. Homework drill is essential. Tests should be cumulative.

The preparation of our young people in English and mathematics ... is pathetic.

- Let us now agree once and for all that not all students have equal abilities and introduce some stratification (honors sections, sections which coordinate with elementary physics or business courses, etc.) into the organization of classes.
- Above all, we must put our best teachers, the most caring and inspiring, in the first year courses.
- A program that emphasizes mathematical thinking and word problems will require more instructor time per student. Consequently, it will be necessary to restructure the incentive and reward system for university professors. Research is vital, no one questions that. But there is presently far too much recognition given in American universities for mediocre research and far too little for excellence in teaching. This attitude has not only led to an imbalance in distribution of personal time and effort, but also has skewed the internal distribution of faculty among departments within a university.

Changes in Curriculum

We continue with a similar list of fundamental principles directed at the calculus curriculum itself:

- We must understand that the calculus sequence is the basic set of courses in mathematical thinking to which a student is exposed. Although the sequence must be responsive to the university community as a whole and must prepare students to handle the mathematics used in the courses of other disciplines, the teaching of mathematics at this level should not have a purely service connotation.

- It is important that students be exposed to logical thought processes, problem solving and other areas of mathematics such as probability and number theory which stress these activities. The actual material covered is not as important as the confidence induced by gaining understanding through mathematics and simple logical arguments.

Above all, we must put our best teachers, the most caring and inspiring, in the first year courses.

- We must use to the fullest extent the products of modern technology. Homework can be monitored and corrected and help can be given on computers. The prerequisites of a given course can be introduced on VCR tapes which can be used on home televisions, or on discs which can be used in most standard personal computers. Much of the material which a student must do alone (lots of examples, homework, etc.), all the repetition which is so necessary in learning mathematics, can be done better by the student with technological aids.
- Many students are computer literate (they can operate the computer, play games, etc.), but very few can write a good program or use packages effectively. All summing operations (series, integrals, etc.) should be done numerically. There is no harm in introducing students to exact methods; it increases their literacy and fluency; but it should be stressed that there is no difference in principle between calculating $\int_0^1 \sin^2 \pi x\, dx$ and $\int_0^1 \sqrt{1+x^2} \sin^2 \pi x\, dx$. Each is a number and the latter number is no less good than the former just because it cannot be exactly calculated but must be approximated.

Mathematics education cannot be egalitarian.

- The curriculum should stress word problems, the art of judicious approximation, and the importance of converting conservation laws and problems in optimization into mathematical language. After all, practically all laws used in engineering and science are derived from looking at balances of mass, concentration, momentum, heat, energy. Moreover, many of the challenges in modern business involve choosing optimal configurations of the coordinates to achieve certain desirable ends.

Before writing down an outline of a curriculum, it is important to return to the idea that mathematics education cannot be egalitarian. We suggest therefore that the best students be separated into classes which cover an expanded curriculum in the traditional three courses. However, the extra depth and amount of material introduced should be recognized and so we recommend that each of these three courses receive four hours credit.

In order to coordinate the curriculum between the brighter and the less qualified student, we suggest that the vast majority take calculus as a sequence of either three or four three-credit hour courses. The first option would cover less material but still emphasize mathematical thinking, problem solving, approximation techniques, and the importance of computers. The four course option would allow the less qualified student to have an equal preparation should he or she decide to continue in a course of studies where all the calculus material is required.

In addition, it should be possible for a student majoring in engineering or the physical sciences (or any other interested student for that matter) to choose a calculus track which coordinates calculus topics with relevant material in the parallel science sequence (e.g., motion problems, Newton's laws, calculation of chemical compositions). After all, most major universities run over twenty sections of each calculus course each semester. Diversity and choice should have their place in the education system.

Real learning requires work.

We want to stress again, however, that except for vital prerequisite material, the exact topics introduced are not crucial. We see the role of the national mathematics leadership in suggesting a curriculum as limited to preparing a readable framework about which individual instructors and teachers can build and which an average student can understand. What is important is that we structure the curriculum so as to nurture the students' abilities to develop clear thinking, do word problems and be willing to use modern technology with confidence. Former students who have some work experience are far more apt to say that it is their ability to think and learn rather than their precise knowledge of a given topic which is important to them in later life.

The Honors Curriculum

For honors work we do not suggest any radical changes from present practice. However, we do recommend inclusion of a wider range of topics in the calculus sequence. In particular, the mathematical background of almost all students studying the quantitative sciences should include

- writing programs and using computer packages effectively,
- basic ideas in probability and statistics,
- the solution of linear equations, matrices, linear inequalities,
- game theory and the choice of strategies,
- a little number theory.

The last member of this list is included, not because it has immediate application but because many of the problems and ideas can be understood without a great deal of background, results can be numerically checked, and many tools required in constructing logical proofs (e.g., induction) can be readily introduced.

What is important is that we structure the curriculum so as to nurture the students' abilities to develop clear thinking, do word problems and be willing to use modern technology with confidence.

We now introduce what we call the Honors Curriculum consisting of three substantial courses, each with 4 hours credit. Specific comments will follow each semester's outline; general comments will be given at the end.

1ST SEMESTER: Number systems, functions, limits, continuity, differentiation, finite differences, indefinite integrals, special functions.

Comments: Little change from the present curriculum for the first 1 and 1/3 semesters. Introduces a little number theory. At least one-sixth of the teacher's time should be spent on the mathematization of word problems. Emphasize pictures and curve sketching.

2ND SEMESTER: A mathematical laboratory emphasizing computer programming and use of packages. Definite integration by Riemann sums. Introduction of numerical methods. Techniques of exact integration. Improper integrals. Series, Taylor series, maximum and minimum problems. Solution of systems of linear equations. Matrices, determinants. Linear inequalities and programming.

Comments: Introduction to the idea and implementation of approximations. Techniques of exact integration are exhibited only once; the student using exercises and drills, is responsible for mastering this material out of class. Numerical verification of all results should be encouraged. A microcomputer laboratory is essential and the student should expect to spend at least three homework hours per week working on the computer. If time permits, a little game theory and choice of strategies might be introduced here. A knowledge of solutions of linear equations, matrices and determinants is important for the differential equations course, which is often taken in the third semester.

3RD SEMESTER: An introduction to probability and elementary concepts in statistics (mean, variance, least squares fitting). Analytic geometry. Coordinate systems and the differences between Euclidean and other manifolds. Tangent lines, arc length. Vectors and the description of curves surfaces. Partial derivatives, Taylor series, maxima and minima in higher dimensions. Multiple integrals. Line and surface integrals. The notions of circulation and flux. The theorems of Gauss, Green, and Stokes.

Comments: Except for the inclusion of probability, this program is not a great deal different from the present curriculum. We suggest, however, that proofs be replaced by a demand that the student understand the results sufficiently well to carry out nontrivial computations of circulation, heat flux, etc.

The Regular Curriculum

The regular curriculum, which would consist of four three-credit hour courses, would omit special functions and number theory from semester one, would include special functions in semester two but exclude solutions of linear equations. Semester three would be new and include much of the discrete mathematics, probability, statistics, systems of linear equations, matrices, linear programming, game theory and strategies and a little number theory. Semester four, which would be required of students following certain of the engineering and science majors, would be the same as the honors section without probability and statistics.

Logical Sequence

One of the primary aims of a mathematical education is to teach students to think logically and correctly about the problems at hand. The problem need not be mathematical in origin or nature, but the solution will require logic. It is therefore important that this logical attack be seen in action. For example, in the standard calculus course we should *not* be teaching limits, then differentiability, then continuity—as do so many texts (and hence courses). We should, instead, be stressing the logic involved. For example:

Here is the definition of a limit, but who wants to "dwell in Hell"? Here are a few useful results involving limits which can be obtained from that definition. Here are a few good theorems which also can be proved from that definition. Here is how we can combine these results and these theorems to produce more results, without going back to the initial definition. However, if new functions appear (and they do) we will then need results associated with them (presumably derived from the original definition of limit) to use these theorems.

OK, so some functions have limits and some don't. Let's concentrate on those that do. What other properties might these functions have? Ah, continuity. Now repeat the above paragraph with "limit" replaced by "continuity." Are there any important functions that are not continuous?

OK, so some functions are continuous and some are not. Let's concentrate on those that are. What other properties might these functions have? Ah, differentiability. Now repeat the earlier paragraph with "limit" replaced by "differentiability." Are there any important functions that are not differentiable?

One of the primary aims of a mathematical education is to teach students to think logically and correctly.

Those "theorems" that are missing from the known theorems should also be mentioned (no formula for limits of composite functions until we have continuity, no result for the integral of products or quotients).

Checking Answers

An important aim of mathematics is to solve problems we haven't seen before. However, it is equally important to develop the expertise to decide whether the results we get are right. We should be emphasizing habits that promote this. For example:

- While it is true that a curve can be sketched correctly using only a few results, every available result and theorem should be used to ensure that the curve "hangs together." Asymptotes, first and second derivative texts, x and y intercepts, symmetry, regions where the function is positive and negative, while frequently giving overlapping information, also lead to inconsistencies if an error has been made. But more importantly, they give the student confidence that the result is correct without looking at the back of the book.
- Check constructed formulae by using special cases. For example, in calculating volumes via integrals in terms of a cross-sectional area, the formula for the area, $A(x)$, must be constructed. Before integrating, check $A(0)$ and A at any other value where the result is known independently of the formula.
- Habitually check indefinite integrals by differentiation.
- Does the answer make sense? The age-old example of this (from absolute extrema) has someone rowing from an island and then walking along a straight shoreline to a destination in minimum time. Faulty mathematics suggests that the person should row past the destination and walk back.
- What does the answer tell us? The problem (again from absolute extrema) of finding the dimensions of a cylindrical can, holding 1 liter, with minimum surface area should be interpreted in terms of the *shape* of the resulting cylinder. (Who cares that the radius is approximately 5.4 cms?) Much more important is the conclusion that the cross section of the cylinder is a square, and even more important is the question "Why don't manufacturers make them that way?"

Mathematical Experimentation

That mathematics is an experimental subject needs to be emphasized more. Contrary to classroom demonstrations, most real problems are not done correctly, or even the right way, the first time. We should spend time explaining our intuition as well as our knowledge. This means doing things that don't work! For example, some time should be spent trying to integrate $\sec(x)$, and then when it is finally done (by writing $\sec(x) = \cos(x)/(1 - \sin^2 x)$) we should point out how most texts do it (by a trick substitution) presumably based on knowing the answer. We shouldn't be encouraging the idea that mathematics is just a bunch of tricks.

Related to this is the fact that mathematicians are not above thinking like physicists when necessary (all functions are infinitely differentiable and have convergent Taylor series). One of the main reasons for doing this is that, as a result, we might be able to guess the answer. Knowing a tentative answer frequently helps in proving it. (To a student, knowing what the answer is means looking in the back of the book!)

As an example, consider the following problem: Find all functions that satisfy the conditions of the mean value theorem, for which c is always the mid-point, i.e., find all functions $f(x)$ whose derivative at $(a+b)/2$ is always $(f(b)-f(a))/(b-a)$ for all a, b. If we could guess the answer, we might then define $g(x) = f(x)-$guess, and show that $g(x) = 0$. So let's assume that the answer has a Taylor series expansion and substitute it in the condition on f to see what we get. Lo and behold, a quadratic! So that would now be our guess, which then won't depend on the Taylor expansion assumption.

We need to get the student away from the notion that mathematics is just a collection of formulae that need to be memorized. This can be done in at least two ways, by giving interpretations and applications of the formulas, and by stressing that memory does not a good mathematician make.

For example: The interpretation of the mean value theorem in terms of average velocity, or the formula for $1+4+9+\ldots$ being used by spies to count how many cannonballs were in a pile, by knowing how many layers the (square) pyramid has. We should stress the use of tables of integrals, and tables of infinite series, and point out, that faced with the problem of integrating powers of $\sin(x)$ and $\cos(x)$ we look it up! We should be stressing, in techniques of integration, not how to do the integrals which appear in a table of integrals, but how to get from the integral at hand to the integral in the table.

We need to stress that "little" things make a big difference in mathematics, and attention must be paid to them. The results involving the existence of absolute extrema on closed intervals, as opposed to open intervals, is one example. Just as dramatic, and an important topic which is seldom covered, is sketching one-parameter families of curves, such as $x^2 + c/x^2$, and discussing the three cases for c (positive, negative, zero). This also stresses that a mathematical problem is frequently solved by looking at different cases.

A good place to demonstrate the different case concept is in the previously mentioned extreme value problem involving rowing and walking. That is actually an infinite domain problem which can be split into three cases, land between the closest point to the island and the destination, land past the destination, and land before the closest point. All *three* cases should be analyzed, not just one as is often the case.

Common Sense

We must emphasize how nature, symmetry, and common sense can frequently help in understanding a problem or its solution. For example, knowing that the circle is the shape which has a maximum area for a fixed perimeter helps make sense of the solution to the problem of cutting a piece of string into the perimeters of a circle and a square, the sum of whose areas is to be a maximum. After solving the problem involving the shortest distance from a point to a curve, we should explain that common sense says that the answer could be obtained by considering a small circle centered at the point. Then imagine increasing its radius until it just touches the curve. Intuition suggests that at this point, the tangent to the curve should be at right angles to the radius. Is it? Snell's laws should also be mentioned in this context (why does nature behave the way it does?) to build up some intuition in this area.

Infinite Series

The standard treatment of infinite series leaves much to be desired. If the course follows the usual pattern, the good student comes away with the ability to decide whether a series converges or diverges, with the idea that divergent series are diseased and all convergent series are healthy, but no feeling for what the illness is. They leave with no real idea why convergence series are categorized as absolutely or conditionally convergent, and no understanding why we care about convergence or divergence. The main justification for emphasizing the various tests for convergence seems to be that they make good exam questions.

The good student comes away ... with the idea that divergent series are diseased and all convergent series are healthy, but no feeling for what the illness is.

We must de-emphasize these tests. Who cares? We must stress why a knowledge of convergence or divergence is important. We must stress why the distinction between conditional convergence and absolute convergence is important. To introduce conditional convergence without mentioning Riemann's rearrangement theorem seems pointless. We should point out that divergent series are not useless, by doing the "brick problem" (given bricks of the same size, is it possible to place one brick on top of another, in a vertical plane, in such a way that no part of the top brick is above the starting brick?). But overall we must give examples on the use of Taylor series. The previously mentioned example involving the mean value theorem is one such application. Evaluating definite integrals is another. Solving $y'' - y = 0$ is a third.

Changing Attitudes

- It should be repeatedly explained to students that the answer alone is not good enough. What is required is the correct logic leading to the solution. There are plenty of examples where the answer is right, but the technique is flawed. Don't check the answer by looking at the back of the book: find some other way of convincing yourself that you are right or wrong.
- Some material should be covered in class even though it does not lend itself to testing by examination. For example, the proof that $\sin(x)/x$ goes to 1 as x goes to 0 is the entire justification for using radians in calculus, but it is not usually important enough to examine.
- Some material should not be covered in class even though it does lend itself to testing by examination. Covering every single technique of integration generates nice exam questions and kills time, but is it justified?
- Like it or not, computers are here (and we believe to stay), so let's learn how to use them effectively.

New Material

There are a number of topics, absent from a standard calculus sequence, to which students should be exposed.

Mathematical induction. Somehow students never seem to get this in any course. Apart from the usefulness of the technique, it also puts the ability to create a proof within the grasp and expertise of most students. It also stresses the experimental aspect of mathematics, assuming the actual conjecture is not given, just the problem (i.e., "what is $1+4+9\ldots$," as opposed to "prove that $1+4+9\ldots=\ldots$").

Approximations. That some integrals, such as $\operatorname{erf}(x)$, have no closed form representation should be discussed within the context of the approximation of integrals. Various arc length integrals should also be done numerically. With the easy access to computers, standard problems that lead to uncontrived, non-standard algebra should be done as a matter of routine.

For example, finding the shortest distance from $(0,1)$ to $\sin(x)$ (as opposed to the textbook classic, x^2), leads to solving the equation $x-\cos(x)+\sin(x)\cos(x)=0$. In the past we have avoided such problems like the plague, as though they didn't exist. We should no longer shrink from such tasks.

Problems with no solution. Most students are under the impression that all problems have solutions. ("I have a problem, what are you going to do about it?") Non-closed-form integrals come as a shock. Posing problems which have no solutions is important. For example, find the area of a four-sided figure given the four side-lengths. Trying to decide what information is relevant in solving a problem is also a skill that requires fostering. The standard is to give exactly the right amount of information to solve the problem, which is fine when learning a technique.

However, we should also give problems where there is overlapping relevant information (in addition to giving the rower/walker's velocity, mention that he can walk so far in so many seconds), too little relevant information (no means of calculating the rower/walker's velocity from the data supplied), or irrelevant information (the rower/walker has red hair, has recently signed a recall petition, has a girl friend at the destination, etc.).

Problems using many techniques. Calculus has become very compartmentalized. Problems using diverse techniques should be tackled. For example, after surface area of solids of revolution, find the shape of the cone of fixed volume and minimum surface area. Since most students (and faculty!) don't know the formula for the surface area of a cone, this problem is usually avoided earlier when extrema are covered. Now, with the benefit of integration, this formula can be derived, and the extrema problem re-attempted.

It should be repeatedly explained to students that the answer alone is not good enough. What is required is the correct logic leading to the solution.

After arc lengths has been completed, one can do the following problem (based on Greenspan and Benney's *Calculus*). A plane is flying horizontally 5 miles from the end of the runway, and is 1 mile high. It plans to land horizontally following a cubic equation. (Before and after, it follows a horizontal straight line.) What is the maximum slope of the curve it follows? How far does it fly? If its horizontal speed is 150 mph, what is its maximum vertical acceleration?

Finite difference calculus. Many students never hear of finite difference calculus, functional equations or recurrence relations, let alone see them. They bring together a number of different ideas and techniques. Some students believe that $f(x+h) = f(x) + f(h)$. Why not solve it? Show that $ln(x)$ is essentially a consequence of $f(xy) = f(x) + f(y)$.

Euler's formula. By introducing cis(x) in the context of a little complex arithmetic, at the end of the first semester (after exponentials) many of the techniques of integration can be bypassed, and many of the trigonometric identities can safely be forgotten!

Non-mathematical techniques. We should always point out other ways of doing things. For example:

a. A gardener has a kidney-shaped lawn to sod. He needs to know the area to order the correct amount of sod. (Solution: Cut out kidney-shaped piece of card to scale. Cut out what 1 square yard is. Weigh both on a chemical balance!)
b. Someone wants the volume of some strange shaped container. (Solution: Fill it with liquid, measure amount of liquid used.)
c. Derive Snell's law of reflection. (Solution: Reflect source about mirror. Join eye to source via a straight line, reflect line back.)

Use of Computers

This is not a negotiable item—it is essential. Computer labs must be accessible to mathematics students and faculty, at all levels. These labs should have packages emphasizing demonstrations, drill and practice, experimentation, numerics, and graphics, in addition to program writing. Software such as MACSYMA, Eureka, and Gnuplot should be readily available.

With the increase in the number of such labs around the country, it is inevitable, and highly desirable, that specialized educational computer packages will continue to be developed by faculty, released to mathematics students, and placed in the public domain. We recommend that minimum cost distribution of such software be encouraged. Furthermore, a central library of such software should be established and vigorously maintained. Access to it should be via modem.

DAVID LOVELOCK is Professor of Mathematics at the University of Arizona. A relativist by training and reputation but a computer addict by compulsion, Lovelock has taught at Bristol University, England and the University of Waterloo, Canada. In 1986 Lovelock received two awards from the University of Arizona for distinguished teaching and high scholarly standards.

ALAN NEWELL is presently Head of the Mathematics Department at the University of Arizona and Director of the Arizona Center for Mathematical Sciences, a Center of Excellence supported by the Air Force Office of Scientific Research under the University Research Initiative Program. He also serves as Chairman of the Advisory Committee on the Mathematical Sciences for the National Science Foundation.

Computers and Calculus: The Second Stage

R. Creighton Buck

UNIVERSITY OF WISCONSIN

Computers have finally reached the mathematics classrooms. Recent joint AMS-MAA summer meetings have highlighted talks and workshops dealing with various aspects of educational software. Leading calculus texts offer floppy disks with programs intended to add visual or numerical reinforcement to topics treated in the textbook. The computer magazine *Byte* features four universities identified as having made significant efforts to computerize their instruction in many subjects, including mathematics. It is clear that there is strong interest among college and university mathematicians, yet this interest is restrained by caution, inadequate technical experience, funding problems, and sometimes a degree of skepticism.

For many students, the first course in calculus has been a very difficult transition from the algebra and geometry courses they have taken previously, both in content and style. The routine algebra courses usually focus on formal manipulation of symbols according to specified rules, accompanied by a collection of algorithms which solve related classes of problems.

The geometry course may begin with an extensive nomenclature for plane and solid figures and their component parts, move to a collection of algorithms for certain geometric constructions, and then to a relatively short list of statements that describe various relationships among components parts making up geometric figures. (Sample: "If two sides of a triangle are equal then ....") Some of these may be given formal proofs, based on a list of accepted axioms. Additional topics often covered are right triangle trigonometry and a brief introduction to coordinate geometry with emphasis on lines and circles.

While some of the assigned problems in either course will be interesting and challenging, most are likely to fit into templates given in the text. Only a few will be seen by students as having direct connection with realistic applications.

Complexity of Calculus

In contrast to these courses, calculus deals with a vastly more complex collection of skills and concepts:

1. Scalar and vector functions of one or more variables; differentiation, integration and their properties; definition and properties of the standard transcendental functions.
2. Mathematical models for physical concepts such as velocity, acceleration, center of mass, moments of inertia, work, pressure, gradient, gravitation and planetary motion, ... (and this in spite of the fact that only one out of six high school graduates has taken a course in physics!).
3. A mathematical and computational approach to geometric concepts such as area, volume, length of a curve, area of a surface, curvature.
4. Infinite series, used both as a tool in numerical computation and as a way to define and work with specific functions, or to construct new functions.
5. Polar and spherical coordinates, advanced two and three dimensional analytic geometry, conics and quadric surfaces, vector analysis and its applications.
6. Assignments that require the student to build a mathematical model for a concrete realistic situation involving time-dependent components or other variables, and then use the model to answer specific questions about its behavior or properties; this may involve solving a differential equation or optimizing a related quantity, or perhaps carrying out other mathematical procedures suggested by the model itself.

It is important to keep in mind that, in addition to traditional mathematical knowledge, a growing number of students now arrive at college with various levels of experience with computers and programming, some of which may have involved isolated mathematical concepts such as factoring, sorting, and various graphing techniques. This adds another nonhomogeneous boundary condition for the college instructor who is trying to design a suitable modern calculus course for the incoming students.

Current Uses of Computers

Of the calculus software I have examined, many packages are tied implicitly to familiar texts even if not associated with its author or publisher. In use, many will be run in the classroom with a suitable projection apparatus and screen as a demonstration that replaces or supplements the usual chalk and blackboard treatment of a concept or technique. The use of color and partial animation may give a much more professional touch that the teacher could not have provided.

Some other programs are intended to be used by students in a tutorial mode as a review or drill to test a specific technique (e.g., formal differentiation), usually with immediate feedback on success or failure. These may also replace a portion of routine homework if there is an adequate bookkeeping system for student IDs and performance records.

Other programs will permit a student to enter a function and then ask for its graph or its zeros, or even a display of its successive derivatives given as explicit formulas. Programs are also available that permit a student to ask for the formal solution to any second order constant coefficient homogeneous ordinary differential equation, or the numerical solution of any first order initial value problem. Similar programs now exist in a limited form on hand held calculators, and more sophisticated calculators are just over the horizon.

Basic Questions

This situation obviously poses several crucial questions to those responsible for planning the content and administration of undergraduate mathematics courses. When should a student be allowed or encouraged to turn to the computer for assistance or insight? When should the computer be used in the classroom?

Here is one obvious ground rule: Don't do it with a computer if you can do it better with gestures, pencil and paper, chalk and blackboard, or any other traditional way. However, the word "better" must be interpreted correctly. We know that an uninterested student will seldom learn; thus, one must reach a balance between the role of computers as interactive information sources, and as theatrical attractions.

Don't do it with a computer if you can do it better with gestures, pencil and paper, chalk and blackboard, or any other traditional way.

However, a third question seems to be more central, and much harder to answer: How does the existence and wide availability of computers modify the objectives of the calculus course?

A satisfactory answer to these questions requires the collective work of individual mathematicians, with support and recognition from their departments. These mathematicians must be willing to spend time in a thoughtful analysis of the mathematical content of calculus and then carry out experiments to test whether access to computers results in more effective instruction and better educated students.

Asking the Right Questions

The starting point in this project may be to pose the right questions. Perhaps several samples will clarify this.

- Some students who have difficulty with calculus seem to be deficient in spatial intuition. Is it possible to help these students with special graphics programs? Or is this a case where the use of tactile wire or plaster models is better?
- At present "curve tracing" is a standard topic in calculus, ostensibly for the purpose of producing a reasonably accurate graph of a specific function or equation, but also to reinforce the techniques of differentiation. Now that a computer (or even a hand-held calculator) will display the graph immediately, is this still important? What do we really mean when we say "skill in curve tracing is an essential component of calculus?"

  (One possible answer: We want students, based on their mathematical knowledge and computer experimentation, to be able to look at an equation of the form $y = f(x)$, where the formula for f contains one or more parameters, be able to say: "The graph of this sort of function has one of the following shapes, depending on the values of the parameters $a, b, \ldots$.")
- When a student has access to programs that do formal differentiation, should we continue to stress routine differentiation exercises? (After all, we now allow students to use calculators to assist with their arithmetic on homework and tests!) Clearly, a student ought to know how and when to use the formulas for differentiating the sum, product, or reciprocal of functions, and understand the use of the "chain rule" for composite functions, since these supply essential skills and insight in many other mathematical areas. But at what point in this learning process do we let the student replace paper and pencil manipulation by a canned production-style differentiation machine?
- It is important that the student be able to describe the geometrical meaning of a statement such as:

  $$f'(x)f''(x) > 0 \quad \text{for all } x \text{ between -4 and 4.}$$

  Is ability to answer this helped if the students have seen appropriate classroom demonstrations and carried out experimentation on their own?
- Integration presents a different problem, since we must deal both with the definite integral, and with anti-differentiation. Accurate numerical integration programs obviously should be available. (Students already have access to these on some hand calculators.) However, substitution (change of variable) in either indefinite or definite integrals remains a useful and very important technique in many areas of analysis, and so must be retained. When should this approach to the evaluation of a definite integral be required, and when is a numerical integration acceptable?
- Simple max/min problems provide some of the more interesting elementary applications of calculus. However, programs can quickly do a search to locate the absolute maximum and minimum of a function of one variable on an interval. Shall such programs be permitted? Can good problems be selected which require students to locate *local* extrema?

  Problems of this type in more than one variable require finding critical points; this in turn often leads to the solution of complicated non-linear systems. Do we provide students with the sophisticated black box Newton-type programs that are appropriate here? If so, can we make an honest attempt to explain their nature?
- What about advanced analytic geometry? Certainly, lines and planes in space must be covered in order to work with normals, tangent planes, and properties of curves in space. But what about the usual treatment of general conics and quadric surfaces? Is this

something that is now best replaced by a well chosen picture gallery or labeled samples, while all the rest is left to be covered in a linear algebra course as illustrations of quadratic forms and eigenvalue computation?

- Is there no value for today's calculus student in working through the proofs of the optical properties of the parabola and ellipse, or any of the other classical geometric theorems?
- In the study of infinite series, which computer programs should be demonstrated in class and which made available to students? For example, should they have access to a program that takes an explicit function as input, and supplies a stipulated number of coefficients of its MacLaurin series?
- Calculus courses usually include some introduction to differential equations. In the past, this has often been a compendium of special techniques, tied to specific classes of equations. Should all this be replaced by a discussion of initial valued problems, both ordinary and systems, followed by experiments with a black-box differential equations solver? Would this rob the student of useful basic mathematical experiences and skills? Would such a student be less able to deal with certain applied problems that do not immediately fit the standard patterns?

A New Syllabus

I hope that it is clear that there are more questions to be asked about the factors that should be examined, as one moves toward computerization of the calculus sequence. The end product might be a radically different topic sequence for the calculus course.

At what point ... do we let the student replace paper and pencil manipulation by a canned production-style differentiation machine?

Here is an outline for the first four days in one such course. It's purpose is provocative rather than descriptive. (Since it was devised during a canoe trip, it certainly has not been tested in class, nor researched exhaustively!)

The course starts by discussing familiar functions: polynomials and rational functions. These are identified as specific algorithms, accepting a numerical input and delivering a numerical output. Several different formats or conventions are suggested for describing a function. For example, one might write $P() = 6()^2 + () - 2$ as well as $P(var) = 6 * (var\hat{\ }2) + var - 2$. The natural domain of a function is the set of admissible inputs. Sequences are presented as functions on the integers; here it is appropriate to include sequences defined recursively.

Each student receives a disk containing the programs PLOT, TABLE, and ZEROS and an assortment of others. Some are demonstrated in class, and the simple procedures for use and function input are explained. The assigned problems require students to use the computer programs to answer questions about specific functions, and then to carry out experiments on their own.

Among the demonstration functions is a polynomial of degree 5 with one parameter. As different values for the parameter are chosen, the effects are seen in the graph and in the nature and location of the roots; as the number of visible zeros change, the existence of complex roots is mentioned, and related to the graph.

Further exploration of functions; experimentation with combinations of functions and discussion of the architecture of TABLE, PLOT, and ZEROS. Introduction of the functions sin and cos, defined by their series. Brief discussion of how series are "summed" and the sources of error. Intuitive explanation of the concept of convergence of series. Discussion of time and position functions that model motion on a line ("dog on a road").

The assigned problems deal with the new types of functions; use of PLOT and ZEROS to solve equations such as $3 + 2t - t^2 = \sin(4t + 1)$.

The ZOOM command is used to examine a polynomial plot in neighborhoods of a point on the graph. Conclusion: "In a microscope, curves look like straight lines." Counterexamples: $(t^2 - .1)\sin(1/(t^2 - .1))$. Conclusion: "Some functions are smooth everywhere and some are not."

Recall "slope of a line;" definition of: "local slope of a curve." Conclusion: "Smooth functions have local slopes at each point on their graph." Return to functions that model motion, and interpret slope as velocity.

Demonstrate program SLOPE, a program that gives the approximate local slopes of smooth functions. Discussion of nature of the algorithm, as compared with "the exact slope;" formulation of several definitions of "limit."

DERIVATIVE introduced as an algorithm that takes a smooth function f as an input and delivers a function $f' = D(f)$. Sample calculations for linear and quadratic functions; graphs of f and f' are compared.

Looking to the Future

I am sure this is enough of a sample. My choice of topics was not quite off-the-cuff. The six objectives I listed earlier for the calculus course are some of those that have been traditionally cited by textbook authors, publishers, and curriculum designers. However, as we move toward the next century, I believe that other objectives should be mentioned. In the last 40 years, the rate of change in all sciences has accelerated, and the level of mathematical sophistication used in these sciences has increased dramatically.

In the last 40 years, the rate of change in all sciences has accelerated, and the level of mathematical sophistication used in these sciences has increased dramatically.

The calculus sequence seems to offer an ideal opportunity for us to give students their first exposure to certain simple tools and unifying viewpoints of modern analysis that will make their future path smoother, whatever their ultimate interests. I would hope that computers can be used intelligently in calculus instruction in such a way that room could be found for such topics.

My first candidates are function spaces and linear operators. As students gain experience with specific functions, and are more at home with the use of formulas using symbols that represent arbitrary functions in some specific class, they will find that the concept "function" has begun to acquire the same concreteness held by "number."

The moment when a student finds it natural to think of an arbitrary continuous function as a point-like object having a specific location with respect to other similar point-like objects represents a major insight! (Is it perhaps the mathematical equivalent of a "rite of passage," with attendant maturity implications?) Such students then find it easy to move on to the idea of a geometric linear space whose "points" are functions.

As students gain experience ... using symbols that represent arbitrary functions ... they will find that the concept "function" has begun to acquire the same concreteness held by "number."

Linear operators are even easier to bring in. In my outline, the calculus instructor has already introduced the concept of a linear operator by showing that "differentiation" is merely another type of function that accepts a suitable numerical function as its input and then delivers a numerical function as its output.

I own an elementary calculus text from 1831 that uses fluxions and fluents. I am willing to predict that our current elementary texts may look equally strange and out-of-date to an undergraduate in 2020.

R. CREIGHTON BUCK is Hilldale Professor of Mathematics at the University of Wisconsin at Madison. He has served as Vice President of both the Mathematical Association of America and of the American Mathematical Society. He is a former Chairman of the Committee on the Undergraduate Program in Mathematics (CUPM), and is a former Director of the Mathematics Research Center at the University of Wisconsin. Buck has served on many government advisory committees on science and education.

Present Problems and Future Prospects

Gail S. Young

NATIONAL SCIENCE FOUNDATION

In 1935, the year I took calculus, the content and spirit of the course were essentially the same as now. But there were major differences in the environment.

Then a college education was still primarily for an elite—a social elite, not an intellectual elite. The students knew they were expected to graduate, that there was a stigma for failure.

Some mathematics was a usual requirement for graduation, as "mind-training," but calculus—a sophomore course—was taken almost exclusively by students in the physical sciences, engineering, and mathematics, a small part of the student body, a homogeneous group.

Almost all were concurrently taking physics. One of our problems is that this homogeneity has disappeared.

Since it was a sophomore course, the preparation of the calculus student was controlled by the department. When polar coordinates came up, the teacher knew precisely what the students had been taught before. That preparation was usually a year course covering college algebra, trigonometry, and plane and solid analytic geometry. With the placing of calculus in the freshman year, starting in the '60s, control over the preparation passed to the secondary schools.

In the '60s that was not bad. Whatever one thinks of its content, the New Math had changed the entire atmosphere of secondary school mathematics. In fact, in the first CBMS Survey, 1965, 75% of reporting departments said it was the New Math that made the change to freshman calculus possible.

That enthusiasm, that push for change, has long ended, stopped by the crushing problems of our schools. We are left with the heterogeneous preparations, but in students who no longer have the same enthusiasm and confidence.

Calculus in Context

Since before Sputnik few students took calculus, for staffing purposes it could be considered to be essentially an upper-division course. In the first few years of teaching, a Ph.D. rarely taught a course above calculus, and often did not even teach calculus. Only three times in my first five years did I teach a course above calculus, though I got pretty good at "Mathematics for Home Economics."

Now 31% of our undergraduate mathematics students take calculus, and hardly anyone regards it as a privilege to teach it, nor as a compliment to one's ability.

Calculus is our most important course, and the future of our subject ...depends on improving it.

These changes are all for the worse and it is hard to see what can be done about some of them. Calculus is our most important course, and the future of our subject as a separate discipline depends on improving it.

Students no longer enter calculus with enthusiasm. One reason, I think, is that many students, particularly the better ones, have had experience with the computer, and expect that their college and university mathematics courses will use computers heavily. However, the 1985 CBMS Survey showed that only 7% of the courses used computers at all. From anecdotal evidence, much of that usage seems to be as a large calculator, for such topics as Simpson's Rule. I do not believe the main reason for bringing the computer into calculus is to make the students happier, but that would be a desirable result.

I am one of the people who believe that the computer will revolutionize our subject as greatly as did Arabic numerals, the invention of algebra, and the invention of calculus itself. All these were democratizing discoveries; problems solvable only as research and by an elite suddenly became routine. The computer will do the same for our mathematics, and calculus is the place to begin.

Our calculus course (as well as differential equations) comes from the British curriculum of the last part of the last century with the British emphasis on hard problems and little theory. It is a course not in the spirit of contemporary mathematics, and needs change for that reason alone. What is left of the British tradition now is an emphasis on working many problems, finding, say, 25 centroids, without really understanding what a centroid is. That in itself is reason to change.

Using Computers

To my mind, the most important change will be the introduction of symbol manipulation (SM). Perhaps this should come before calculus. The present calculus is really a collection of algorithms with a little theory and a few applications. It is not easy for a student to see that. After learning the common substitutions for changing variables in indefinite integrals, I thought that given some function to integrate, my task was to invent some substitution to make it integrable. That the whole topic could be covered by one algorithm never entered my head. I had no concept of an algorithm. But practically every method in a first calculus course is an algorithm, never clearly stated, never explicit.

Once a topic is reduced to an algorithm, we are in the realm of the computer, and the problem becomes one of optimal programming. Once programmed, there is no need to repeat it. Except for its possible effect on moral character, I see no point in mastering hand computation of square roots, with calculators available at $5.95 that will do it with one key punch. Nor do I see any point in finding the partial fraction expansion of a rational function if a computer can do it.

The intellectual merits of the above discussion, however, are not relevant. The fact is that we can now do all the manipulation in calculus on the computer cheaply enough for classroom use. I mention MAPLE and mu-MATH as examples. One can also consider the *HP-28C*, a hand-held calculator that can do algebra, and that is clearly a precursor of more powerful SM calculators. These developments make the introduction to SM inevitable, if not by us, by our customer departments.

The computer will revolutionize our subject as greatly as did Arabic numerals, the invention of algebra, and the invention of calculus itself.

Are there problems in introducing SM? Yes, of course. Here is one I can't handle. In research papers using SM, one sees a hand calculation, then SM, then more hand calculation, then more SM, ... How do people learn the necessary hand techniques? Is that a real problem? I've seen indications that students learn hand techniques better with SM.

At a different level, how do you get the necessary training for the faculty? What equipment do you need, and where does the money come from? Do you have the necessary additional space? It will probably not be possible to use as many graduate students and part-time faculty as one can in the present course. How do you handle that? Where are texts? Etc.

Besides these local problems, there are ones that will require national effort, to carry out necessary experimentation. For example, how much hand computation will still be needed? Such questions seem to me to require research at a number of schools.

Numbers and Graphs

But SM is not the only use of the computer in calculus. One could say that the objective of most real-world problems in calculus is to give a foreman some directions. For that, one needs numbers. In most of my career, we have not been able to give calculus students numerical problems of the sort they might actually meet, because of the time required. One gave a few simple problems with the Trapezoidal Rule or Simpson's Rule, or approximated the sum of a Taylor's Series at some point, and that was it.

We can do better.

We can do better. In addition, there are other choices. I suspect, for example, that to evaluate a definite integral with the computer, SM is not the way to go. One should instead immediately turn to numerical integration, and the question of indefinite integration becomes wildly irrelevant for that purpose.

Computer graphics must have a place. For the other two uses, I know that it is rather easy to learn the necessary techniques; I don't know about graphics. But the ignorance of geometrical techniques by our students could be largely remedied by graphics. For example, one could sketch a volume determined by several equations. Is it too much to ask for all three techniques?

In my opinion computer methods will let us cut in half the time required to teach our present course, that is, the current set of algorithms done by the computer and not by the student, and the same amount of theory and applications. If so, what do we do with the extra time? One choice is to use it for topics that do not now usually get into the first two years. This possibility was discussed extensively in the report of a conference at Williams College [1]; Kemeny's talk is particularly relevant. I believe that for many types of students (social science majors, for example), this would be best. But for students going into fields where analysis is much used, why not use the time to teach a better understanding of calculus and how to apply it?

Teaching Applications

Let me discuss applications first. Consider the standard topic of force exerted by water on the face of a dam. When I took calculus, the necessary physics had been taught all of us earlier in the year (in Physics), and we all had done simple problems. We had only to learn how to set up sums by appropriate partitions that would converge on the one hand to the total force on the dam, and on the other hand to the definite integral of a function.

We had a good basis for an intuitive understanding of the first, and there were results in the text for the second. There were many exercises devoted to finding the force on improbable dam faces, which dams, however, all had the common property that the resultant integral could be calculated by indefinite integration.

The only major change I see from then to now is that most current students have no knowledge of fluid pressure, and must use not physical intuition, but faith as justification. In such topics we are not teaching the application of mathematics, and our good students know that.

Our applications should be taught as modelling. We are dealing with continuous models; fluids, gases, electrical currents are continuous functions in calculus. Reasons for the use of continuous models instead of discrete ones should be given, and some examples where discrete models are better should be given. There should be enough explanation of the underlying physics, economics, whatever, so that the student can understand the basis for the model. The fact that the model is not reality should be explained over and over. Should there be some purely computer treatments, perhaps a simulation?

Computer methods will let us cut in half the time required to teach our present courses. ... Why not use the time to teach a better understanding of calculus and how to apply it?

How much of such treatments can be done is a subject for experiment, but a few good models will beat our present array of poorly understood problems apparently intended mostly for civil engineers.

Teaching Theory

We do a poor job of theory. I am not here calling for rigor; I have long been convinced that epsilon-delta proofs won't work with students who have no great talent or interest in mathematics *per se*. (There is the possibility of approaching epsilontics from numerical analysis, as error and control of error. That might work.) What *is* possible, I believe, is the sort of clear explanation found, e.g., in Courant's *Calculus,* at times being close to a rigorous proof.

It is difficult to persuade most students that a proof that the limit of a sum is the sum of the limits is necessary or interesting. On the other hand, the "proof" that a function continuous and monotone on a closed interval is integrable there is perceived differently, as a clarification of what is going on in integration. At present, time pressures force most teachers to rush through such a topic, to be understood only by the best students.

With a course of this nature and with freedom from the necessity of manipulative skills, more students will go on to take higher courses, more students will be able to actually use calculus in their major fields, and more graduates will be mathematically competent.

Reference

[1] Ralston, Anthony and Young, Gail S. (eds.) *The Future of College Mathematics.* New York: Springer-Verlag, 1983.

GAIL S. YOUNG is Program Director for Mathematics in the Directorate of Science and Engineering Education at the National Science Foundation. Young has served as Professor of Mathematics at the University of Michigan, Tulane University, Rochester University, Case Western Reserve, and the University of Wyoming. In 1969-70 he served as President of the Mathematical Association of America and in 1984 he chaired the Mathematics Section of the American Association for the Advancement of Science. Young is a recent recipient of the Distinguished Service Award of the Mathematical Association of America.

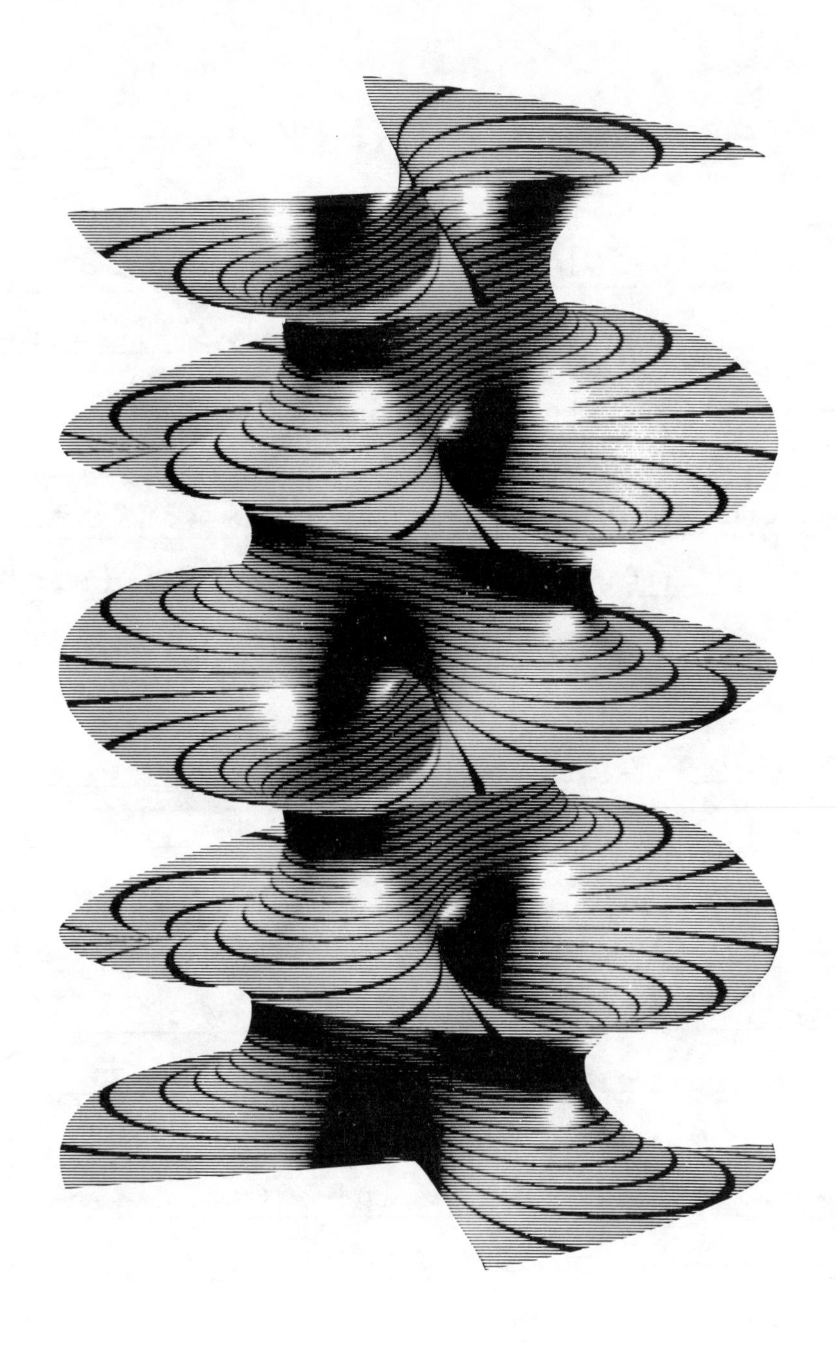

EXAMINATIONS

PARTICIPANT RESPONSES

Question: What can be done to improve calculus instruction?

Answers

- Reward the instructor
- Invite complaining professors from "client" disciplines to teach calculus.
- Use lots of pictures.
- Emphasize the beauty, the big ideas.
- Honestly require that prerequsites be met.
- Make students read the textbook.
- Small classes with motivated instructors
- Reward good teaching at the same level as good research.
- Recognize that not everyone can pass calculus.
- A national newsletter or database that shares ideas
- Use the best faculty to teach calculus, not the worst.

Final Examinations for Calculus I

The following thirteen examinations represent a cross section of final examinations given in 1986–87 for students enrolled in the first calculus course. The examinations come from large universities and small colleges, from Ph. D. granting institutions and two year colleges.

Brief descriptive information preceding each examination gives a profile of the institution and circumstances of the examination. Unless otherwise noted, no books or notes are allowed during these examinations; calculator usage is explicitly noted where the information was provided. Except for information on problem weights and grading protocol, the texts of the examinations are reproduced here exactly as they were presented to the students.

INSTITUTION: *A public Canadian university with nearly 20,000 students that in 1987 awarded 15 bachelor's degrees, 4 master's degrees, and 6 Ph.D. degrees in mathematics.*

EXAM: *A two-hour exam (calculators not allowed) for all students (except engineers) taking first term calculus. Of the 2500 students who enrolled in the course, 50% passed, including 15% who received grades of A or B.*

1. a. Use the definition of continuity to show that the given function is discontinuous at $x = 1$.

$$f(x) = \begin{cases} 2x - 1 & \text{if } x < 1 \\ 0 & \text{if } x = 1 \\ x^2 & \text{if } x > 1 \end{cases}$$

   b. Evaluate $\lim\limits_{x\to 1} \left[\dfrac{1}{x-1} - \dfrac{2}{x^2-1}\right]$.

   c. Evaluate $\lim\limits_{x\to 0} \tan(3x)/x$.

2. a. Use the limit definition of derivative to find $f'(0)$ if $f(x) = x/(x^2+1)$.

   b. Find $\frac{d}{dx}\left[\sin\left(\cos(e^{1/x})\right)\right]$.

   c. Find $\dfrac{d}{dx}\left[\dfrac{x \ln x}{\sqrt{1+x^2}}\right]$.

3. a. Find the equation of the line tangent to the curve $2(x^3+y^2)^{3/2} = 27xy$ at the point $(2, 1)$.

   b. If $g(-1) = 1$, $g'(-1) = 2$ and $f'(1) = -1$, evaluate $\frac{d}{dx}\left[f\left(g(-\frac{1}{2}x)\right)\right]$ at $x = 2$.

4. A spot light located on the ground shines on the wall 12m. away. If a man 2m. tall, walks from the spotlight toward the wall at a speed of 96m./min., how fast is his shadow on the wall decreasing when he is 4m. from the wall?

5. Given $f(x) = x/(x-2)^2$.

   a. Find

      i. the domain of f

      ii. the asymptotes of f

      iii. intervals where f is increasing and decreasing

      iv. relative extrema of f.

   b. Further, find the intervals where f is concave up and concave down and the points of inflection. And sketch an accurate graph of f.

6. If 1200 cm^2 of material is available to make a box with a square base, rectangular sides and no top, find the volume of the largest possible box.

7. a. Find $\int (x^3/\sqrt{x^2+1})\,dx$

   b. Evaluate $\int_1^4 (1/\sqrt{x}) \cos\left(\frac{\pi}{2}\sqrt{x}\right)\,dx$

   c. Show that $\int_{-1}^{1} a^x/(1+a^x)\,dx = 1$, where a is a positive constant.

INSTITUTION: *A state university in the Midwest with 20,000 undergraduate and 6,000 graduate students that in 1987 awarded 45 bachelor's degrees, 14 master's degrees, and 7 doctor's degrees in mathematics.*

EXAM: *A two-hour exam (calculators allowed) on elementary calculus for business students. Of the 1440 students who enrolled in the course, approximately 74% passed, including 40% who received grades of A or B.*

1. Find each of the following limits which exist.

   a. $\lim\limits_{x\to -4} \dfrac{x^2+2x-8}{x^2-6x+8}$ b. $\lim\limits_{x\to 2} \dfrac{x^2+2x-8}{x^2-6x+8}$

   c. $\lim\limits_{x\to\infty} \dfrac{3x^2-4x+7}{-5x^2+9x-11}$ d. $\lim\limits_{x\to\infty} \dfrac{16}{1+4e^{-3x}}$

   e. $\lim\limits_{x\to -\infty} \dfrac{16}{1+4e^{-3x}}$

2. Find the derivative of each of the following functions.
   a. $f(x) = \sqrt{9+4x^2}$
   b. $f(x) = (3x+2)\sqrt{9+4x^2}$
   c. $f(x) = 3e^{(-4x^2+2x+3)}$
   d. $f(x) = -2\ln\left(2x^3 - 3x + \frac{1}{x}\right)$
   e. $f(x) = \dfrac{2x-1}{x^2+4}$
   f. $f(x) = e^{(-4x^2+2x+3)}(4x-1)$

3. Let $f(x) = x^4 - 8x^3 + 216$.
   a. Determine those intervals of the x-axis on which $f(x)$ is increasing and those intervals of the x-axis on which $f(x)$ is decreasing.
   b. Determine where $f(x)$ has a relative maximum value and where $f(x)$ has a relative minimum value.
   c. Determine those intervals of the x-axis on which the curve $y = f(x)$ is concave upward and those intervals of the x-axis on which the curve $y = f(x)$ is concave downward.
   d. Find all inflection points on the curve $y = f(x)$.
   e. Sketch the curve $y = f(x)$ carefully and neatly.
   f. Determine where $f(x)$ has an absolute maximum value on the closed interval $[2,8]$.

4. Find each of the following.
   a. $\int 3x\sqrt{9+4x^2}\,dx$ b. $\int \dfrac{-x^2+4x}{x^3-6x^2-9}\,dx$
   c. $\int 16x\,e^{(-4x^2+1)}\,dx$ d. $\int 4\left[\ln(x^2)\right]^3 \frac{1}{x}\,dx$
   e. $\int \dfrac{e^{-4x}}{1+3e^{-4x}}\,dx$ f. $\int_0^4 \sqrt{16+5x}\,dx$

5. Set up the trapezoidal rule ready for calculating an approximation to $\int_1^5 3e^{-\sqrt{4+9x^2}}\,x\,dx$ using 8 subintervals.

6. Determine which of the following improper integrals converges, and find the value of each convergent one.
   a. $\int_4^\infty \frac{1}{\sqrt{9+4x}}\,dx$ b. $\int_0^\infty x^2 e^{-x^3}\,dx$

7. Determine where the graph of each of the following functions has a relative maximum point, a relative minimum point, a saddle point.
   a. $f(x,y) = 3x^2 - 4xy + 5y^2 - 14x + 24y - 17$
   b. $f(x,y) = x^2 - 4xy + 2y^4 + 11$

8. Use the method of Lagrange multipliers to find the maximum value of $f(x,y) = -32x^2 + 3xy - 2y^2 + 45$ subject to the constraint $g(x,y) = 16x^2 + y^2 = 32$.

INSTITUTION: *A southern urban two-year college with 5,000 students.*

EXAM: *A two-hour exam (calculators allowed) for the first term of calculus for science and business students. Of the 28 students who enrolled in the course, 64% passed, including 40% who received grades of A or B.*

1. Answer either a or b but *not* both:
   a. If $f(x) = 2x^2 - 3x + 1$, find $f'(\sqrt{3}/2)$ by using the definition of the derivative.
   b. By means of an appropriate Riemann Sum, find $\int_{-1}^3 (2x^2 - 3x + 1)\,dx$.

2. Answer a or b but *not* both:
   a. Find the area of the region bounded by $x = y^2$ and $x + y = 2$.
   b. Find the volume of the solid of revolution generated when the region bounded by $x = y^2$ and $x + y = 2$ is revolved about the line $y = 1$.

3. Do either a or b but *not* both:
   a. Use Newton's Method to estimate $\sqrt[3]{6}$, starting with $a_1 = 2$ and finding a_3.
   b. Estimate $(.94)^3 - 1/(.94)^3$ by using differentials.

4. Answer either a or b below, but *not* both:
   a. Sketch the graph of $y = x^4/4 - (4x^3)/3 + 2x^2 - 1$ by finding the local max and local min; point(s) of inflection; intervals where the graph is increasing; decreasing; concave upward; concave downward.
   b. Estimate $\int_1^5 x^4\,dx$ using the Trapezoidal Rule with $n = 4$.

5. Do either part a or b, but *not* both:
   a. A drinking glass is in the shape of a truncated cone with a base radius of 3cm, a top radius of 5cm, and an altitude of 10cm. A beverage is poured into the glass at a constant rate of 48cm^3/sec. Find the rate at which the level of the beverage is rising in the glass when it is at a depth of 5cm. [Note: Volume of a cone = $(1/3)\pi(\text{radius})^2(\text{altitude})$.]

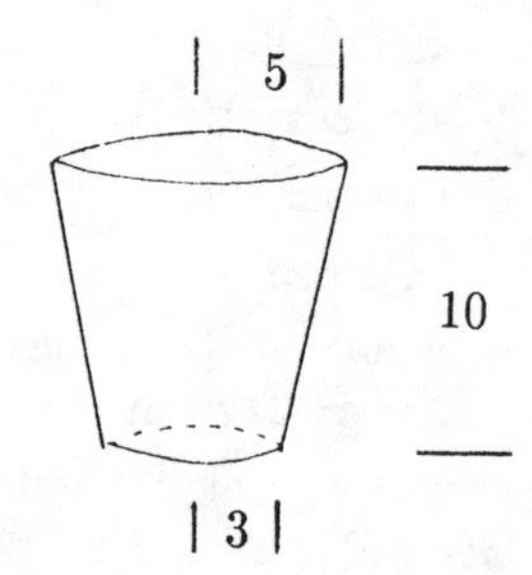

b. Two straight roads intersect at right angles, one running north and south, while the other runs east and west. Bill jogs north through the intersection at a steady rate of 6 miles per hour. Sue pedals her bicycle west at a steady rate of 12 miles per hour and goes through the intersection one half hour after Bill. Find the minimum distance between Bill and Sue.

6. For each of the following functions, determine both $f'(x)$ (you need *not* simplify $f'(x)$) and $\int f(x)\,dx$.

   a. $f(x) = 21x^6 + 15x^2 - 3x + e^{\pi}$

   b. $f(x) = 3x\left(e^{x^2}\right) + x\sqrt{9 - x^2}$

   c. $f(x) = \dfrac{1}{\sqrt{1 - x^2}} - \dfrac{1}{4 + x^2}$

   d. $f(x) = e^{\ln(x^4 + 2x^2 + 1)}$

   e. $f(x) = (\sin x)\left(e^{\cos x}\right)$

INSTITUTION: *A midwestern liberal arts college with 1750 students that in 1987 awarded 24 bachelor's degrees in mathematics.*

EXAM: *A two-hour exam (calculators allowed) from one section of first-term calculus taken all students whose intended course of study requires calculus. Of the 160 students who originally enrolled in the course, approximately 88% passed, including 50% who received grades of A or B. 30 students were in the section that took this exam.*

1. Differentiation.

   a. Define $f'(x)$.

   b. On the pictures below, indicate Δy and dy.

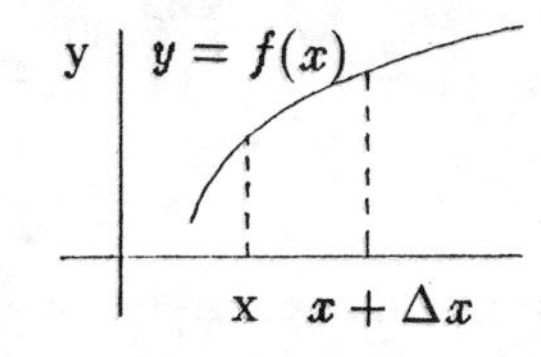

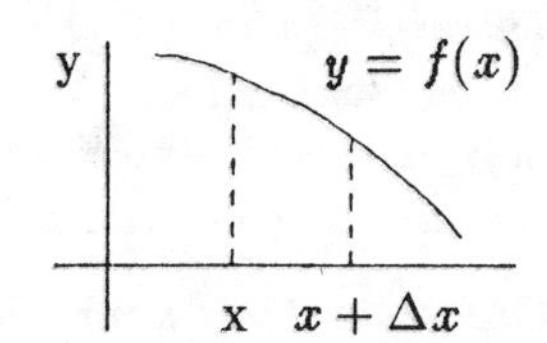

   c. According to Newton's law of heat transfer, the rate at which the temperature of a body changes is proportional to the difference between the temperature of the body (call the temperature of the body u) and the temperature T of the surrounding medium. Express this law as a mathematical equation.

   d. Given $f(2) = e$, $f'(2) = -1$, $g(2) = 4$, $g'(2) = 3$, $r(x) = [f(x)]^{g(x)}$. Find $r'(2)$.

   e. Find $\frac{dy}{dx}$ if $y/x + x/y = 4$.

2. Integration.

   a. Consider the function $f(t) = 1/t$ defined on the interval $[1,2]$. Let P_n be the partition of $[1,2]$ into n equal subintervals, and in each subinterval $[x_{i-1}, x_i]$, choose $t_i = x_{i-1}$. Then the sum S_n is defined to be

$$S_n = f(t_0)(x_1 - x_0) + f(t_1)(x_2 - x_1) + \cdots + f(t_{n-1})(x_n - x_{n-1}).$$

   Find the two sums, S_2 and S_3.

   b. If you had the use of the tables in your text or a calculator, you should be able to find the number to which the sums in part a above will converge (that is, the number which will be approached) as n increases. What number would you look up (or enter into your calculator)?

   c. Let $G(x) = \int_0^x \sqrt{16 - t^2}\,dt$

   i. Which is correct?

   $G(2) = -G(-2)$ or $G(2) = G(-2)$

   ii. What is $G(0)$?

   iii. What is $G'(2)$?

   iv. What is $G(4)$?

   d. Find the indicated antiderivatives.

   i. $\int (t+1)/\sqrt{t}\,dt$ *ii.* $\int \sqrt{t}/(t^{3/2} + 1)\,dt$

3. Graphs and derivatives. For each of the following functions, find $f'(x)$. Give the answer, then find the critical points, if any, and draw a graph of $y = f(x)$.

   a. $f(x) = 3x^4 - 8x^3 + 6x^2 - 5$

   b. $f(x) = 4x/(x^2 + 2)$

   c. $f(x) = \ln x/x$

   d. $f(x) = x^{2x - x^2}$

   e. $f(x) = x^{-x}$

4. The following two questions both refer to the function $f(x) = \ln x/x$ graphed in part c of question 3 above.

   a. Write the equation of the line tangent to the graph at $x = \sqrt{e}$.

   b. Find the area under the graph from $x = \sqrt{e}$ to $x = e^2$.

INSTITUTION: *A western land-grant university with 15,000 students that awarded 29 bachelor's, 9 master's, and 3 doctoral degrees in mathematics in 1987.*

EXAM: *A three-hour exam (calculators not allowed) given to all students in a first term calculus for engineering, science, and mathematics students. Of the 435 students who enrolled in the course, 46% withdrew before the final exam, 6% failed the course, and 24% received grades of A or B.*

1. Find the equation of the line tangent to the graph of $y = f(x)$ at the point $(0, -4)$ where $f(x) = \tan x - 4 + 3x$.

2. Determine all relative extreme values of $y = x^5 - 20x$. (Justify your results.)

3. Find the derivatives of the following functions:
   a. $f(x) = (x^2 - 1)/(x^2 + 1)$
   b. $f(x) = (1/x \sin x)^{2/3}$
   c. $f(x) = e^{x^3}$
   d. $f(x) = (\ln x)^{\ln x}$
   e. $f(x) = \arctan((x + 1)/(x - 1))$

4. Evaluate the following limit in two distinct ways:

$$\lim_{x \to \frac{1}{2}} \frac{8x^3 - 1}{2x - 1}$$

5. Sketch the graph of $f(x) = x^3 - 6x^2 + 12x - 4$, showing all principal features (such as asymptotes, extrema, and inflecton points).

6. Approximate $\sqrt[4]{17}$ by using differentials.

7. A crate open at the top has vertical sides, a square bottom, and a volume of 4 cubic meters. If the crate has the least possible surface area, find its dimensions.

8. Compute the following:
   a. $\int_{-1}^{1} \left[\frac{d}{dx}\sqrt{2 + x^3}\,\right] dx$
   b. $\int x^2/(1 - x^3)\,dx$
   c. $\int_1^2 \left(t^2 - (1/t^2)\right)^2 dt$
   d. $\int x\sqrt{x - 1}\,dx$
   e. $\int_0^2 1/\sqrt{16 - x^2}\,dx$

9. Evaluate $\lim_{x \to x_0} (\sin 3x)/x$ for
   a. $x_0 = \pi/3$
   b. $x_0 = \pi/6$
   c. $x_0 = 0$

10. Suppose that the equation $y = (2x)/(1 + 2^y)$ defines y as a differentiable function of x. Use implicit differentiation to determine $\frac{dy}{dx}$.

11. A board 5 feet long slides down a vertical wall. At the instant the bottom end is 4 feet from the base of the wall, the top end is moving down the wall at the rate of 2 feet per second. At that moment, how fast is the bottom sliding along the horizontal ground?

12. Assume that the half-life of radon gas is 4 days. Determine a formula which gives the time required for 10% of a given quantity of this gas to become harmless.

13. Evaluate the area under the curve $y = 2^{-x}$ on the interval $[\frac{1}{2}, 1]$.

14. Find a formula for the inverse of $f(x) = \dfrac{3x + 5}{x - 4}$.

15. Sketch the graph of $f(x) = \ln(2 - x - x^2)$. [Note that x must be such that $2 - x - x^2 > 0$.]

16. Determine whether the following statements are true or false:
   a. The function $f(x) = 1/x$ is its own inverse.
   b. Defining $f(-2) = 1$ makes $f(x) = \dfrac{x^2 + 5x + 6}{x + 2}$ continuous at $x = -2$.
   c. $\lim_{x \to 1^-} (\sqrt{x^2 - x} + x) = 1$
   d. $\int_0^{\pi} \tan x\,dx = -\ln|\cos x|\Big|_0^{\pi} = -(\ln 1 - \ln 1) = 0.$
   e. If $\lim_{x \to a^-} f(x) = \lim_{x \to a^+} f(x)$, then $f(x)$ is continuous at the point $x = a$.

INSTITUTION: *A highly selective private northeastern university with 4,500 undergraduate and 4,500 graduate students. In 1987, 68 students received bachelor's degrees in mathematics, 2 received master's, and 11 received mathematics Ph.D. degrees.*

EXAM: *A three-hour exam for all students enrolled in first term calculus. Course enrollment rose from 230 to 275 during the term, and 94% of the enrolled students passed; 62% received grades of A or B.*

1. If $f(x) = x^2 \cos x$, then $f'(\pi/2) =$
   (A) π (B) $-\pi^2/4$ (C) $\pi^2/4$
   (D) $-\pi$ (E) $\pi - \pi^2/4$

2. For what value of the constant c will the tangent line to the curve $y = x^4 + cx^2 - cx + 1$ at the point $(x, y) = (1, 2)$ intersect the x-axis at $x = 2$?
(A) -1 (B) 3 (C) -6 (D) 2 (E) 15

3. Which of the following gives the best approximation for $f(1.01)$ if $f(x) = 2/x^7 + 3x^4$?
(A) 5.00 (B) 5.02 (C) 5.01
(D) 4.98 (E) 4.99

4. The equation of the tangent line to the curve $x^2 + y^2 + 3xy^3 + 4x^3y = 1$ at the point $(1, 0)$ is
(A) $y = -(1/2)x + 1/2$ (B) $y = (1/2)x - 1/2$
(C) $y = x - 1$ (D) $y = 2x - 2$
(E) $y = -2x + 2$

5. If $f(x) = \cos \pi\,(x^2 - (1/2)x)$, then $f'(1) =$
(A) $3/2$ (B) $(2/3)\pi$ (C) $-(3/2)\pi$
(D) $-3/2$ (E) $(3/2)\pi$

6. The curve $y = 4x^3 - 3x^2 + 2x - 1$ has a point of inflection at $x =$
(A) 1 (B) 1/2 (C) 1/3 (D) 1/4 (E) 1/5

7. The curve $y = x^2/3 + 12/x^2$ has a
(A) relative minimum at $x = 6$
(B) relative minimum at $x = \sqrt{6}$
(C) relative maximum at $x = 6$
(D) relative maximum at $x = \sqrt{6}$
(E) point of inflection at $x = 0$

8. If a solid right circular cylinder of volume V is made in such a way as to minimize its surface area then the ratio of its height to its base radius is
(A) π (B) 1/2 (C) 0 (D) 1 (E) 2

9. A ladder of length 15 ft. is leaning against the side of a building. The foot of the ladder is sliding along the ground away from the wall at 1/2 ft./sec. How fast is the top of the ladder falling when the foot of the ladder is 9 ft. from the wall?
(A) 3/2 ft./sec. (B) 3/4 ft./sec.
(C) 3/8 ft./sec. (D) 1/3 ft./sec.
(E) 1/2 ft./sec.

10. Consider the following two statements:
P. If a function is continuous then it is differentiable.
Q. If a function is differentiable then it is continuous.
(A) P and Q are both true.
(B) P and Q are both false.
(C) P is true and Q is false.
(D) P is false and Q is true.

11. For which values of the constants a and b will the function $f(x)$ defined below be differentiable?
$$f(x) = \begin{cases} ax & x \le 1 \\ bx^2 + x + 1 & x > 1 \end{cases}$$
(A) $a = b = 1$ (B) $a = 1$, $b = 2$
(C) $a = 1$, $b = -1$ (D) $a = 2$, $b = 0$
(E) $a = 3$, $b = 1$

12. The total area enclosed between the curves $y = x$ and $y = x^3$ is
(A) 1 (B) 1/2 (C) 1/3 (D) 1/4 (E) 1/5

13. The area bounded by the curve $y = \pi \sin(\pi x)$ and the x-axis between two consecutive points of intersection is
(A) 1 (B) 2 (C) 3 (D) π (E) $\pi/2$

14. If the acceleration of a particle moving in a straight line is given by $a = 10t^2$ and the particle has a velocity of 2 ft./sec. at $t = 0$, then the distance traveled by the particle between $t = 0$ and $t = 3$ sec. is
(A) 27π ft. (B) 147/2 ft. (C) 135/2 ft.
(D) 41 ft. (E) 83 ft.

15. If $f(x) = \int_0^x \sin^2 t \cos^6 t\,dt$ then $f'(\pi/4)$ is
(A) 1/2 (B) 1/4 (C) 1/8 (D) 1/16 (E) 1/32

16. $\int_0^{\pi/2} \sin 2x\,dx =$
(A) 0 (B) 1/2 (C) 1 (D) 2 (E) π

17. The region bounded above by the curve $y = x - x^2$ and below by the x-axis is rotated about the x-axis. The volume generated is
(A) $\pi/6$ (B) $\pi/30$ (C) $\pi/12$ (D) $\pi/15$ (E) $\pi/3$

18. The region in problem 17 is rotated about the y-axis. The volume generated is
(A) $\pi/6$ (B) $\pi/30$ (C) $\pi/12$ (D) $\pi/15$ (E) $\pi/3$

19. The length of the curve $y = (2/3)x^{3/2}$ between $x = 1$ and $x = 2$ is
(A) 1 (B) $\sqrt{2} - 1$
(C) $3\sqrt{3}$ (D) 1/2
(E) $(2/3)(3\sqrt{3} - 2\sqrt{2})$

20. The average value of the function $f(x) = x^3$ between $x = 0$ and $x = 2$ is
(A) 1/4 (B) 1/3 (C) 1/2 (D) 1 (E) 2

21. The following table lists the known values of a certain function f

x	1	2	3	4	5	6	7
$f(x)$	0	1.1	1.4	1.2	1.5	1.6	1.1

If the trapezoid rule is used to approximate $\int_1^7 f(x)\,dx$ the answer obtained is

(A) 7.9 (B) 7.5 (C) 6.8
(D) 7.35 (E) 7.05

22. If Simpsons's rule is used to estimate $\int_1^7 f(x)\,dx$ for the function given in problem 21, the answer obtained is

(A) 7.9 (B) 7.5 (C) 6.8
(D) 7.35 (E) 7.05

23. If $f(x) = e^{\sin 2x}$, then $f'(\pi/2) =$

(A) 2 (B) 1 (C) 0 (D) -1 (E) -2

24. If $f(x) = \ln(\sin^{-1}(x/2))$ then $f'(1) =$

(A) $1/\pi$ (B) $(2\sqrt{3})/\pi$ (C) $(\pi\sqrt{2})/4$
(D) $\pi\sqrt{2}$ (E) $\pi/2$

25. If $\lim_{x\to\infty}(e^x+1)/(x^4+\ln x) = A$ and $\lim_{x\to\infty}\dfrac{\ln x}{x^3+1} = B$ then

(A) $A = 0,\ B = \infty$ (B) $A = 0,\ B = 0$
(C) $A = \infty,\ B = 0$ (D) $A = \infty,\ B = \infty$
(E) $A = 1,\ B = 1$

26. $\lim_{x\to\infty}(x\ln x + x^2)/(x^2+2) =$

(A) 0 (B) ∞ (C) 1 (D) 2 (E) 1/2

27. The temperature, T, of a body satisfies the equation $dT/dt = kT$, t in seconds. If $T(0) = 30^\circ$ and $T(5) = 5^\circ$, then $k =$

(A) -1 (B) $-\ln 2$
(C) $-(1/5)\ln 6$ (D) $-(1/3)\ln 3$
(E) $-(1/7)\ln 5$

28. $\displaystyle\int_0^3 \frac{x}{\sqrt{x+1}}\,dx =$

(A) 1/3 (B) 2/3 (C) 4/3 (D) 8/3 (E) 16/3

29. $\displaystyle\int_0^{\pi/4} \sin^2 x\cos^3 x\,dx =$

(A) $(4\sqrt{2})/81$ (B) $(5\sqrt{2})/7$ (C) $(3\sqrt{2})/17$
(D) $\frac{127\sqrt{2}}{1391}$ (E) $7\sqrt{2}/120$

30. $\displaystyle\int_1^2 xe^x\,dx =$

(A) 1 (B) 0 (C) e (D) e^2 (E) $1/e$

31. $\displaystyle\int_1^2 (2x+3)/(x^3-2x^2-3x)\,dx =$

(A) $2\ln 3 + \ln 2$ (B) $(1/4)\ln 3 - 2\ln 2$
(C) $(1/2)\ln 3 + (1/2)\ln 2$ (D) $2\ln 3 - (1/3)\ln 2$
(E) $\ln 3 - \ln 2$

32. The integral $\displaystyle\int_0^\infty \frac{dx}{x^p}$

(A) diverges for $p > 1$, converges for $p \le 1$
(B) diverges for $p \ge 1$, converges for $p < 1$
(C) converges for $p > 1$, diverges for $p \ge 1$
(D) converges for $p \ge 1$, diverges for $p > 1$
(E) diverges for all p

INSTITUTION: *A southwestern public university with 12,000 students that in 1987 awarded 24 bachelor's and 5 master's degrees in mathematics.*

EXAM: *A three-hour exam (calculators not allowed) given to 18 students in one section of first term calculus for engineering, mathematics, and science majors. Of the 269 students who enrolled in the course, 42% completed the course with a passing grade; 60% of those who passed received grades of A or B.*

1. An object shot upward has height $x = -5t^2 + 30t$ m after t seconds. Compute its velocity after 1.5 sec., its maximum height, and the speed with which it strikes the ground

2. Two cars leave an intersection P. After 60 sec., the car traveling north has speed 50 ft/sec. and distance 2000 ft. from P, and the car traveling west has speed 75 ft/sec. and distance 2500 ft. from P. At that instant, how fast are the cars separating from each other?

3. Find the maximum and minimum of $f(x) = x\sqrt{1-x^2}$ on the interval $[-1, 1]$.

4. Find the coordinates of all local maxima and minima of $f(x) = x^3/(1+x^4)$.

5. Find the point on the graph of the equation $y = \sqrt{x}$ nearest to the point $(1, 0)$.

6. An athletic field of 400-meter perimeter consists of a rectangle with a semi circle at each end. Find the dimensions of the field so that the area of the rectangular portion is the largest possible.

7. Differentiate:
   a. $f(x) = e^{5x}(x^2 - 3x + 6)$
   b. $g(x) = 1/(1 + e^{-x})$
8. In a certain calculus course, the number of students dropping out each class day was proportional to the number still enrolled. If 2000 started out and 10% dropped after 12 classes, estimate the number left after 36 classes.
9. Differentiate:
   a. $f(x) = \ln(x^2 + x)$
   b. $g(x) = \ln(\sec x + \tan x)$
10. Use the logarithm function to differentiate:
    a. $y = (2 + \sin x)^x$ b. $y = x^{1/x}$
11. Evaluate $\arccos(1/2) - \arccos(-1/2)$.
12. Differentiate $f(x) = x \arcsin x + \arccos(2x + 1)$.
13. Differentiate $f(x) = x \cosh x - \sinh x$.
14. Find
    a. $\lim_{x \to 1} ((1 - x)/(e^x - e))$
    b. $\lim_{x \to \pi/2} (x - \pi/2) \tan x$
15. Assume the population of a certain city grows at a rate proportional to the population itself. If the population was 100,000 in 1940 and 150,000 in 1980, predict what it will be in year 2000.
16. Evaluate:
    a. $\int_1^3 (x^2 - 4x + 4)\,dx$ b. $\int_0^{\pi/2} (\sin x \cos x)\,dx$
17. Find $\frac{dy}{dx}$
    a. $y = \int_0^{\sin x} \sqrt{t}\,dt$ b. $y = \int_0^x \sec t\,dt$
18. Compute the area under the graph $f(x) = x + \sin x$ over the interval $[0, \pi/2]$.
19. Show that $2 < \int_0^4 1/(1 + \sin^2 x)\,dx < 4$. [HINT: If $f(x) \leq g(x)$ on $[a, b]$ then $\int_a^b f(x)\,dx \leq \int_a^b g(x)\,dx$.]

INSTITUTION: *A midwestern community college with approximately 8500 full time students.*

EXAM: *A two-hour exam (calculators allowed) given to 60% of the students who completed a first term calculus course for students specializing in business or liberal arts. Of the 142 students who enrolled in the course, 30% withdrew, 1% failed, and 37% received grades of A or B.*

1. A manufacturer has been selling lamps at a price of \$15 per lamp. At this price, customers have purchased 2,000 lamps per month. Management has decided to raise the price to p dollars in an attempt to improve profit. For each \$1 increase in price, it is expected that sales will fall by 100 lamps per month.

   If the manufacturer produces the lamps at a cost of \$8 each, then express the monthly profit for the sale of lamps as a function of price p. Then maximize the profit by finding the price p at which greatest profit occurs.

2. Find each derivative.
   a. $f'(2)$ if $f(x) = \sqrt{x^3 + 4x}$
   b. $g(x) = \frac{1}{3}x^6 + 3x^2$
   c. $h(x) = xe^x$
   d. $j(x) = \ln(x^2 + 4x)$
   e. $n(x) = (x + 1)/(x - 1)$

3. A fine restaurant has purchased several cases of wine. For a while, the value of the wine increases; but eventually, it passes its prime and decreases in value. In t years, the value of a case of the wine will be changing at the rate of $62 - 12t$ dollars per year. If the annual storage costs remain fixed at \$2 per case, then when should the restaurant sell the wine in order to maximize profit? [HINT: Profit depends upon value "minus" storage costs.]

4. Use implicit differentiation to find $\frac{dy}{dx}$ if $x^2y + 2y^2 = 30$.

5. It is estimated that t days from now a citrus farmer's crop will be increasing at the rate of $0.3t^2 + 0.6t + 1$ bushels per day. By how much will the value of the crop increase during the next 6 days if the market price remains fixed at \$7 per bushel?

6. Evaluate these integrals:
   a. $\int_0^4 (6x^2 + 3\sqrt{x})\,dx$
   b. $\int 2e^{2x+1}\,dx$
   c. $\int (3x^2)/(x^3 + 1)\,dx$
   d. $\int (x^2 + 1)(x^3 + 3x + 1)^5\,dx$
   e. $\int xe^x\,dx$ [Use parts.]
   f. $\int_1^\infty 3/x^2\,dx$

7. At a certain factory, the output is $Q = 120K^{1/2}L^{1/3}$ units, where K is the capital investment in $1000 units and L is the size of the labor force measured in worker-hours. Currently, capital investment is $400,000 and labor force is 1000 worker-hours. Use the total differential of Q to estimate the change in output which results if capital investment is increased by $2000 and labor is increased by 4 worker-hours.

8. A sociologist claims that the population of a certain country is growing at the rate of 2% per year. If the present population (1987) of the country is 200 million people, then what will the population be in the year 2000? [HINT: Solve the separable differential equation $\frac{dp}{dt} = .02P$.]

9. Sketch carefully the graphs of $f(x) = 2x - x^2$ and $g(x) = 2x - 4$. Then find the area of the region enclosed by the two graphs.

10. Determine the maxima and minima for the function $f(x, y) = x^2 + xy + y^2 - 3x$.

INSTITUTION: *A western public university with 23,000 undergraduate and 10,000 graduate students that in 1987 awarded 350 bachelor's degrees, 19 master's degrees, and 13 Ph.D. degrees in mathematics.*

EXAM: *A three-hour exam (calculators not allowed) given to 71 students in one section of a first term calculus course for students intending to major in the life sciences and economics. Of the 914 students who enrolled in the course, 79% passed the course, 44% of whom received grades of A or B.*

1. Differentiate the following:

   a. $\frac{d}{dx} \sec\left(e^{3x}\right)$

   b. $\frac{d}{dx} 3^{\cos x}$

   c. $\frac{d}{dx} \arcsin\left(x^{\pi}\right)$

   d. $\frac{d}{dx} \ln(\arctan x)$

2. Find the area of the 3-sided region in the first quadrant enclosed by the straight lines $y = x$ and $y = x/8$ and the curve $y = 8/x^2$. Draw a picture.

3. Consider the region enclosed between the curves $y = 2x^2$ and $y = x^2 + 1$ as shown:

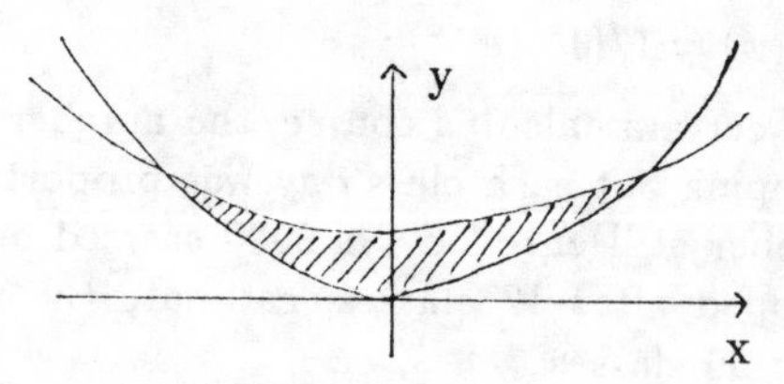

   a. Suppose the region is rotated about the axis $x = 5$. Write down a definite integral which would give the volume. *Do not evaluate* this integral.

   b. Same as part (a), except let the axis be the line $y = 7$.

4. A spring, whose natural length is 10 ft., requires 40 ft-lbs. of work to be stretched from a length of 13 ft. to a length of 17 ft. How much work would it require to stretch it from a length of 15 ft. to a length of 21 ft.?

5. An object is removed from an oven and left to sit in a 70° room. After 2/3's of an hour it is 170°, and after 2 hours it is 120°. Assuming that Newton's law of cooling applies, how hot was the object when it was removed from the oven? (Note: You may leave your answer in any reasonable form; the actual numerical answer will contain a $\sqrt{\ }$.)

6. Find the following indefinite integrals, using substitution if necessary:

   a. $\int (e^{3x})/(1 + e^{3x})^5 \, dx$

   b. $\int \sqrt{x}/(\sqrt{x} - 1) \, dx$

7. Using integration by parts twice, find $\int x^2 \sin x \, dx$. (Recall you can check your answer by differentiation.)

8. Solve the following differential equation, making use of an integrating factor. Then evaluate your constant using the initial condition.

$$\frac{dy}{dx} = 3 - y/(2x + 1); \quad y(4) = 5$$

(Assume $2x + 1 > 0$ always.)

9. At time $t = 0$ a tank contains 13 gallons of water in which 2 pounds of salt are dissolved. Water containing 6 pounds of salt per gallon is added to the tank at the the rate of 8 gallons per minute. The solutions mix instantly and the mixture is drained at the rate of 3 gallons per minute. Write down a differential equation which relates the amount of salt s (in pounds) to the time t. *Do not solve* this equation.

10. Using Theorem 19.3, find the solutions of the following two differential equations, use the initial data to evaluate your constants, and identify the appropriate graphs.

a. $y'' + 6y' + 9y = 0$; $y(0) = 0$ and $y'(0) = 2$.

b. $y'' + 5y' + 4y = 0$; $y(0) = 2$ and $y'(0) = -5$.

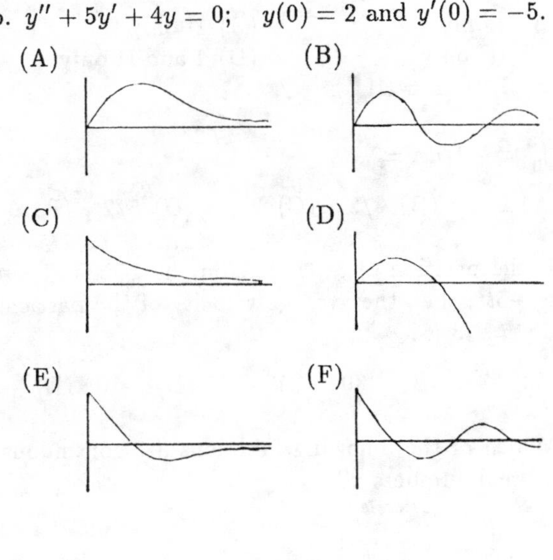

EXAM: *A three-hour exam (calculators not allowed) administered in 1985 to 35,000 high school students by the advance placement (AP) program. This is the "AB" exam for students who have studied the prescribed curriculum. 35% of those who took the exam received grades of 4 or 5, the highest two grades available.*

1. $\int_1^2 x^{-3}\,dx =$

(A) $-7/8$ (B) $-3/4$ (C) $15/64$
(D) $3/8$ (E) $15/16$

2. If $f(x) = (2x+1)^4$, then the 4th derivative of $f(x)$ at $x = 0$ is

(A) 0 (B) 24 (C) 48 (D) 240 (E) 384

3. If $y = 3/(4+x^2)$, then $\frac{dy}{dx} =$

(A) $-6x/(4+x^2)^2$ (B) $3x/(4+x^2)^2$
(C) $6x/(4+x^2)^2$ (D) $-3/(4+x^2)^2$
(E) $3/2x$

4. If $\frac{dy}{dx} = \cos(2x)$, then $y =$

(A) $-\frac{1}{2}\cos(2x) + C$ (B) $-\frac{1}{2}\cos^2(2x) + C$
(C) $\frac{1}{2}\sin(2x) + C$ (D) $\frac{1}{2}\sin^2(2x) + C$
(E) $-\frac{1}{2}\sin(2x) + C$

5. $\lim_{n\to\infty} (4n^2)/(n^2 + 10,000n)$ is

(A) 0 (B) 1/2,500 (C) 1
(D) 4 (E) nonexistent

6. If $f(x) = x$, then $f'(5) =$

(A) 0 (B) 1/5 (C) 1 (D) 5 (E) 25/2

7. Which of the following is equal to $\ln 4$?

(A) $\ln 3 + \ln 1$ (B) $\ln 8/\ln 2$
(C) $\int_1^4 e^t\,dt$ (D) $\int_1^4 \ln x\,dx$
(E) $\int_1^4 1/t\,dt$

8. The slope of the line tangent to the graph of $y = \ln(x/2)$ at $x = 4$ is

(A) 1/8 (B) 1/4 (C) 1/2 (D) 1 (E) 4

9. If $\int_{-1}^{1} e^{-x^2}\,dx = k$, then $\int_{-1}^{0} e^{-x^2}\,dx =$

(A) $-2k$ (B) $-k$ (C) $-k/2$
(D) $k/2$ (E) $2k$

10. If $y = 10^{(x^2-1)}$, then $\frac{dy}{dx} =$

(A) $(\ln 10)10^{(x^2-1)}$ (B) $(2x)10^{(x^2-1)}$
(C) $(x^2-1)10^{(x^2-2)}$ (D) $2x(\ln 10)10^{(x^2-1)}$
(E) $x^2(\ln 10)10^{(x^2-1)}$

11. The position of a particle moving along a straight line at any time t is given by $s(t) = t^2 + 4t + 4$. What is the acceleration of the particle when $t = 4$?

(A) 0 (B) 2 (C) 4 (D) 8 (E) 12

12. If $f(g(x)) = \ln(x^2+4)$, $f(x) = \ln(x^2)$, and $g(x) > 0$ for all real x, then $g(x) =$

(A) $1/\sqrt{x^2+4}$ (B) $1/(x^2+4)$
(C) $\sqrt{x^2+4}$ (D) x^2+4
(E) $x+2$

13. If $x^2 + xy + y^3 = 0$, then, in terms of x and y, $\frac{dy}{dx} =$

(A) $-(2x+y)/(x+3y^2)$
(B) $-(x+3y^2)/(2x+y)$
(C) $-2x/(1+3y^2)$
(D) $-2x/(x+3y^2)$
(E) $-(2x+y)/(x+3y^2-1)$

14. The velocity of a particle moving on a line at time t is $v = 3t^{1/2} + 5t^{3/2}$ meters per second. How many meters did the particle travel from $t = 0$ to $t = 4$?

(A) 32 (B) 40 (C) 64 (D) 80 (E) 184

15. The domain of the function defined by $f(x) = \ln(x^2 - 4)$ is the set of all real numbers x such that

(A) $|x| < 2$ (B) $|x| \le 2$ (C) $|x| > 2$
(D) $|x| \ge 2$ (E) x is a real number

16. The function defined by $f(x) = x^3 - 3x^2$ for all real numbers x has a relative maximum at $x =$

(A) -2 (B) 0 (C) 1 (D) 2 (E) 4

17. $\int_0^1 xe^{-x}\,dx =$

(A) $1-2e$ (B) -1 (C) $1-2e^{-1}$
(D) 1 (E) $2e-1$

18. If $y = \cos^2 x - \sin^2 x$, then $y' =$

(A) -1 (B) 0
(C) $-2\sin(2x)$ (D) $-2(\cos x + \sin x)$
(E) $2(\cos x - \sin x)$

19. If $f(x_1) + f(x_2) = f(x_1 + x_2)$ for all real numbers x_1 and x_2, which of the following could define f?

(A) $f(x) = x + 1$ (B) $f(x) = 2x$
(C) $f(x) = 1/x$ (D) $f(x) = e^x$
(E) $f(x) = x^2$

20. If $y = \text{Arctan}(\cos x)$, then $\frac{dy}{dx} =$

(A) $-\sin x/(1+\cos^2 x)$ (B) $-(\text{Arcsec}(\cos x))^2 \sin x$
(C) $(\text{Arcsec}(\cos x))^2$ (D) $1/((\text{Arccos}\, x)^2 + 1)$
(E) $1/(1+\cos^2 x)$

21. If the domain of the function f given by $f(x) = 1/(1-x^2)$ is $\{x : |x| > 1\}$, what is the range of f?

(A) $\{x : -\infty < x < -1\}$

(B) $\{x : -\infty < x < 0\}$

(C) $\{x : -\infty < x < 1\}$

(D) $\{x : -1 < x < \infty\}$

(E) $\{x : 0 < x < \infty\}$

22. $\int_1^2 (x^2 - 1)/(x+1)\,dx =$

(A) 1/2 (B) 1 (C) 2 (D) 5/2 (E) ln 3

23. $\frac{d}{dx}\left(1/x^3 - 1/x + x^2\right)$ at $x = -1$ is

(A) -6 (B) -4 (C) 0 (D) 2 (E) 6

24. If $\int_{-2}^2 (x^7 + k)\,dx = 16$, then $k =$

(A) -12 (B) -4 (C) 0 (D) 4 (E) 12

25. If $f(x) = e^x$, which of the following is equal to $f'(e)$?

(A) $\lim_{h\to 0}(e^{x+h})/h$ (B) $\lim_{h\to 0}(e^{x+h} - e^e)/h$
(C) $\lim_{h\to 0}(e^{e+h} - e)/h$ (D) $\lim_{h\to 0}(e^{x+h} - 1)/h$
(E) $\lim_{h\to 0}(e^{e+h} - e^e)/h$

26. The graph of $y^2 = x^2 + 9$ is symmetric with respect to which of the following?

I. The x-axis
II. The y-axis
III. The origin

(A) I only (B) II only
(C) III only (D) I and II only
(E) I, II, and III

27. $\int_0^3 |x-1|\,dx =$

(A) 0 (B) 3/2 (C) 2 (D) 5/2 (E) 6

28. If the position of a particle on the x-axis at time t is $-5t^2$, then the average velocity of the particle for $0 \le t \le 3$ is

(A) -45 (B) -30 (C) -15 (D) -10 (E) -5

29. Which of the following functions are continuous for all real numbers x?

I. $y = x^{2/3}$
II. $y = e^x$
III. $y = \tan x$

(A) None (B) I only (C) II only
(D) I and II (E) I and III

30. $\int \tan(2x)\,dx =$

(A) $-2\ln|\cos(2x)| + C$ (B) $-\frac{1}{2}\ln|\cos(2x)| + C$
(C) $\frac{1}{2}\ln|\cos(2x)| + C$ (D) $2\ln|\cos(2x)| + C$
(E) $\frac{1}{2}\sec(2x)\tan(2x) + C$

31. The volume of a cone of radius r and height h is given by $V = \frac{1}{3}\pi r^2 h$. If the radius and the height both increase at a constant rate of 1/2 centimeter per second, at what rate, in cubic centimeters per second, is the volume increasing when the height is 9 centimeters and the radius is 6 centimeters?

(A) $\frac{1}{2}\pi$ (B) 10π (C) 24π (D) 54π (E) 108π

32. $\int_0^{\pi/3} \sin(3x)\,dx =$

(A) -2 (B) $-2/3$ (C) 0 (D) 2/3 (E) 2

33. The graph of the *derivative* of f is shown in the figure at the right. Which of the following could be the graph of f?

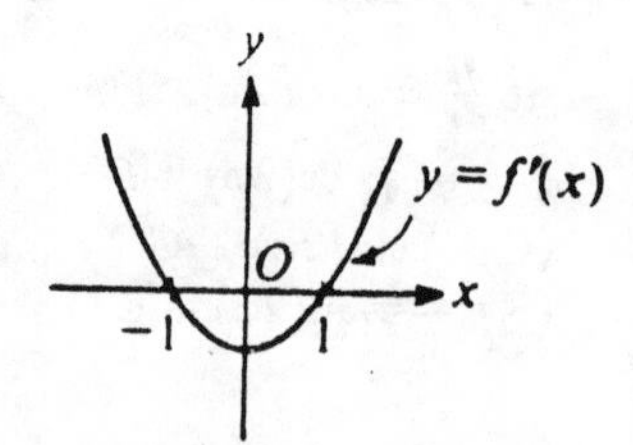

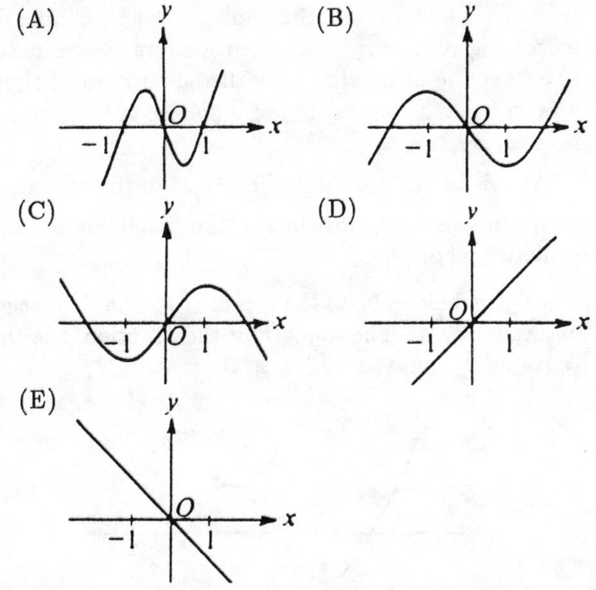

34. The area of the region in the *first quadrant* that is enclosed by the graphs of $y = x^3 + 8$ and $y = x + 8$ is

(A) 1/4 (B) 1/2 (C) 3/4 (D) 1 (E) 65/4

35. The figure at the right shows the graph of a sine function for one complete period. Which of the following is an equation for the graph?

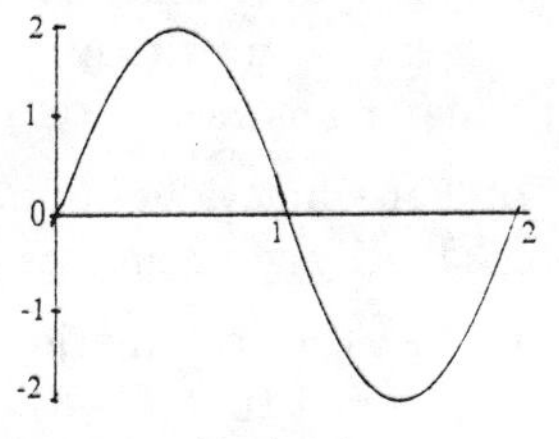

(A) $y = 2\sin\left(\frac{\pi}{2}x\right)$ (B) $y = \sin(\pi x)$
(C) $y = 2\sin(2x)$ (D) $y = 2\sin(\pi x)$
(E) $y = \sin(2x)$

36. If f is a continuous function defined for all real numbers x and if the maximum value of $f(x)$ is 5 and the minimum value of $f(x)$ is -7, then which of the following must be true?

I. The maximum value of $f(|x|)$ is 5.
II. The maximum value of $|f(x)|$ is 7.
III. The minimum value of $f(|x|)$ is 0.

(A) I only (B) II only
(C) I and II only (D) II and III only
(E) I, II, and III

37. $\lim_{x \to 0}(x \csc x)$ is

(A) $-\infty$ (B) -1 (C) 0 (D) 1 (E) ∞

38. Let f and g have continuous first and second derivatives everywhere. If $f(x) \leq g(x)$ for all real x, which of the following must be true?

I. $f'(x) \leq g'(x)$ for all real x
II. $f''(x) \leq g''(x)$ for all real x
III. $\int_0^1 f(x)\,dx \leq \int_0^1 g(x)\,dx$

(A) None (B) I only
(C) III only (D) I and II only
(E) I, II, and III

39. If $f(x) = \ln x/x$ for $x > 0$, which of the following is true?

(A) f is increasing for all x greater than 0.

(B) f is increasing for all x greater than 1.

(C) f is decreasing for all x between 0 and 1.

(D) f is decreasing for all x between 1 and e.

(E) f is decreasing for all x greater than e.

40. Let f be a continuous function on the closed interval $[0, 2]$. If $2 \leq f(x) \leq 4$, then the greatest possible value of $\int_0^2 f(x)\,dx$ is

(A) 0 (B) 2 (C) 4 (D) 8 (E) 16

41. If $\lim_{x \to a} f(x) = L$, where L is a real number, which of the following must be true?

(A) $f'(a)$ exists.

(B) $f(x)$ is continuous at $x = a$.

(C) $f(x)$ is defined at $x = a$.

(D) $f(a) = L$.

(E) None of the above

42. $\frac{d}{dx}\int_2^x \sqrt{1+t^2}\,dx =$

(A) $x/\sqrt{1+x^2}$ (B) $\sqrt{1+x^2} - 5$
(C) $\sqrt{1+x^2}$ (D) $x/\sqrt{1+x^2} - 1/\sqrt{5}$
(E) $1/(2\sqrt{1+x^2}) - 1/(2\sqrt{5})$

43. An equation of the line tangent to $y = x^3 + 3x + 2$ at its point of inflection is

(A) $y = -6x - 6$ (B) $y = -3x + 1$
(C) $y = 2x + 10$ (D) $y = 3x - 1$
(E) $y = 4x + 1$

44. The average value of $f(x) = x^2\sqrt{x^3+1}$ on the closed interval $[0, 2]$ is

(A) 26/9 (B) 13/3 (C) 26/3 (D) 13 (E) 26

45. The region enclosed by the graph of $y = x^2$, the line $x = 2$, and the x-axis is revolved about the *y-axis*. The volume of the solid generated is

(A) 8π (B) $\frac{32}{5}\pi$ (C) $\frac{16}{3}\pi$ (D) 4π (E) $\frac{8}{3}\pi$

FREE-RESPONSE QUESTIONS:

1. Let f be the function given by $f(x) = \frac{2x-5}{x^2-4}$.
   a. Find the domain of f.
   b. Write an equation for each vertical and each horizontal asymptote for the graph of f.
   c. Find $f'(x)$.
   d. Write an equation for the line tangent to the graph of f at the point $(0, f(0))$.

2. A particle moves along the x-axis with acceleration given by $a(t) = \cos t$ for $t \geq 0$. At $t = 0$ the velocity $v(t)$ of the particle is 2 and the position $x(t)$ is 5.
   a. Write an expression for the velocity $v(t)$ of the particle.
   b. Write an expression for the position $x(t)$.
   c. For what values of t is the particle moving to the right? Justify your answer.
   d. Find the total distance traveled by the particle from $t = 0$ to $t = \pi/2$.

3. Let R be the region enclosed by the graphs of $y = e^{-x}$, $y = e^x$, and $x = \ln 4$.
   a. Find the area of R by setting up and evaluating a definite integral.
   b. Set up, but *do not integrate*, an integral expression in terms of a single variable for the volume generated when the region R is revolved about the *x-axis*.
   c. Set up, but *do not integrate*, an integral expression in terms of a single variable for the volume generated when the region R is revolved about the *y-axis*.

4. Let $f(x) = 14\pi x^2$ and $g(x) = k^2 \sin(\pi x)/(2k)$ for $k > 0$.
   a. Find the average value of f on $[1, 4]$.
   b. For what value of k will the average value of g on $[0, k]$ be equal to the average value of f on $[1, 4]$?

5. The balloon shown at the right is in the shape of a cylinder with hemispherical ends of the same radius as that of the cylinder. The balloon is being inflated at the rate of 261π cubic centimeters per minute. At the instant the radius of the cylinder is 3 centimeters, the volume of the balloon is 144π cubic centimeters and the radius of the cylinder is increasing at the rate of 2 centimeters per minute. (The volume of a cylinder with radius r and height h is $\pi r^2 h$, and the volume of a sphere with radius r is $(4/3)\pi r^3$.)

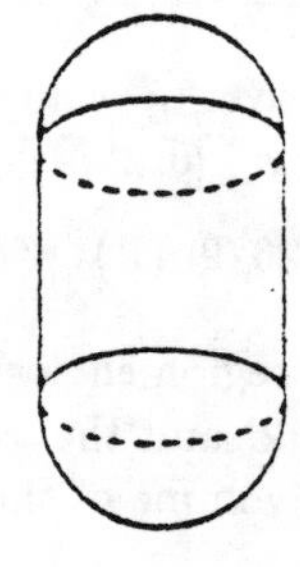

   a. At this instant, what is the height of the cylinder?
   b. At this instant, how fast is the height of the cylinder increasing?

6. The figure below shows the graph of f', the derivative of a function f. The domain of the function f is the set of all x such that $-3 \leq x \leq 3$.

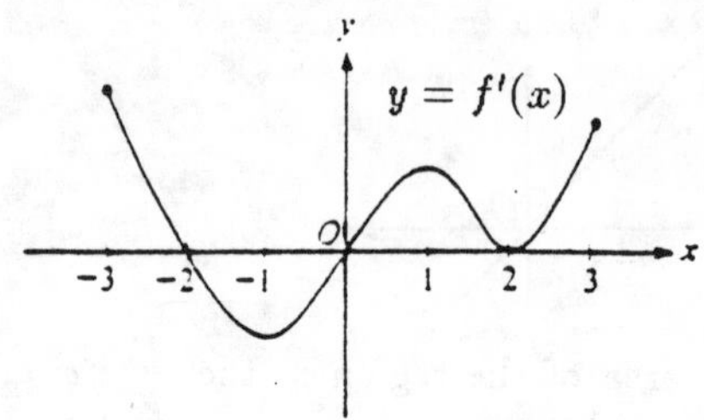

[NOTE: This is the graph of the *derivative* of f, *not* the graph of f.]

   a. For what values of x, $-3 < x < 3$, does f have a relative maximum? A relative minimum? Justify your answer.
   b. For what values of x is the graph of f concave up? Justify your answer.
   c. Use the information found in parts a and b and the fact that $f(-3) = 0$ to sketch a possible graph of f.

INSTITUTION: *A major midwestern research university with approximately 40,000 students. In 1987, 75 students received bachelor's degrees, 30 received master's degrees, and 11 received Ph.D. degrees in mathematics.*

EXAM: *A two-hour exam (calculators not allowed) on first term calculus for science and engineering students. Of the 1650 students who enrolled in the course, approximately 85% passed, including 40% who received grades of A or B.*

1. For each of the following find $\frac{dy}{dx}$.
   a. $y = (\cos(5x))^{3/2}$
   b. $y = \dfrac{x^3}{\sqrt{2x+1}}$
   c. $y = x \tan x$

d. $y = \dfrac{x^4}{4} - \dfrac{4}{x^4}$

e. $\dfrac{x}{y} + y^2 = 2x^3 + 1$

2. Find the following limits. If the limit is a finite number, write that number. If there is no finite limit, write $+\infty$, $-\infty$, or "none."

a. $\displaystyle\lim_{x\to 1^-} \frac{x-1}{x^2-2x+1}$

b. $\displaystyle\lim_{x\to +\infty} \frac{7x^5-3x^2+1}{x^6+1}$

c. $\displaystyle\lim_{x\to 3} \frac{\sin(x-3)}{x-3}$

d. $\displaystyle\lim_{x\to 2} \int_0^x (3t^2-1)\,dt$

3. For the function $y = x^3 - 6x^2 + 9x - 3$ answer the following questions.

a. Give the coordinates of all relative minima and relative maxima (if any). Justify your answer.

b. Give the coordinates of all inflection points (if any). Justify your answer.

c. For what interval(s) is the function concave up? For what intervals(s) is it concave down?

d. Sketch the graph using the information from a–c.

4. Let $f(x) = 3x^2 - 5x + 1$.

a. State carefully what properties of f guarantee that the Mean Value Theorem applies to f on the interval $[1, 2]$.

b. This theorem asserts the existence of a certain number c. Find this c for f on $[1, 2]$.

5. A flood light is positioned on the ground shining on a white wall 30 feet away. A woman 5 feet tall starts at the wall and walks towards the light at a rate of 4 ft./sec. How fast is the height of her shadow on the wall increasing when she is 10 feet from the light?

6. Find the following indefinite integrals. Do *not* simplify your answers.

a. $\displaystyle\int \frac{3x^4+2x}{x^3}\,dx$

b. $\displaystyle\int \frac{1}{t^2}\sqrt{3-\frac{1}{2t}}\,dt$

c. $\int \sin(2x+1)\,dx$

d. $\int (4x^4 - 2x + 1)\,dx$

7. Find the solution to the differential equation $\frac{dy}{dx} = (x^3+1)y^2$ subject to the initial condition $y = 1$ when $x = 2$.

8. Suppose that $f(x)$ is continuous for all real values of x and symmetric about the y-axis, F is an antiderivative of f, and F has values $F(0) = 3$, $F(1) = 5$, and $F(2) = 9$. Find

a. $\int_0^1 f(x)\,dx$

b. $\int_{-1}^2 f(x)\,dx$

9. Evaluate the following definite integrals.

a. $\int_{-1}^0 x(x^2+1)^5\,dx$

b. $\displaystyle\int_8^{34} \frac{1}{\sqrt[3]{7-x}}\,dx$

10. Let $f(x)$ be the function with graph as shown. Answer the following questions from the graph. You should *not* try to find an equation for $f(x)$.

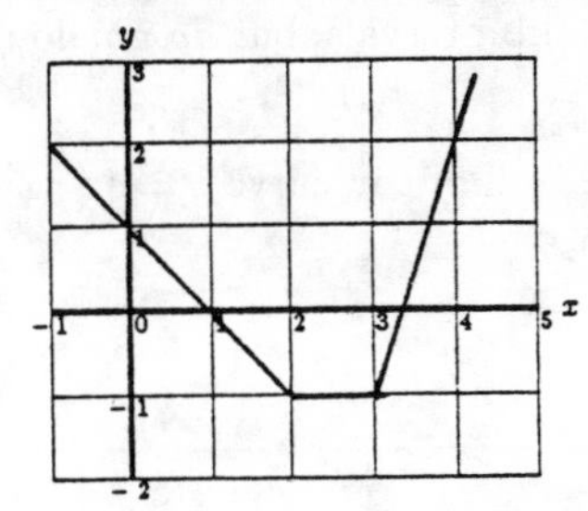

a. Find $\int_0^4 f(x)\,dx$

b. For what number a, $0 \le a \le 4$, does $\int_0^a f(x)\,dx$ have the smallest value?

11. Find the area enclosed by the two curves $y = 2x^2 + 1$ and $y = x^2 + 5$.

12. Suppose that $\int_0^x f(t)\,dt = x + \int_1^x t\,f(t)\,dt$. Give an explicit algebraic expression (not involving integrals) for $f(x)$.

13. For each part set up *do not evaluate* the integral(s) which give(s) the result. Both parts relate to the given graph of $y = \cos x$.

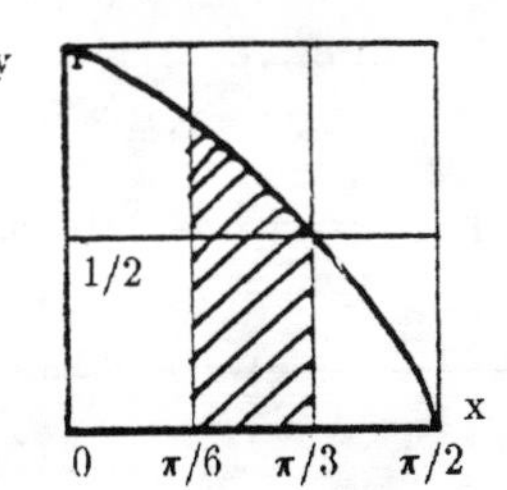

a. Find the volume of the solid formed by rotating the shaded region about the x-axis.

b. Find the volume of the solid formed by rotating the shaded region about the y-axis.

14. An object starts from rest and moves with a velocity in m/sec given by $v = t^2 - 2t$, where t is time in seconds.

    a. Find the displacement of the object between the times $t = 0$ and $t = 3$.

    b. Find the total distance travelled by the object between the times $t = 0$ and $t = 3$.

    c. What is the average acceleration between times $t = 0$ and $t = 3$?

15. For what value(s) of x does the slope of the tangent line to the curve $y = -x^3 + 3x^2 + 1$ have the largest value? Justify your answer.

16. Set up the computation to find an approximate value for the integral $\int_1^3 (1/x)\,dx$, using the trapezoidal rule with $n = 4$ subintervals, but *do not* do the final arithmetic.

17. Find the length of the curve $y = \left(\frac{2}{3}\right) x^{3/2}$ from $(0,0)$ to $(3, 2\sqrt{3})$.

INSTITUTION: *A private western university with about 4,000 undergraduates that in 1987 awarded 50 bachelor's degrees in mathematics.*

EXAM: *A three-hour exam (calculators allowed) on first tem calculus for a mixture of engineering, business, and science students. Of the 277 students who enrolled in the course, 41 were in the particular section that was given this exam. Approximately 32% of the students in the course received grades of A or B.*

1. Find formulas (in split-rule form) for the function graphed below.

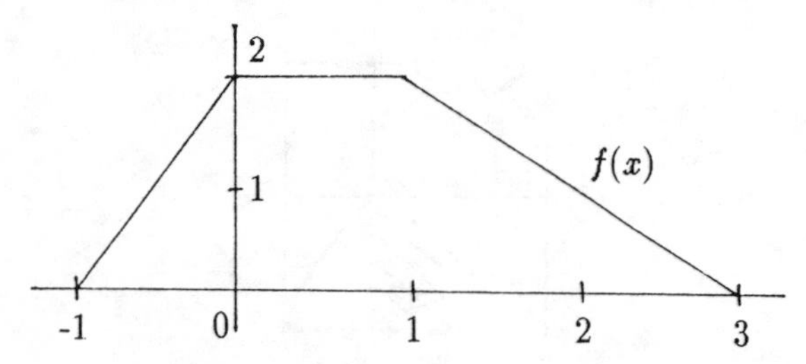

2. Find $\frac{dy}{dx}$ for each of the following:

    a. $y = \frac{1}{\sqrt[3]{4x+5}}$

    b. $y = \sin^2(3t)$, $x = \cos(3t)$

    c. $\tan(1 + xy) = 86$.

3. Suppose the position of a moving object at time t (seconds) is given by $s = (t^2 + 1)^5$ feet. Find each of the following:

    a. The average rate of change of s between $t = 0$ and $t = 1$.

    b. The instantaneous velocity at $t = 1$.

    c. The acceleration at $t = 1$.

4. Evaluate each of the following limits:

    a. $\lim_{x\to 0} \frac{\sin^2(5x)}{5x^2}$

    b. $\lim_{x\to 0} \frac{\sin(5x^2)}{5x^2}$

    c. $\lim_{x\to 0} \frac{\sqrt{5x+1}-1}{x}$

5. Suppose $y = f(x) = \frac{2x+3}{x-2}$

    a. State the equations of any horizontal or vertical asymptotes to the graph of $f(x)$.

    b. Find $\frac{dy}{dx}$ (the derivative of the inverse function) and evaluate it at $x = 3$.

    c. Sketch the graph of $f(x)$

    d. Find $\frac{d}{dx} f(f(x))$

6. Sketch the graph of $y = 3x^4 + 16x^3 + 24x^2 - 2$, after completing tables for the sign of y' and y''. Label turning points. Also, find the values of a and b such that the curve is concave down for $a < x < b$.

7. A printer is to use a page that has a total area of 96 in^2. Margins are to be 1 in. at both sides and $1\frac{1}{2}$ in. at the top and bottom of the page. Find the (outer) dimensions of the page so that the area of the actual printed matter is a maximum.

8. A variable line through the point $(1, 2)$ intersects the x-axis at the point $A(x, 0)$ and the y-axis at the point $B(0, y)$, where x and y are positive. How fast is the area of triangle AOB changing at the instant when $x = 5$, if x increases at a constant rate of 16 units/sec.?

9. Let $f(x) = \sqrt{3x-2}$. Find a point $x = c$ between $a = 1$ and $b = 2$ at which the slope of the tangent line equals the slope of the line connecting $(1, f(1))$ and $(2, f(2))$. Also, what is the name of the theorem that guarantees the existence of the point $x = c$?

10. a. Find all functions $y(x)$ such that $\frac{dy}{dx} = x^2 + \frac{1}{x}$

    b. $\int \frac{dx}{\sqrt{1-x}}$

    c. $\int x^3(2 - 3x^4)^7\,dx$

INSTITUTION: *A large midwestern state university with approximately 25,000 students. In 1987, 50 students received bachelor's degrees, 6 received master's degrees, and 3 received Ph.D. degrees in mathematics.*

EXAM: *A two-hour exam (calculators not allowed) on elementary calculus for business students. Students are allowed one 5/8 "crib card." Of the 640 students who enrolled in the course, approximately 60% passed, including 27% who received grades of A or B.*

1. Find the equation of the tangent line to the curve $y = 4x^3 - 3x + 30$ at the point where $x = -2$.

2. Evaluate the following limits:

   a. $\lim\limits_{x\to 3} \dfrac{x^2+4x-21}{x-3}$ b. $\lim\limits_{h\to 0} \dfrac{(x+h)^2 - x^2}{h}$

   c. $\lim\limits_{x\to\infty} \dfrac{3x^4 - x}{1+5x^4}$

3. a. If $y = \ln\left(x^4 + x^2\right) + e^{x^3}$, find $\frac{dy}{dx}$.

   b. If $F(x) = (x+1)^4(x+2)^3$, find $F'(x)$.

   c. If $y = \dfrac{x+1}{x^3+2}$, find $y'(1)$.

   d. If $x^2 + xy + y^3 = 3$, use implicit differentiation to find $\frac{dy}{dx}$ in terms of x and y. Evaluate $\frac{dy}{dx}$ at the point $(1,1)$.

4. Find the absolute maximum and the absolute minimum of $F(x) = x^2 - 4x + 1$ in the interval $1 \le x \le 4$.

5. a. If $y = \sqrt{x}$, find the differential dy.

   b. Use differentials to estimate $\sqrt{18}$.

6. Sketch the graph of $y = x^3 - 6x^2$ and use it to identify

   a. x-intercepts

   b. critical points

   c. inflection points

   d. intervals in which y is decreasing

   e. intervals in which y is concave up

7. Evaluate the following integrals:

   a. $\int x^2 e^{x^3}\, dx$

   b. $\int \dfrac{x\, dx}{x^2+1}$

   c. $\int x e^x\, dx$ (by parts)

   d. $\int_1^4 \left(3\sqrt{x} + \frac{1}{x^2}\right) dx$

   e. $\int (2x^{-1} + 3e^x)\, dx$

8. Sketch the curves $y = x^2$ and $y = 2x$ and compute the area between them.

9. The Pullman company knows that the cost of making x spittoons is $C(x) = 50 + x + \frac{1600}{x}$.

   a. Find the number x of spittoons that must be made in order to minimize the cost.

   b. What is the minimum cost?

10. Let $F(x,y) = 8x^2 + 6y^2 - 8x + xy + 10$. Find

    a. $F_x(1,2)$ b. $F_y(1,2)$

    c. $F_{xx}(1,2)$ d. $F_{xy}(1,2)$

11. a. Find the critical points of $F(x,y) = \dfrac{x^3}{3} + \dfrac{y^2}{2} - xy$.

    b. Determine any relative maximum, relative minimum, or saddle points.

12. Use the method of Lagrange multipliers to find the minimum of $F(x,y) = x^2 + y^2$ subject to the constraint $4x + 2y = 10$.

Final Examinations for Calculus II

The following twelve examinations represent a cross section of final examinations given in 1986–87 for students enrolled in the second calculus course. The examinations come from large universities and small colleges, from Ph. D. granting institutions and two year colleges.

Brief descriptive information preceding each examination gives a profile of the institution and circumstances of the examination. Unless otherwise noted, no books or notes are allowed during these examinations; calculator usage is explicitly noted where the information was provided. Except for information on problem weights and grading protocol, the texts of the examinations are reproduced here exactly as they were presented to the students.

INSTITUTION: *A southeastern land-grant university with 12,500 students that in 1987 awarded 65 bachelor's degrees, 26 master's degrees, and 5 doctor's degrees in mathematics.*

EXAM: *A final exam (calculators allowed) from one section of second term calculus for engineering students. Of the 390 students who originally enrolled in the course, 72% passed, including 38% who received grades of A or B. 32 students were in the section that took this exam.*

1. Find the area of the region bounded by the curves $y = 6 - x^2$ and $y = x$.

2. Consider the region R bounded by the x-axis, the curve $y = 1/x$, and the lines $x = 1$, $x = 4$. Find the volume, V, of the solid obtained when R is

   a. revolved around the x-axis.

   b. revolved around the y-axis.

3. Find the length of the curve $y = x^{3/2}$ from $(0,0)$ to $(5, \sqrt{5})$.

4. In 1977 the world's population was 4.3 billion persons and growing at a rate of 2.12% per year. If this continues,

   a. how long will it take for the world's population to double?

   b. what will the population be in 2000 AD?

5. Find y':

   a. $y = \ln x^{2/3}$ b. $y = \ln(\sec x)$

   c. $y = e^{\sqrt{x^2+1}}$ d. $y = e^{\tan 3x}$

   e. $y = x^x$ f. $y = 2^{x^2+1}$

   g. $y = \csc \sqrt{x}$ h. $y = \log_2 e^{x^2}$

   i. $y = x^2 \sin^{-1} x$ j. $y = \tan^{-1}(\pi x)$

6. Antiderivatives

   a. $\int y e^{y^2}\,dy$

   b. $\int (3/e^t)\sqrt{1+e^{-t}}\,dt$

   c. $\int \sec^2 z/(1-\tan z)\,dz$

   d. $\int x\pi^{x^2}\,dx$

   e. $\int \tan(e/x)/x^2\,dx$

   f. $\int \frac{1}{x\sqrt{9x^2-1}}\,dx$

   g. $\int \frac{1}{t^2+4t+5}\,dt$

   h. $\int x^{-3}\ln x\,dx$

   i. $\int \cos^3(2x)\,dx$

   j. $\int \frac{1}{(2+x^2)^{3/2}}\,dx$

   k. $\int \frac{x}{(x+1)(x^2+1)}\,dx$

7. Determine whether the following series are absolutely convergent, conditionally convergent, or divergent; show your work.

   a. $\sum_{k=0}^{\infty} (-1)^k \frac{k^2}{k+1}$ b. $\sum_{k=1}^{\infty} (-1)^k \frac{\cos k}{3^k}$

   c. $\sum_{k=1}^{\infty} \frac{(-1)^{k+1}}{\sqrt{k}}$

8. Write the Taylor series expansion for $c = 0$ and

   a. e^x b. $\sin x$

   c. $\cos x$

9. Find *all* x's for which the following power series converge.

   a. $\sum_{k=1}^{\infty} \frac{(-1)^k x^k}{k^2}$ b. $\sum_{k=0}^{\infty} \frac{(x-4)^k}{3^k}$

INSTITUTION: *A midwestern church-related college with 400 undergraduates that in 1987 awarded 5 bachelor's degrees in mathematics.*

EXAM: *A two-hour exam (calculators allowed) given to students in a second term calculus course. Of the 7 students who enrolled in the course, 6 passed, 3 of whom received a grade of A or B.*

1. a. Sketch (carefully label) the region bounded by the two curves, $y = x^2$ and $y = x + 2$.
   b. Find the volume of the solid generated by revolving the above region about the x-axis.

2. a. Determine the eigenvalues of $A = \begin{pmatrix} 2 & -2 \\ -2 & 5 \end{pmatrix}$.
   b. Determine the eigenvectors for A.
   c. Carefully and thoroughly discuss the graph of the equation $2x^2 - 4xy + 5y^2 = 6$.

3. a. Carefully sketch (with labels) the graphs of $r = 1 + \cos\theta$, $r = 2 - \cos\theta$.
   b. Identify by name these curves.
   c. Determine the points of intersection of these curves.
   d. Determine the area inside $r = 2 - \cos\theta$ that is outside $r = 1 + \cos\theta$.

4. Determine the following:
   a. $\int (xe^x)/(1 + x^2)\, dx$
   b. $\int (2x^2 + 3)/(x(x-1)^2)\, dx$
   c. $\lim_{x \to 0^+} (x+1)^{\ln x}$
   d. $\int_e^\infty 1/(x(\ln x)^2)\, dx$

5. Determine the most general form of the solution of the system of equations

$$\begin{bmatrix} 1 & 4 & 2 \\ 2 & 5 & 1 \\ 1 & 1 & -1 \end{bmatrix} \begin{bmatrix} x_1 \\ x_2 \\ x_3 \end{bmatrix} = \begin{bmatrix} 1 \\ 5 \\ 4 \end{bmatrix}$$

6. a. Determine the equation of the plane through the points $P(5, 2, -1)$, $Q(2, 4, -2)$, $R(1, 1, 4)$.
   b. Determine a unit vector that is perpendicular to the above plane.

7. Assume $M_{2\times3}$ is a vector space. Which of the following are subspaces of $M_{2\times3}$?
   a. $W = \left\{ \begin{pmatrix} a & b & c \\ d & 0 & 0 \end{pmatrix} \mid \text{where } b = a + c \right\}$
   b. $U = \left\{ \begin{pmatrix} a & b & c \\ d & 0 & 0 \end{pmatrix} \mid \text{where } b + d = 1 \right\}$

8. a. Write the equations of the tangent(s) to the curve $x(t) = t^2 - 2t + 1$, $y(t) = t^4 - 4t^2 + 4$, at the point $(1, 4)$.
   b. Is there any point at which the curve has a vertical tangent? Explain.

INSTITUTION: *A public Canadian university with 14,000 students that in 1987 awarded 52 bachelor's degrees, 5 master's degrees, and 2 Ph.D. degrees in mathematics.*

EXAM: *A three-hour exam (calculators allowed) for all students in a year-long course for social science students. Of the 820 students who enrolled in the course, 69% passed, including 32% who received grades of A or B.*

1. INVENTORY CONTROL PROBLEM: The success of "Le declin de l'empire americain" *inside* Canada has given hope to the Memphrémagog Music company located in Saint-Benoît du-Lac, Québec and they are now optimistic that Canadians might even start buying Canadian music. They have predicted that they can sell 20,000 records in the next year and thus a decision must be made as to the number of production runs. Since each run involves a production set-up cost of \$200 they do not want too many runs. On the other hand there are storage costs involved with keeping records on hand so they do not want the production runs to be too big. The total storage cost for the year is \$.50 times the average number of records on hand during the year.
   a. Let x be the number of records produced in one run. Assume that the records are sold at a constant rate in the period from the beginning of one run until the beginning of the next run and that a run is sold out just as the next one starts. Draw a diagram to indicate the production process throughout the year.
   b. What will be the average number of records on hand from the beginning of one run until the beginning of the next? Why is this number also the average number on hand during the year?
   c. Find the number of runs that the Memphrémagog Music company should have so as to minimize its total set up and storage costs.
   d. Show that the value obtained does indeed correspond to minimum cost.

e. [Optional cultural question.] What other famous musical organization, aside from the Memphrémagog Music company, is located in Saint-Benoît du-Lac, Québec?

2. The Camrose (Alberta) Souvenir corporation specializes in the manufacture of models of oil drilling rigs. Production involves a fixed cost of $5000 for market research and tooling up, but experience has shown that the marginal cost is constant at $2 per oil rig. All items produced are sold. The demand equation for the rigs is $x = 10000 - 1000p$.

   a. Find the cost function.

   b. Sketch the cost and revenue functions on the same graph using the production level as the independent variable (abscissa).

   c. What range of production levels should the company analyze?

   d. For what levels of production does the company make a profit?

3. The economists at the Bank of New Brunswick have discovered a new semi-cyclic law governing mortgage rates; if t is time, in years, measured from April 12, 1964 then the interest rate r is given by $r(t) = \sin(\ln[x^3 + \ln(\sin x^3)])$. On the other hand the New Brunswick board of realtors has noticed that housing starts varies with the mortgage rate according to the formula $N(r) = 4.591 \cdot (10)^3 \cdot e^{-(5r^2+2.71)}$. At what rate was the number of housing starts changing seven years after the April 12, 1964 date? Do not simplify your answer; at certain points in your calculation you may simply indicate which quantities you would evaluate numerically.

4. Because of a drop in demand for salmon the minister of fisheries has decided to reduce the allowable catch for British Columbia fishermen. The demand has been dropping according to an exponential function. The records for the month of January 1986 were lost, but the demand for March and May were 400 and 100 metric tonnes respectively.

   a. Find a numerical relationship between the rate at which the demand is dropping and the demand at any point in time.

   b. Tell the minister what the demand was in January 1986. Simplify your answer as much as possible.

   c. Tell the minister when the demand for salmon will be zero.

5. In order to predict how many branch plants it should set up in northern Saskatchewan to produce its unique brand of junk food, the Aunt Sarah corporation has made a population study of the Beuval region. In 1970 the population was 25. It was found that starting in 1970 people entered the region (this included births and immigration) at the rate of $20 + 200t$ people per year where t is years measured from January 1, 1970. With the passage of time people also leave the region (due to deaths and emigration) and it was found that if there are p people in the region at a given time, then t time units later $pe^{-.1t}$ of these people will still be there.

   a. Out of the 25 people in the Beuval region in 1970 how many will be there in 2000?

   b. Approximately how many people entered the region in the first two months of 1972?

   c. Of the people who entered the region in the first two months of 1972 how many will still be there in the year 2000?

   d. Find an expression for the population of the Beuval region in the year 2000. Do not evaluate this expression.

6. The CCCC (Canadian Calligraphy and Clacker Company) produces two kinds of typewriters, manual and electric. It sells the manual typewriters for $100 and the electric typewriters for $300 each. The company has determined that the weekly cost of producing x manual and y electric typewriters is given by $C(x, y) = 2000 + 50x + x^2 + 2y^2$. Assuming that every typewriter produced is sold, find the number of manual and electric typewriters that the company should produce so as to maximize profits. Show that the value obtained does indeed correspond to maximum profit.

7. Different items that a consumer can purchase have different "utility" values, i.e., how much the item is "worth." If the consumer considers two items at a time, then they will have a joint utility value. Suppose that a consumer has $600 to spend on two commodities, the first of which costs $20 per unit and the second $30 per unit. Further let the joint utility of x units of the first commodity and y units of the second commodity be given by $U(x, y) = 10x^{.6}y^{.4}$. Use the method of Lagrange multipliers to determine how many units of each commodity the consumer should buy in order to maximize utility. [Note: a joint utility function which is of the form $U(x, y) = C \cdot x^a y^b$ is called a Cobb-Douglas utility function.]

8. Evaluate:

   a. $\int_2^3 \dfrac{x^3}{7x^4 + 5}\,dx$ b. $\int \cos(4x + 2)\,dx$

   c. $\int x \ln x\,dx$

9. Use an appropriate 2nd Taylor Polynomial to estimate $\sqrt[4]{17}$. Do *not* simplify the arithmetical quantities that you obtain.

INSTITUTION: *A mid-Atlantic suburban community college with 3500 students that awarded 750 associate degrees in 1987.*

EXAM: *A two and one-half hour exam (calculators allowed) for the second term of calculus for engineering, science, and mathematics students. Students are allowed to use a handbook of tables, together with an instructor-supplied set of basic formulas. Of the 16 students who enrolled in the course, 93% passed, including 63% who received grades of A or B.*

1. Find the area under the curve $r(x) = \sqrt{x+1}$ from $x = -1$ to $x = 8$. Sketch the region.

2. Find the area between the curves $f(x) = x^3$ and $g(x) = x$. Sketch the region.

3. Find the volume of the solid of revolution generated by revolving the region bounded by $y = 1/x$, the x-axis, $x = 1$, and $x = 6$ about the x-axis. Sketch.

4. Find all points of intersection of the curves $r = 2 - 2\cos\theta$ and $r = 2\cos\theta$; sketch the system.

5. Using the curves in problem 4, find the area between the curves.

6. Find an equation of the curve where $d^2y/dx^2 = 18x - 8$ if the curve passes through $(1, -1)$ and the slope of the tangent line is 9 at $(1, -1)$.

7. Find four derivatives:

   a. $y = \ln^3 5x$ b. $y = \ln(5x)^3$
   c. $f(t) = t^2 e^{-5t}$ d. $r(\theta) = 2\cos(2\theta^2)$
   e. $x^2 y = \tan^{-1}(x/y)$

8. Evaluate five integrals (without using tables):

   a. $\int_0^1 \frac{6x^2}{(x^3+2)^4}\,dx$
   b. $\int_0^4 x\,e^x\,dx$
   c. $\int 3^{2t-1}\,dt$
   d. $\int \frac{x+2}{x^3+x^2}\,dx$
   e. $\int \frac{\cos 3\theta}{\sin^4 3\theta}\,d\theta$
   f. $\int \sin^4 2x\,dx$

9. Simplify:

   a. $\lim_{n\to+\infty}\left(1+\frac{1}{n}\right)^{n-3}$ b. $\lim_{n\to+\infty}\left(1+\frac{1}{2n}\right)^{n}$

10. Find the angle of rotation for the conic $\sqrt{3}x^2 + 3xy = \sqrt{3}/2$.

11. For the conic $x^2 + 4y^2 + 8y + 6x = 3$, determine the coordinates (h, k) for the origin in the $x'y'$ plane.

12. Identify each conic and sketch:

    a. $x^2 - 6x + 12y = 3$ b. $y^2 = 4x^2 + 64$

13. Find each limit, if it exists:

    a. $\lim_{\theta\to 0}(1-\cos\theta)/\theta^2$ b. $\lim_{x\to+\infty} x^2/e^x$
    c. $\lim_{x\to 0+} \cot x/\ln x$ d. $\lim_{x\to\pi/2+}(\sec x - \tan x)$

14. Write the Taylor polynomial for $n = 4$ at $c = 1$ for $f(x) = e^{-x}$.

15. Determine whether each of the following converges or diverges:

    a. $\int_0^1 \frac{dx}{1-x^2}$ b. $\int_{-\infty}^{+\infty} \frac{du}{u^2+4u+5}$
    c. $\int_{-\infty}^1 \frac{x}{\sqrt{2-x}}\,dx$

INSTITUTION: *A publicly-funded Canadian university with approximately 30,000 students that in 1987 awarded 125 bachelor's degrees, 15 master's degrees, and 6 Ph.D. degrees in mathematics.*

EXAM: *A three-hour exam (calculators allowed) given to all students completing a year-long calculus course for students of finance, commerce, business, and economics. Students are allowed one $8\frac{1}{2} \times 11$ "aid sheet" in the student's own handwriting. Of the 750 students who enrolled in the course, 69% passed, including 19% who received grades of A or B.*

1. You have \$1500 to invest for 120 days. Which of the following investment strategies is better, and by how much?

   (A) 8% annual interest compounded daily.
   (B) 8.25% simple annual interest.

2. Calculate $\int_{-2}^{2}(x+1)\,dx$.

3. If $y = f(x)$, find $\int (x + yy')\,dx$.

4. If $y^3 = x + 1$, find $\lim_{x\to 0}(y-1)/x$.

5. Let $g(x,y) = xyf(x,y)$, where $f(2,1) = 1$, $\frac{\partial f}{\partial x}(2,1) = -1$, $\frac{\partial f}{\partial y}(2,1) = -1$. Find the value of $\frac{\partial g}{\partial y}(2,1)$.

6. Compute $\begin{bmatrix} 1 & 1/2 \\ 1/3 & 1/4 \end{bmatrix}^{-1}$.

7. Find a saddle point and a local minimum point for $f(x,y) = x^3 + y^3 - 3xy$.

8. Find (graphically or otherwise) the maximum value of $6x + 4y$, subject to $5x + 4y \leq 100$, $2x + y \leq 30$, $7x + 4y \leq 108$, $x \geq 0$, $y \geq 0$.

9. Find the volume obtained by rotating $y = x^2$ about the y-axis, $0 \leq x \leq 2$.

10. If $e^{xy} = 10$, find $\frac{dy}{dx}$.

11. Evaluate $\int_{-1}^{1} e^{-|x|}\,dx$ using Simpson's rule with four intervals.

12. Evaluate:

 a. $\int x^3 \ln(5x)\,dx$

 b. $\int (x^2+3)/(x^2+2x+1)\,dx$

 c. $\int e^x \cosh 2x\,dx$

13. Given $f(x) = |x|e^{-x}$

 a. Graph $f(x)$ for $-2 \leq x \leq 4$, showing all local extrema and inflection points.

 b. Find the absolute maximum and absolute minimum of $f(x)$ for $-2 \leq x \leq 4$.

 c. Evaluate $\int_0^4 f(x)\,dx$.

14. A factory assembles sedan cars with 4 wheels, 1 engine and 4 seats, trucks with 6 wheels, 1 engine and 2 seats, and sports cars with 4 wheels, 1 engine and 2 seats.

 a. How many wheels, engines and seats are required to assemble x sedans, y trucks and z sports cars?

 b. How many sedans, trucks and sports cars respectively can be assembled using exactly 218 wheels, 47 engines and 134 seats?

 c. If sedans sell for $10,000, trucks for $12,000 and sports cars for $9,000, find the total value of output in part b.

 d. If an extra pair of wheels, an extra engine, and an extra pair of seats become available, can the output value in part c be increased? Is so, how much?

15. A rectangular 3-story building is to be constructed of material costing $100 per square metre for floors, $200/m^2 for roofing, side and back walls, and $400/m^2 for the front wall. If $144,000 is available for materials, then, neglecting all other costs, find the dimensions of the building that will enclose the greatest volume.

INSTITUTION: *A southwestern public university with 45,000 students that in 1987 awarded 36 bachelor's degrees, 10 master's degrees, and 1 Ph.D. in mathematics.*

EXAM: *A three-hour, closed book final exam (calculators allowed) from one section of second term calculus for engineering and science students. Students are allowed to use pre-approved outline notes during the exam. Of the 1150 students who originally enrolled in the course, 77% passed, including 35% who received grades of A or B—4% A's, 31% B's. 113 students were in the section that took this particular exam.*

1. Using limit theorems (like "the limit of a sum is the sum of the limits"), prove that

$$\lim_{n\to-\infty} \frac{2n}{3n^2+1} \sin n = 0.$$

2. State (with reasons) whether the infinite series

$$\sum_{k=2}^{\infty} (-1)^k/((k+2)\ln k)$$

is absolutely convegent, conditionally convergent, or divergent.

3. Evaluate the improper integral $\int_0^\infty f'(x)\,dx$ if $f(x) = x^3e^{-x}$. (Note that f' appears in the integral).

4. Find the interval of convergence of

$$\sum_{k=3}^{\infty} (-x)^k/(2^k \ln k)$$

5. Let Ω denote the banana-shaped region between the graph of $y = x^2$ and that of $y = x^3$. Find $\int_\Omega\int xy^2\,dx\,dy$.

6. Write down (but do not evaluate) a formula using only a single integral of a function of one variable for the area of the region inside the graph of $r = 3\cos\theta$ but outside the graph of $r = \cos\theta$, where $0 \leq \theta \leq \pi$.

7. For each limit below, find the limit if its exists and call the limit non-existent otherwise:

   a. $\lim_{h\to 0}(1-2h)^{1/h}$ b. $\lim_{n\to\infty} n^{2/n}$

   c. $\lim_{n\to-\infty} \sin n/n^{1/3}$

8. Find the Taylor series expansion in powers of $x-2$ for $\cos 3x$.

9. Find the Taylor series in powers of x for an antiderivative of $x^{-1}\ln(1+x)$.

10. Show that the point $(x,y)=(9\pi^2,0)$ is in the intersection $\theta=\pi$.

11. You're suppose to design a fully enclosed rectangular box (having all six of top, bottom, front, back, left, and right) whose cost is exactly \$60, where the material from which the six faces are made costs \$2 per square foot. Find the dimensions l, w, and h of such a box having the largest possible volume of all such boxes.

12. Find $\lim_{x\to 1}\ln(2/(1+x))/(x-1)$

13. Find $\lim_{x\to\infty}(\sqrt{x}+1)^{1/x}$

INSTITUTION: *A midwestern liberal arts college with 2200 students that in 1987 awarded 8 bachelor's degrees in mathematics.*

EXAM: *Two two-hour exams (calculators not allowed) from two different sections of second term calculus taken primarily by prospective mathematics, computer science, economics, and science majors. Of the 75 students who originally enrolled in the course, approximately 83% passed, including 55% who received grades of A or B. 24 students were in the section that took the first of these exams, 18 in the section that took the second exam.*

1. Find the area bounded by the graphs of $y^2=1-x$ and $2y=x+2$. (Use a horizontal element of area.)

2. Find y' if

   a. $y=\sin^{-1}(5x-1)$

   b. $y=\ln(3x^4+5)^{3/2}$

   c. $xy^2=e^x-e^y$

   d. $y=(2x+3)\sin^3(x^2)$

3. Find:

   a. $\int(2x)/(\sqrt{9-x^4})\,dx$ b. $\int x\ln x\,dx$

4. Graph $y=e^x/x$ and identify the value of x that corresponds to a vertical asymptote, the value of y that corresponds to a horizontal asymptote, and the value of (x,y) at which a minimum occurs. [Note: $y'=(e^x(x-1))/x^2$; $y''=(e^x((x-1)^2+1))/x^3$.]

5. $\int(9x+27)/(x^4+9x^2)\,dx$

6. a. $\lim_{x\to\infty}(1-(3/x))^{2x}$

   b. $\lim_{x\to\infty}(x\ln x)/(x^2+1)$

7. Argue the case for convergence or divergence. Find the value if convergent:

   a. $\int_0^\infty \dfrac{x}{(x^2+9)^2}\,dx$ b. $\int_{-1}^0 \dfrac{1}{\sqrt{1+x}}\,dx$

8. Find the Taylor polynomial for $f(x)=\ln x$ if $a=1$ and $n=4$. Also give R_4.

9. The region in the first quadrant bounded by the graphs of $y=1/\sqrt{x}$, $x=1$, $x=4$ and $y=0$ is revolved about the x-axis. Find the volume of the solid of revolution.

10. a. Explain why $\sum_{n=1}^{\infty}\frac{n}{3n+1}$ is divergent.

    b. Find the exact sum of the geometric series

$$\sum_{n=2}^{\infty}2^{-n+1}$$

    c. If $a_n=(2n+1)/(4n+3)$ then $\lim_{n\to\infty}a_n=?$

---

1. Evaluate the following limits:

   a. $\lim_{x\to 0+}(e^x+3x)^{1/x}$

   b. $\lim_{x\to-3}\left(\dfrac{x}{(x^2+2x-3)}-\dfrac{4}{(x+3)}\right)$

2. Differentiate the following functions. Derivatives need not be simplified.

   a. $\sqrt{\ln\sqrt{x}}$ b. $x^{\ln x}$

   c. $(\cos x)^{x+1}$ d. $\sin^3 e^{-2x}$

3. Evaluate the following integrals:

   a. $\int 1/(x+x^3)\,dx$ b. $\int\sqrt{4-x^2}/x\,dx$

   c. $\int x\,e^{-4x}\,dx$ d. $\int_0^\infty 1/\sqrt[3]{x+1}\,dx$

4. Set up integrals that could be evaluated (but do not evaluate) to find the volume of the solid generated by revolving the region bounded by the graphs of $y = x^3$, $x = 2$, and $y = 0$.

   a. about the x-axis;

   b. about the y-axis;

   c. about the line $x = 3$.

5. An above-ground swimming pool has the shape of a right circular cylinder of diameter 12 ft and height 5 ft. If the depth of the water in the pool is 4 ft., find the work required to empty the pool by pumping the water out over the top.

6. Find the local extrema for $f(x) = x \ln x$, $x > 0$. Discuss concavity, find the points of inflection, and sketch the graph of f.

7. The rate at which sugar decomposes in water is proportional to the amount that remains undecomposed. Suppose that 10 lb. of sugar is placed in a container of water at 1:00 P.M., and one-half is dissolved at 4:00 P.M. How much of the 10 lb. will be dissolved at 8:00 P.M?

8. A person on level ground, $\frac{1}{2}$km from a point at which a balloon was released, observes its vertical ascent. If the balloon is rising at a constant rate of 2 m/sec., find the rate at which the angle of elevation of the observer's line of sight is changing at the instant the balloon is 100 m. above the level of the observer's eyes. [NOTE: 1 km= 1000m.]

9. Find a second degree Taylor polynomial for $f(x) = \sin(x)$ around $a = \pi/4$. Use this polynomial to estimate $\sin(46°)$, but do not simplify. Find an upper bound on the error of your estimate.

INSTITUTION: *A highly selective northeastern private university with 6,500 undergraduates that in 1987 awarded 28 bachelor's degrees, 3 master's degrees, and 10 doctor's degrees in mathematics.*

EXAM: *A three-hour final exam (calculators not allowed) for all students who take second term calculus. Of approximately 160 students who originally enrolled in the course, 92% passed, including 47% who received grades of A or B.*

1. $\int_3^4 (2x)/(x^2 - 3x + 2)\, dx =$

   (A) $\ln 3 - \ln 2$ (B) $\ln(3/4)$
   (C) $\ln 2 - \ln 3$ (D) $6 \ln 2 - 2 \ln 3$
   (E) Does not converge.

2. $\int_0^\infty xe^{-3x^2}\, dx =$

   (A) 1/6 (B) −2/9 (C) 0
   (D) 1 (E) Does not converge.

3. Which integral gives the volume generated when the area under the arc of $y = \sin x$, $0 \le x \le \pi$ is rotated about the vertical line $x = 2\pi$?

   (A) $\int_0^\pi \pi \sin^2 x\, dx$ (B) $\int_0^\pi 2\pi x \sin^2 x\, dx$
   (C) $\int_0^\pi 2\pi x \sin x\, dx$ (D) $\int_0^\pi 2\pi(2\pi - x) \sin x\, dx$
   (E) $\int_0^2 2\pi x \sin x\, dx$

4. $$\int_{2/\sqrt{3}}^{2} \frac{1}{x^2\sqrt{x^2-1}}\, dx$$

   (A) 1 (B) $(\sqrt{3}-1)/2$ (C) $\pi/6$
   (D) $9/2^6$ (E) $\sqrt{3} - \frac{1}{\sqrt{3}}$

5. The area of the region R shaded below can be represented by which of the following? (R is the region outside the circle $r = 3\cos\theta$ but inside $r = 2 - \cos\theta$ and above the x-axis.)

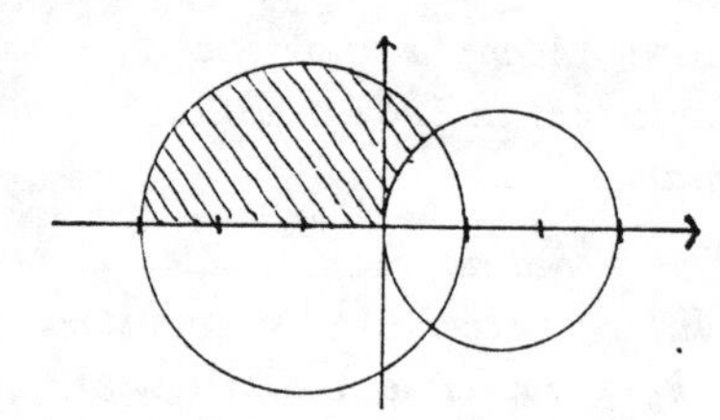

   (A) $\displaystyle\int_{\pi/3}^{\pi/2} \frac{(2-\cos\theta)^2}{2}\, d\theta + \int_{\pi/2}^{\pi} \frac{(2-\cos\theta)^2}{2}\, d\theta$

   (B) $\displaystyle\int_{\pi/3}^{\pi} \frac{1}{2}(2-\cos\theta)\, d\theta - \int_{\pi/3}^{\pi/2} \frac{1}{2}(3\cos\theta)\, d\theta$

   (C) $\displaystyle\int_{\pi/3}^{\pi} \frac{1}{2}(2-\cos\theta)^2\, d\theta - \int_{0}^{\pi/3} \frac{1}{2}(3\cos\theta)^2\, d\theta$

   (D) $\displaystyle\int_{\pi/3}^{\pi} \frac{1}{2}(2-\cos\theta)^2\, d\theta - \int_{\pi/3}^{\pi/2} \frac{1}{2}(3\cos\theta)^2\, d\theta$

   (E) $\displaystyle\int_{\pi/3}^{\pi} \frac{1}{2}(2-\cos\theta)^2\, d\theta - \int_{\pi/3}^{\pi} \frac{1}{2}(3\cos\theta)^2\, d\theta$

6. $\int_0^{.6} \frac{\sin x}{x}\, dx$ is closest to:

   (A) .302 (B) .282 (C) .588
   (D) .602 (E) .988

7. The first few terms of the Taylor series for $f(x) = x/(1+x)$ about $x = 0$ are:
(A) $x + x^2 + x^3 + x^4 + x^5 \cdots$
(B) $x - x^2 + x^3 - x^4 + x^5 \cdots$
(C) $x - x^2 + 2x^3 - 3x^4 + 4x^5 \cdots$
(D) $x - x^2 + 2!x^3 - 3!x^4 + 4!x^5 \cdots$
(E) $x - 2!x^2 + 3!x^3 - 4!x^4 - 5!x^5 \cdots$

8. Let x be a positive but small quantity. Use the first few terms of the Taylor series for the functions x, $\sin x$, and $e^x - 1$ to determine which of the following inequalities is correct:
(A) $x < \sin x < e^x - 1$
(B) $\sin x < x < e^x - 1$
(C) $e^x - 1 < \sin x < x$
(D) $e^x - 1 < x < \sin x$
(E) $\sin x < e^x - 1 < x$

9. A medication has a half life of 6 hours. A person takes a first dose of the medication now, the second dose in six hours, and continues to take the medicine every six hours. If the dose is 10 grams, the amount of medication in the blood immediately after the n-th dose will be:
(A) $20(1 - (\frac{1}{2})^n)$ grams
(B) $20(10(\frac{1}{2})^{n-1})$ grams
(C) $20(1 - (\frac{1}{2})^{n+1})$ grams
(D) 20 grams
(E) $20(\frac{1}{2})^n$ grams

10. If $\frac{dy}{dx} = \cos x/e^y$ and $y(0) = 0$, then $y\left(\frac{\pi}{2}\right) =$
(A) 0 (B) $\ln 2$ (C) 1 (D) 1/2 (E) -1

11. A boat is driven forward by its engine, which exerts a constant force F. It encounters water resistance which exerts a force proportional to its velocity. Let $v = f(t)$ be the velocity of the boat at time t, m be the boat's mass, and k be a positive number. Which of the following equations describes the situation?
(A) $dv/dt = \frac{1}{m}(F - kv)$ (B) $dv/dt = \frac{1}{m}kv$
(C) $dv/dt = \frac{1}{m}(F + kv)$ (D) $dv/dt = \frac{1}{m}Fkv$
(E) $dv/dt = \frac{1}{m}(Ft - kv)$

12. Find the solution to $y'' + y' - 6y = 0$ which satisfies the following conditions: i) as $t \to \infty$, $y \to 0$; ii) $y'(0) = 6$.
(A) $y = e^{-t}(2\sin 3x + 3\cos 2x)$
(B) $y = 6e^{2t} + 2e^{-3t}$
(C) $y = -2e^{-3t}$
(D) $y = -3e^{-2t}$
(E) $y = e^{-t}(3\sin 2x + 2\cos 3x)$

13. Which of the following is the graph of the solution to $\frac{d^2y}{dt^2} - \frac{dy}{dt} + y = 0$ and with the initial condition $y(0) = 0$, $y'(0) > 0$?

(A) (B)

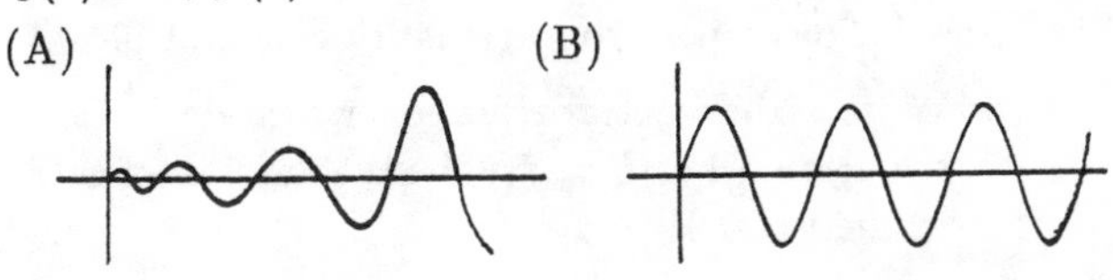

(C) (D)

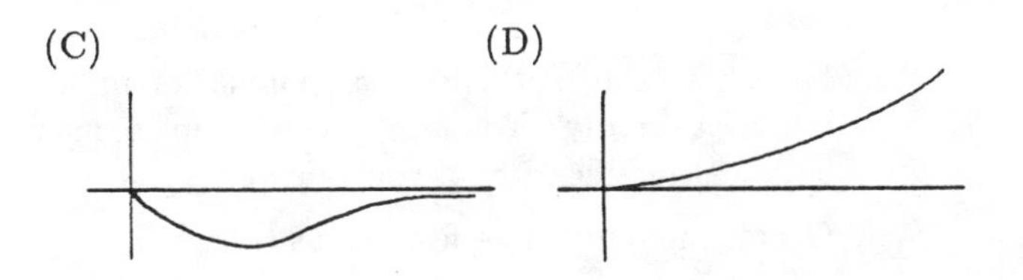

(E)

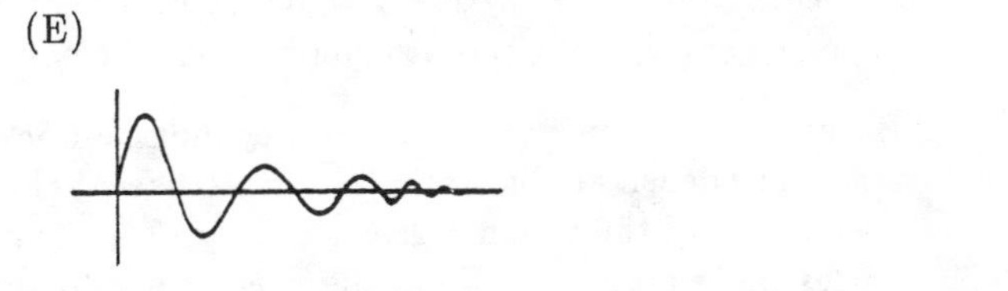

14. $e^{(\pi/2)i} =$
(A) $\sqrt{e^\pi}$ (B) i (C) 1
(D) $1 + i$ (E) $\cos\left(\frac{\pi}{2}i\right)$

15. To approximate the value of $\ln 5$, use the fact that $\ln 5 = \int_1^5 \frac{1}{t}\,dt$ and estimate the integral by dividing the interval of integration into four equal pieces and using the trapezoidal rule. Which of the following sums will you get?
(A) $\frac{1}{2}[\ln 2 + \ln 3 + \ln 4 + \ln 5]$
(B) $\frac{1}{2}\left[1 + 2\left(\frac{1}{2}\right) + 2\left(\frac{1}{3}\right) + 2\left(\frac{1}{4}\right) + \frac{1}{5}\right]$
(C) $\frac{1}{2}\left[1 + \frac{1}{2} + \frac{1}{3} + \frac{1}{4} + \frac{1}{5}\right]$
(D) $\frac{5}{2}\left[1 + 2\left(\frac{1}{2}\right) + 2\left(\frac{1}{3}\right) + 2\left(\frac{1}{4}\right) + \frac{1}{5}\right]$
(E) $\left[1 + \frac{1}{2} + \frac{1}{3} + \frac{1}{4}\right]$

16. a. Find the second degree Taylor polynomial (i.e., up to and including the quadratic term) for the function $f(x) = \sqrt{x}$ about the point $x = 16$.
b. What approximation does part (a) give for $\sqrt{15}$? (You may leave your answers a sum.)
c. Estimate the error in the above approximation. You may use the fact that $3.8 < \sqrt{15} < 3.9$. You must justify all steps. You may leave powers of 3.8 or 3.9 unsimplified in your answers.

17. A part manager plans to put 100 goldfish in a man-made pond in her park. The birthrate (goldfish per week) of goldfish is proportional to the population of goldfish, with proportionality constant 10. Given the size and the ecological conditions of the pond, the manager calculates that the death rate (goldfish per week) is proportional to the interactions between the fish (i.e., is proportional to the square of the goldfish population) with proportionality constant .04.

    a. Write a differential equation which describes this situation. (Let $P = f(t)$ be the number of goldfish in the pond.)

    b. Will the number of fish reach an equilibrium? If so, what is it?

    c. For what value of P is the fish population increasing most rapidly? You must give a complete mathematical justification of your answer.

    d. Graph the number of fish versus time.

    e. Describe in words what would happen if the park manager originally put 350 goldfish into the pond.

18. Barnacles grow on the outside of a cylindrical dock-post. The density of barnacles at a depth d feet below the surface of the water is given by $25 - (3/4)d$ barnacles per square foot, from the ocean floor to the surface of the water. It has a radius of 1/3 ft.

    a. What Riemann sum approximates the total number of barnacles on the dockpost? Please justify your answer carefully.

    b. Use your answer to part (a) to find an integral which represents the total number of barnacles on the dockpost.

    c. Find the total number of barnacles on the dock-post.

19. A contagious fatal disease is spreading through a growing population. Let I denote the number of infected individuals and let S denote the number of susceptible individuals. Assume that in the absence of the infection, the growth of S would be exponential: S grows at a rate proportional to itself with proportionality constant 2. Assume that if everyone were infected, the population would die out exponentially according to the law $I = I_0 e^{-3t}$ and hence the death rate of the infected population is proportional to I. Now suppose that the spread of the infection is directly proportional to the product of S and I with factor of proportionality .5.

    a. Write a system of differential equations for dS/dt and dI/dt describing the above situation.

    b. Is there any pair of positive values of S and I at which both S and I are constant for all time? If so, what are these values of S and I?

    c. Sketch various trajectories of the system of equations in the first quadrant of the $S - I$ plane. (Please draw arrows indicating direction.)

    d. What is the average value of S? You need *not* justify your answer.

INSTITUTION: *A private Midwest university with 7,500 undergraduate and 2,000 graduate students that in 1987 awarded 33 bachelor's degrees, 6 master's degrees, and 6 Ph.D. degrees in mathematics.*

EXAM: *A two-hour exam (calculators not allowed) given to all students in a second term calculus course for students enrolled in the colleges of Business and Arts and Letters. Of the 263 students who enrolled in the course, 87% passed the course, including 66% who received grades of A or B.*

1. The limit of sequence $1/2, 2/3, 3/4, 4/5, \ldots$ is

   (A) 5/8 (B) $1/e$ (C) 1
   (D) $\sqrt{2}$ (E) diverges

2. In the partial fraction decomposition of $(3x^2 - 10x + 12)/(x-1)(x^2-4)$, the numerator of the term with denominator $x - 2$ is

   (A) -2 (B) -1 (C) 0 (D) 1 (E) 2

3. The sum of the infinite series $\sum_{n=1}^{\infty}(5^{-n} - 5^{-(n+1)})$ is

   (A) 4/5 (B) ∞ (C) 0
   (D) 4/25 (E) 1/5

4. The graph of $y = 3\sin(\pi/2 - x)$ most closely resembles

   (A)

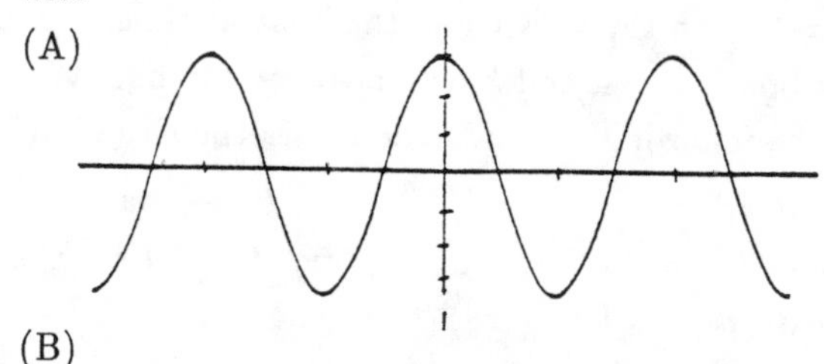

   (B)

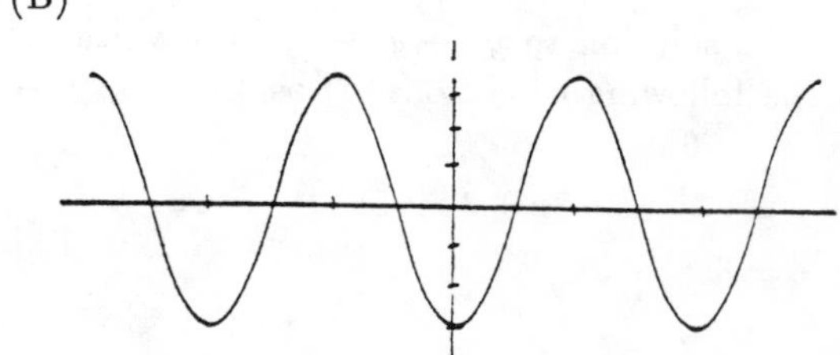

(C)

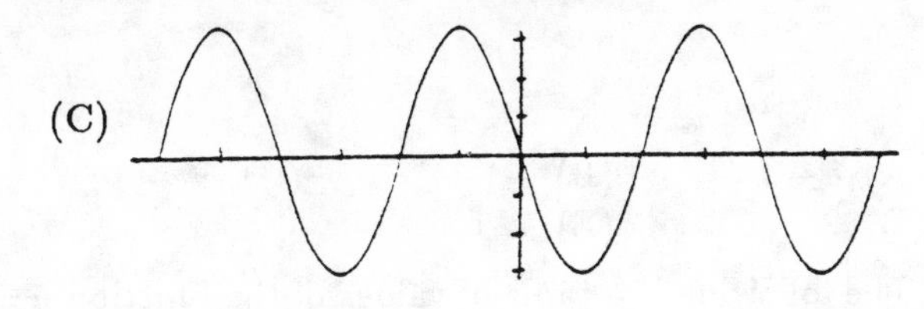

(D)

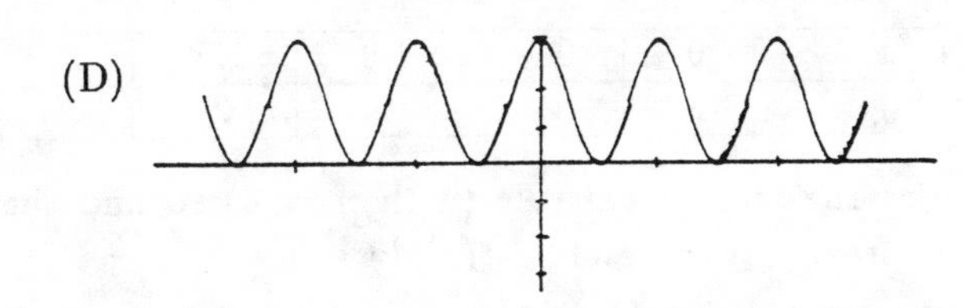

(E)

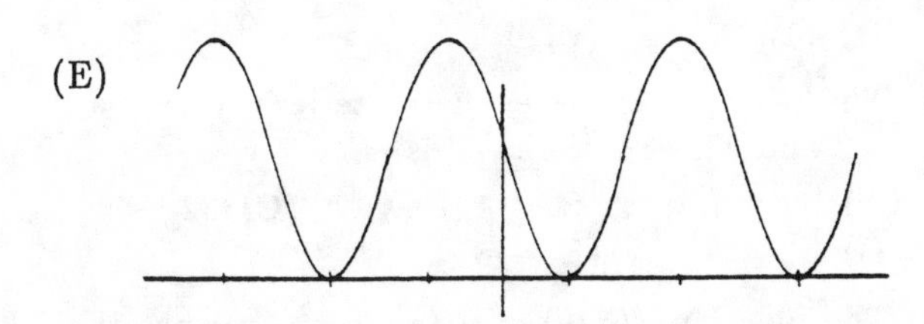

5. The coefficient of x^{10} in the Taylor series for $\sin(2x^2)$ is

(A) 1/20 (B) 4/15 (C) 0
(D) −1/6 (E) −1/10

6. Let z be defined by the implicit equation $\ln(xy+yz+zx) = 1$. Then $\frac{\partial z}{\partial x}$ is

(A) $\dfrac{-(y+z)}{x+y}$ (B) $\dfrac{x+z}{x+y}$
(C) $-\dfrac{x+z}{y+z}$ (D) $-\dfrac{x+y}{x+z}$
(E) $-\dfrac{x+y}{y+z}$

7. If $f(x,y) = x\sin(x+y)$, then $\frac{\partial^2 f}{\partial x \partial y} =$

(A) $\sin(x+y) + x\cos(x+y)$

(B) $x\cos(x+y)$

(C) $\cos(x+y) - x\sin(x+y)$

(D) $2\cos(x+y) - x\sin(x+y)$

(E) $-x\sin(x+y)$

8. Let $f(x) = \sqrt{x}$. The 3rd degree Taylor polynomial about 1 for $f(x)$ is

(A) $1 - \sqrt{x} + \frac{1}{2}(\sqrt{x})^2 - \frac{1}{3}(\sqrt{x})^3$

(B) $1 + (x-1) - \frac{1}{2}(x-1)^2$

(C) $1 + \frac{1}{2}(x-1) - \frac{1}{8}(x-1)^2 + \frac{1}{16}(x-1)^3$

(D) $1 + x + \dfrac{x^2}{2} - \dfrac{x^3}{4}$

(E) $1 + (x-1) + \frac{1}{4}(x-1)^2 + \frac{3}{4}(x-1)^3$

9. The sum of the infinte series $(4/5)^4 - (4/5)^8 + (4/5)^{12} - (4/5)^{16} + \cdots$ is

(A) 256/881 (B) 369/125
(C) 5/4 (D) 256/369
(E) series diverges

10. The improper integral $\int_1^\infty 1/x^2\,dx$ equals

(A) diverges (B) 1 (C) −1
(D) 0 (E) e

11. The function $f(x,y) = y^3 - x^3 - 3y + 12x + 5$ possesses the critical points $(2,-1)$, $(2,1)$, $(-2,-1)$, and $(-2,1)$. The point $(2,-1)$ is

(A) a saddle point.

(B) a local maximum.

(C) a local minimum.

(D) an absolute maximum

(E) an absolute minimum.

12. If $y = \ln|\cos 2x|$ then $\frac{dy}{dx} =$

(A) $-2\tan 2x$ (B) $2\sec 2x$
(C) $\sin 2x$ (D) $-\csc 2x$
(E) $1/(|\cos 2x|)$

13. $d(x\ln(x+y)) =$

(A) $(\ln(x+y) + x/(x+y))\,dx + x/(x+y)\,dy$

(B) $(\ln(x+y) + x/(x+y))\,dy + x/(x+y)\,dx$

(C) $\ln(x+y)\,dx + x/(x+y)\,dy$

(D) $\ln(x+y)\,dy + x/(x+y)\,dx$

(E) $x/(x+y)\,dx + 1/(x+y)\,dy$

14. The volume generated by revolving about the x-axis the area bounded by the graphs of $y = x^2$, $x = 1$, and $x = 2$ is

(A) $32\pi/5$ (B) $31\pi/5$ (C) $8\pi/3$
(D) $7\pi/3$ (E) 4π

15. How many of the following infinite series converge?
i) $1 + 1/7 + 1/7^2 + \cdots + 1/7^n + \cdots$
ii) $3/2 - (3/2)^2 + (3/2)^3 - (3/2)^4 + \cdots$
iii) $1 - 1 + 1 - 1 + 1 - 1 + \cdots$
iv) $1 + 1/2 + 1/3 + 1/4 + \cdots + 1/n + \cdots$
v) $1 + 1/2! + 1/3! + 1/4! + \cdots + 1/n! + \cdots$

(A) 0 (B) 1 (C) 2 (D) 3 (E) 4

16. $\int xe^{-x}\,dx =$

(A) $xe^{-x} + e^{-x}$ (B) $-xe^{-x} + e^{-x}$
(C) $-xe^{-x} - e^{-x}$ (D) $(x^2/2)e^{-x}$
(E) $(x^2/2) - e^{-x}$

17. The Taylor series for $\int_0^x 1/(1+t^2)\,dt$ is
(A) $1+x+x^2/2+x^3/3+\cdots$
(B) $x^2/2!-x^4/4!+x^6/6!-x^8/8!+\cdots$
(C) $1-x^2/2+x^4/4-x^6/6+\cdots$
(D) $x-x^3/3+x^5/5-x^7/7+\cdots$
(E) $1-2x+3x^2-4x^3+\cdots$

18. The volume obtained by revolving about the y-axis the region bounded by the graphs of $f(x)=x^2$ and $f(x)=x^3$ is
(A) $\pi/6$ (B) $\pi/12$ (C) $\pi/10$
(D) $\pi/20$ (E) $\pi/2$

19. $\int_0^{\pi/2}(\cos x)/(1+\sin x)\,dx$
(A) $\ln 1$ (B) $\ln 2$ (C) 1
(D) 2 (E) $\pi/2$

20. $\lim_{x\to 0}(\csc 2x)/(\cot x) =$
(A) Does not exist (B) 2
(C) 1 (D) 1/2
(E) -1

21. $\int 1/(x\ln x)\,dx =$
(A) $\ln\ln x$ (B) $\ln(x\ln x)$
(C) $\ln x + x$ (D) $-1/(x^2\ln x)$
(E) $\ln x + 1/(\ln x)$

22. The surface $x^2+y^2+z^2-2x+4y-6z+10=0$ is a sphere with center
(A) $(-2,4,-6)$ (B) $(2,-4,6)$ (C) $-1,2,-3)$
(D) $(1,-2,3)$ (E) $(4,8,-12)$

23. The solution of the differential equation $\frac{dy}{dx}=y\cos x$ given that $y=8$ when $x=0$ is
(A) $y=8+\sin x$ (B) $y=4\sqrt{\sin x}$
(C) $y=8e^{\sin x}$ (D) $y=8+e^{\cos x}$
(E) $y=8-\cos x$

24. The area bounded by the graphs of $f(x)=x^2$ and $g(x)=x^3$ is
(A) 1/12 (B) 1/6 (C) 1/4
(D) 1/2 (E) 1

25. Let $N(t)$ denote the size of a population of animals at time t. Suppose the growth rate of the population is directly proportional to $100-N$ and the constant of proportionality is 7. Furthermore, $N(0)=10$. Then $N(t)$ for any time t is given by
(A) $N(t)=10e^{-7t}$
(B) $N(t)=100-90e^{-7t}$
(C) $N(t)=100-7t$
(D) $N(t)=100/(100-90e^{-7t})$
(E) $N(t)=10+90e^{7t}$

26. $\int_0^\pi e^{\ln(\sin x)}\,dx =$
(A) $e-1$ (B) -1 (C) $e^\pi-1$
(D) 2 (E) $1/\pi$

27. The following is a table of values of the function $y=f(x)$

x	0	0.5	1.0	1.5	2
$f(x)$	3	2	1	2	3

Use the trapezoidal rule (with $n=4$) to find the approximate value of $\int_0^2 f(x)\,dx$.
(A) 1/8 (B) 11/4 (C) 2
(D) 11/2 (E) 4

28. The fifth term of the geometric series whose first term is 2 and whose ratio is -2 is
(A) -6 (B) $-1/16$ (C) 32
(D) 0 (E) -8

29. The function $f(x,y)=y/(x^2+y^2+9)$ has the following critical points.
(A) $(0,0)$ only (B) $(0,3)$ only
(C) $(0,3)$ and $(0,-3)$ (D) $(0,-3)$ only
(E) $(0,0)$, $(0,3)$, and $(0,-3)$

30. If $\pi<\alpha<3\pi/2$ and $\tan\alpha=1$ then $\sec\alpha=$
(A) -2 (B) $-\sqrt{2}$
(C) $-2/\sqrt{3}$ (D) -1
(E) does not exist

INSTITUTION: *A public Canadian university with 14,000 students that in 1987 awarded 52 bachelor's degrees, 5 master's degrees, and 2 Ph.D. degrees in mathematics.*

EXAM: *A three-hour exam (calculators allowed) for all students in a year-long course for mathematics, physics and computer science students. Of the 190 students enrolled in the course, 33% passed, including 13% who received grades of A or B.*

ANSWER QUESTIONS 1-4:

1. Find the following limits, if they exist. Explain your reasoning.

a. $\lim_{x\to 1}\dfrac{\sin\left(\frac{\pi}{2}(x-1)\right)}{x-1}$

b. $\lim_{t\to 0+}(1+5/t)^t$

c. $\lim_{z\to\infty}(\sin z+z)/(\cos z-3z)$

d. $\lim_{h\to 0}\frac{1}{h}((2+h)\ln(2+h)-2\ln 2)$

2. Determine convergence or divergence of the following series. Justify your answers.

   a. $1 + \frac{1}{\sqrt{2}} + \frac{1}{\sqrt{3}} + \frac{1}{\sqrt{4}} + \cdots$

   b. $\frac{\ln 3}{3} - \frac{\ln 4}{4} + \frac{\ln 5}{5} - \frac{\ln 6}{6} + \cdots$

   c. $\sum_{k=0}^{\infty} \frac{2^k}{3^k + (-2)^k}$

3. Evaluate these integrals and explain clearly what the improper integral in part (c) means.

   a. $\int_0^1 \frac{x}{x^2 + x - 6}\, dx$

   b. $\int x^2 \arcsin(3x)\, dx$

   c. $\int_0^\infty \frac{2}{3 + t^2}\, dt$

4. a. Find the solution to the differential equation $(\sec x)y' + y = 4$ satisfying the condition $y = 1$ when $x = 0$.

   b. Find the general solution of the differential equation $y'' - 4y' + 5y = x$.

ATTEMPT ANY THREE QUESTIONS:

5. Consider the function $f(x) = (e^{-x})/(x+1)$ for $x \in R$, $x \neq -1$.

   a. Find the local extreme values of f and the intervals where f is increasing and decreasing.

   b. Find the points of inflection of f and the intervals where its graph is concave up and concave down.

   c. Find all asymptotes.

   d. Using this information (and any other information you need), sketch the graph of $y = f(x)$.

6. a. Let $0 < K < 1$. Sketch the region bounded by the curves $y = x^2$, $y = 1/x$, and $y = K$ (only the part where $y \geq K$ should be considered). Find the area of this region as a function of K.

   b. Compute the volume of revolution obtained by revolving the region defined in part (a) about the x-axis.

7. a. Define what is meant by the radius of convergence of a power series. If the radius of convergence is R, what does this tell you about the convergence of the power series? Give examples of power series with $R = 0$, 1, and ∞ (no proofs required).

   b. Write down the power series for e^x and find its radius of convergence. Use this to find a power series for e^{-t^2} and give its radius of convergence.

   c. Using *(b.)*, and explaining the theorems which you use, find a power series for $F(x) = \int_0^x e{-}t^2\, dt$.

   d. Show that $0.09666 \leq F(0.1) \leq 0.09668$.

8. a. State and prove the Intermediate Value Theorem.

   b. Prove that the polynomial $p(x) = x^3 - 3x + 1$ has at least 2 roots in the interval $[0, 2]$.

9. a. Show that the curve given parametrically by $x(t) = e^{-t}\cos t$ and $y(t) = e^{-t}\sin t$ for $0 \leq t < \infty$ has finite length, and find this length.

   b. An object of mass m falling near the surface of the earth is retarded by air resistance proportional to its velocity so that, according to Newton's second law of motion, $m\frac{dv}{dt} = mg - kv$ where $v = v(t)$ is the velocity at time t, and g is the acceleration due to gravity near the earth's surface. Assuming the object falls from rest at time $t = 0$, find the velocity $v(t)$ for any time $t > 0$. Show $v(t)$ approaches a limit as $t \to \infty$. Find an expression for the distance fallen after time t.

INSTITUTION: *A publicly-funded Canadian university with approximately 30,000 students that in 1987 awarded 125 bachelor's degrees, 15 master's degrees, and 6 Ph.D. degrees in mathematics.*

EXAM: *A three-hour exam (calculators allowed) given to all students completing a year-long calculus course for specialist students in mathematics. Of the 100 students who enrolled in the course, 39% passed, including 14% who received grades of A or B.*

ANSWER BOTH QUESTIONS 1 AND 2.

1. State and prove one of the following theorems.

   a. The Fundamental Theorem of Calculus (both parts).

   b. The Intermediate Value Theorem for Continuous Functions.

   c. Taylor's Theorem with Remainder for a function f which is $n + 1$-times differentiable on a neighbourhood of a point a. (Give whatever form of the Remainder you wish.)

2. The graph of a function f defined on $\mathbb{R}$ is given below.

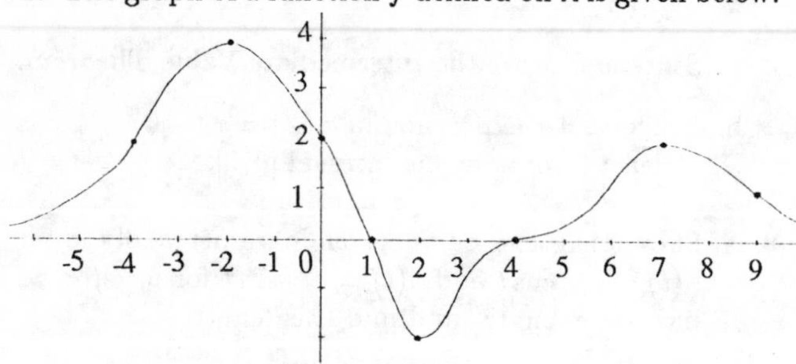

Sketch the graphs of the following functions.

a. $\int_0^x f(t)\,dt$ b. $f'(x)$
c. $1/[f(x)]^2$ d. $f(2-x)$
e. $1+f(x^2)$ f. $f(f(x))$

ANSWER ANY FOUR OF QUESTIONS 3–8.

3. a. Let $f(x) = \begin{cases} 1 & \text{if } x = 0 \\ (\sin x)/x & \text{otherwise.} \end{cases}$
State whether $f'(0)$ and $f''(0)$ exist, and if so, calculate these derivatives.

b. Verify that

$$\int_0^{2x} \frac{\sin t}{t}\,dt = \int_0^x \frac{\sin 2t}{t}\,dt = \frac{\sin^2 x}{x} + \int_0^x \frac{\sin^2 t}{t^2}\,dt$$

and deduce that the improper integral

$$\int_0^\infty \frac{\sin t}{t}\,dt$$

converges.

4. Let $I_n = \int_0^{\pi/4} \tan^n x\,dx$. Show that
a. $I_n = 1/(n+1) - I_{n+2}$ $(n = 0, 1, 2, \ldots)$
b. $\tan x \le (4x)/\pi$ $(0 \le x \le \pi/4)$
c. $\lim_{n\to\infty} I_n = 0$
d. $1 - 1/3 + 1/5 - 1/7 + 1/9 - \cdots = \pi/4$
e. $1/2 - 1/4 + 1/6 - 1/8 + - \cdots = (1/2)\log 2$.

5. Consider the following property (P) of a set S of real numbers:

$$(P) \quad x \in X \Rightarrow \frac{2}{1+x^2} \in S.$$

a. Prove that, if S is bounded and satisfies (P), then $\sup S \ge 1$.
b. Determine, with justification, all sets S that satisfy (P) and for which $\sup S = 1$.
c. Give an example, with justification, of a set S for which i) (P) holds, ii) $\sup S = 2$, iii) $2 \notin S$.

6. a. Find the solution to the equation $(1+x^2)y' + (1-x)^2 y = xe^{-x}$ which satisfies $y = 0$ when $x = 0$.
b. Sketch the graph of the solution obtained in part(a).

7. Consider $I = \int_0^1 \sqrt{x}\sin(x^2)\,dx$.
a. By using a series expansion of $\sin x^2$, evaluate I to 4 decimal places.
b. By dividing the interval $[0,1]$ into four parts, approximate the value of I by using either the trapezoidal rule or Simpson's rule.

8. a. Determine $\lim_{x\to\infty} x \log x$.
b. Determine the volume of the solid of revolution obtained by rotating the region bounded by the curves $y = 0$ and $y = x\log x$ $(0 \le x \le 1)$ about the x-axis.

Final Examinations for Calculus III

The following seven examinations represent a cross section of final examinations given in 1986–87 for students enrolled in the third calculus course. The examinations come from large universities and small colleges, from Ph. D. granting institutions and two year colleges.

Brief descriptive information preceding each examination gives a profile of the institution and circumstances of the examination. Unless otherwise noted, no books or notes are allowed during these examinations; calculator usage is explicitly noted where the information was provided. Except for information on problem weights and grading protocol, the texts of the examinations are reproduced here exactly as they were presented to the students.

INSTITUTION: *A western state university with 16,000 students that in 1987 awarded 36 bachelor's degrees, 9 master's degrees, and 3 Ph.D. degrees in mathematics.*

EXAM: *A two-hour exam (calculators allowed) given to 26 students in one section of a third term calculus course for science, mathematics, and computer science students. Of the 349 students who enrolled in the course, 74% passed the course, including 32% who received grades of A or B.*

1. Find the length of the arc of $x = t^3$, $y = t^2$ from $t = -1$ to $t = 0$.
2. Find the area inside the graph of $r = \cos 2\theta$.
3. Find the limits of these sequences, if they exist:
   a. $a_n = (\sin n)/n$
   b. $a_n = \sin\left(\frac{\pi}{2} + \frac{1}{n}\right)$
   c. $a_n = \sqrt[n]{n^2 + n}$
   d. $a_n = (1 - 2/n)^{-n}$
4. Test and decide if the following series are absolutely convergent, conditionally convergent, or divergent. State the test used.
   a. $\sum_{n=2}^{\infty} (-1)^n \frac{\ln n}{\ln n^2}$
   b. $\sum_{n=1}^{\infty} (-1)^{n+1} \frac{(n!)^2}{(2n)!}$
   c. $\sum_{n=2}^{\infty} (-1)^n \frac{1}{\ln n^3}$
5. Evaluate $\int_0^4 \frac{1}{\sqrt{4-x}}\,dx$ if it converges.
6. Estimate $\int_0^{.1} \frac{\ln(1+x)}{x}\,dx$ to within .001.
7. Find the sum $\sum_{k=0}^{\infty} (-1)^{2k} \frac{\pi^{2k}}{(2k)!}$.
8. Find the Taylor series at $x = 0$ for $f(x) = \frac{e^x + e^{-x}}{2}$ (in powers of x).
9. Find the limit $\lim_{x \to 0} \frac{\sqrt{1+x} - 1 - (x/2)}{x^2}$.
10. Find the third Taylor polynomial $P_3(x)$ for $f(x) = \sqrt{x-1}$ in powers of $(x - 5)$.
11. Find the interval of convergence of $\sum_{k=1}^{\infty} \frac{(2x-5)^k}{k^2}$.

INSTITUTION: *A southwestern public university with 12,000 students that in 1987 awarded 24 bachelor's and 5 master's degrees in mathematics.*

EXAM: *A three-hour exam (calculators not allowed) given to 21 students in one section of third term calculus for engineering, mathematics, and science majors. Of the 138 students who enrolled in the course, 50% passed, including 38% who received grades of A or B.*

1. Given $\vec{x}(t) = (e^t, 2e^t, 3e^t)$. Determine the arclength of $\vec{x}(t)$ for $t \in [0, 1]$.
2. Let $\vec{x}(t) = (x(t), y(t)) = (3\cos t, 4\sin t)$; $t \in (0, \pi)$. Find d^2y/dx^2.
3. Determine the maximum absolute value of the curvature of $y = f(x) = \ln(x)$; $x > 0$.
4. Determine the unit tangent vector and the curvature K of $\vec{x}(t) = (t\cos t, t\sin t)$ and show that $K > 0$ for $t \in \mathbb{R}$.
5. Let $\vec{x}(t) = (\cos t, \sin t, \sin 2t)$. Show that the curvature of $\vec{x}(t)$ vanishes at $t = 1/2 \arccos\left(\pm\sqrt{17/12}\right)$.
6. Find the unit normal vector of $\vec{x}(t) = (3t, \sin t, \cos t)$.

7. Using the chain rule, find $\frac{dW}{dt}$ if $W=\exp(-x)y^2 \sin z$; $x = t$, $y = 2t$, and $z = 4t$.

8. Find the equation to the tangent to $y + \sin(xy) = 1$ at $(0, 1)$ by evaluating the gradient of the respective level curve $f(x, y) = 0$.

9. Compute the directional derivative of $f(x, y, z) = x^2y^2z^2$ at $(1, -1, 2)$ in the direction of $(1, 0, 1)$.

10. Find all unit vectors $\vec{u}$ for which the directional derivative of $f(x, y, z) = x^2 - xy - yz$ at $(-1, -1, 1)$ in the direction of $\vec{u}$ vanishes.

INSTITUTION: *A private midwestern liberal arts college of 2200 students that in 1987 awarded 17 bachelor's degrees in mathematics.*

EXAM: *A three-hour final exam (calculators allowed) from the only section of the third term of a calculus course for prospective mathematics and science majors. Of the 20 students who enrolled in the course, 85% passed, including 33% who received grades of A or B.*

1. Find the cosine of the angle between the two line segments which start at $(0, 0, 0)$ and end at $(2, 3, -4)$ and $(5, -2, 4)$, respectively.

2. Let $z = x^2 - 2y^2$, $x = 3s + 2t$, and $y = 3s - 2t$. Find z_s in two ways.

3. Let $z = (2x^2 - 3x^2)/(xy^3)$. Calculate $xz_x - yz_y$.

4. Let $z = x^2 + xy + y^2$. Find the directional derivative of z at $(3, 1, 13)$ in the direction $\vec{\imath} + \sqrt{3}\vec{\imath}$.

5. Solve $y'' - y' - 2y = e^{2x}$ given $y(0) = y'(0) = 1$.

6. Suppose a particle has position $\vec{r}(t) = t^2\vec{\imath} + (t + t^3/3)\vec{\jmath} + (t - t^3/3)\vec{k}$ at time t. Find the particle's velocity and acceleration, and decompose its acceleration into tangential and normal components.

7. Solve $xy' = y + \sqrt{x^2 + y^2}$.

8. Write two triple integrals—in rectangular and cylindrical coordinates—for the volume bounded below by $z = x^2 + y^2$ and above by $z = 2y$, and evaluate one of them.

9. Find the length of the curve $x = 3t^2$, $y = 3t - t^3$ for $0 \le t \le 2$.

10. Invert the order of integration in

$$\int_1^3 \int_0^{x-1} f(x, y)\, dy\, dx.$$

11. If $(x, y) \neq (0, 0)$, $f(x, y) = (4xy)/(4x^2 + y^2)$. As (x, y) approaches $(0, 0)$ along a line through the origin, $f(x, y)$ approaches 1. What is the slope of the line?

12. Find the y-coordinate of the centroid of the first-quadrant region bounded by $y = x^3$ and $y = \sqrt{x}$.

13. Sketch the first-octant portion of the graph of $x^2 + 2z^2 = 3y^2$ and find the equations of the tangent plane and normal line at $(2, -2, -2)$.

14. Find the minimum distance from $(1, -1, 4)$ to the surface $x^2 + y^2 + z^2 + 2x - 4y + 4z = 16$.

15. Find the equation of the plane passing through $(1, 2, 3)$ and $(3, -2, 1)$ which is perpendicular to the plane with equation $3x - 2y + 4z = 5$.

INSTITUTION: *A private university in the south with 6,600 undergraduates that in 1987 awarded 8 bachelor's degrees, 3 master's degrees, and 3 Ph.D. degrees.*

EXAM: *A four-hour final exam (calculators allowed) from the only section of third term calculus for Arts and Sciences students. (A separate course serves engineering students.) Of the 20 students who enrolled in the course, 50% passed, including 20% who received grades of A or B.*

1. a. Find an equation of the line through $(2, -1, 5)$ parallel to the line $l(t) = (3t, 2 + t, 2 - t)$ where $-\infty < t < \infty$.

   b. Find the parametric equation of the line of intersection of the planes $2x + y + z = 4$ and $3x - y + z = 3$.

   c. Find an equation of the plane passing through $(1, 0, -1)$, $(3, 3, 2)$ and $(4, 5, -1)$.

   d. Find the distance between the parallel planes $2x - y + 2z = 4$ and $2x - y + 2z = 13$.

2. Find (a) the unit tangent, (b) the unit normal, (c) the unit binormal, (d) the curvature, (e) the torsion at each point of the curve

$$\gamma(t) = (t, \frac{\sqrt{2}}{2}t^2, \frac{t^3}{3}).$$

Also, if γ gives the path of a moving particle, what are the (f) tangential and (g) normal accelerations of this particle at time t?

3. Write down parameters equations for the two dimensional torus centered at the origin obtained by rotating $(x-2)^2 + z^2 = 1$ about the z-axis.

4. a. What is the partial differential equation obtained from the equation

$$5\frac{\partial^2 u}{\partial x^2} + 2\frac{\partial^2 u}{\partial x \partial y} + 2\frac{\partial^2 u}{\partial y^2} = 0$$

by substituting $x = 2s - t$, $y = s - t$?

b. Suppose that $y = g(x, z)$ satisfies the equation $F(x, y, z) = 0$. Find $\frac{\partial y}{\partial x}$ in terms of the partials of F.

5. a. An open topped rectangular box is to have a total surface area of 300 in^2. Find the dimensions which maximize the volume.

b. Find the minimum distance between the circle $x^2 + y^2 = 1$ and the line $2x + y = 4$.

6. a. Find and classify all critical points of the function $f(x, y) = 2y^2 - x(x-1)^2$.

b. Find the maximum and minimum values of $f(x, y, z) = x - y + 2z$ on the ellipsoid $M = \{(x, y, z) : x^2 + y^2 + 2z^2 = 2\}$.

7. Find the total mass of an object with density $\delta(x, y, z) = z + 3$ bounded by $x^2 + y^2 = 4$ and $z = \sqrt{x^2 + y^2}$ and the xy-plane.

8. Evaluate $\int\int_R \cos(y-x)/(y+x)\, dx\, dy$ where R is the region bounded by the lines $x + y = 2$, $x + y = 4$, $x = 0$, $y = 0$.

9. Write the parametric equation of the surface given by the equation $z = 4 - (x^2 + y^2)$ and then find the area of the part of this surface which lies above the xy-plane.

10. Let $F(x, y) = \int_{x+y}^{xy} \ln(y/t)\, dt$ and compute $\frac{\partial F}{\partial x}$ and $\frac{\partial F}{\partial y}$.

11. Compute the line integral of $F = (-y, x)$ around the cardiod $r = 1 + \sin\theta$ where $0 \le \theta \le 2\pi$ and find the area enclosed by this curve.

12. Compute

a. $\oint_\gamma (-y\, dx + x\, dy)/(x^2 + y^2)$, if γ is a closed curve about the origin.

b. $\oint_\gamma (x\, dx + y\, dy)/(x^2 + y^2)$ if γ is a closed curve about the origin.

13. Let $f(\vec{r}) = q(\vec{r}/||\vec{r}||^3)$ where q is a constant and $\vec{r} = xi + yj + zk$. Let V be a region surrounding the origin and let S be its surface.

a. Compute $\int\int_S f \cdot \vec{n}\, ds$.

b. Compute $\oint_\gamma f \cdot ds$ where γ is a closed curve within the unit sphere.

c. Explain the different effects of the singularity at $(0, 0, 0)$ on the value of the integrals.

14. By first showing that

$$\int\int\int_V ||\nabla f||^2\, dx\, dy\, dz = \int\int_{\partial V} f\frac{\partial f}{\partial \vec{n}}\, d\sigma$$

conclude that the steady state temperatures within a region V are determined by the surface temperatures where the temperatures u satisfy $\nabla^2 u = \frac{\partial^2 u}{\partial x^2} + \frac{\partial^2 u}{\partial y^2} + \frac{\partial^2 u}{\partial z^2} = 0$.

INSTITUTION: *A major midwestern public research university with 27,000 undergraduate and 9000 graduate students. In 1987, this university awarded 184 bachelor's degrees, 32 master's degrees, and 13 doctor's degrees in mathematics.*

EXAM: *A three-hour final exam (calculators allowed) from one section of third term calculus for engineering and science students. Of the 1200 students who originally enrolled in the course, 87% passed, including 55% who received grades of A or B. 150 students were in the section that took this particular exam.*

1. Locate and classify all the critical points on the surface $z = x^3 + y^3 - 3xy$.

2. Let $L_1 : \left\{ \begin{array}{l} x = 2 - t \\ y = -1 + 3t \\ z = 2t \end{array} \right\}$ and $L_2 : \left\{ \begin{array}{l} x = 1 + 2s \\ y = 7 + 4s \\ z = 3 - 2s \end{array} \right\}$.

Find the following:

a. The point where L_2 intersects the xy plane.

b. The point where L_1 and L_2 intersect.

c. The equation of the plane containing L_1 and L_2.

d. The equation of the line through the point of intersection (part b) and perpendicular to the plane (part c).

3. Sketch the region of integration and evaluate

$$\int_0^4 \int_{\sqrt{y}}^2 \frac{ye^{x^2}}{x^3}\,dx\,dy.$$

4. Consider the curve with the parametric equations $x = 3t^2$, $y = 2t^3$.

a. Carefully sketch the graph of the curve.

b. Find the unit tangent vector T and the unit normal vector N at the point $(3, -2)$.

c. Find the equation of the tangent line at the point $(3, -2)$.

d. Find the equation of the normal line at the point $(3, -2)$.

e. Find the curvature at the point $(3, -2)$.

f. Find the equation of the osculating circle at the point $(3, -2)$.

5. Let $w = f(u)$ and suppose $u = x^3 - y^3$. Find an expression for $\frac{\partial w}{\partial x}$ and $\frac{\partial w}{\partial y}$.

6. The circle $(y-1)^2 + z^2 = 1$ (lying in the yz-plane) is revolved around the z-axis.

a. Write the equation of the solid of revolution.

b. Set up but *do not evaluate* a triple integral in cylindrical coordinates to calculate the volume of the solid.

7. Find the equation of the line tangent at $(1, -2, -3)$ to the curve of intersection of the surface $z = 5 - 4x^2 - y^2$ and the plane $x - y + 2z = -3$.

8. Consider the surface $x^2 + y^2 - z^2 = 1$.

a. Sketch the surface.

b. Use Lagrange multipliers to find the point(s) on the surface which are closest to $(4, 0, 0)$.

c. Is there a point on the surface farthest away from $(4, 0, 0)$? Is so, find it; if not, explain why not.

9. Let $w(t) = (t, \frac{2}{3}t^3, t^2)$ be the position vector of a particle moving in three space.

a. Find the distance that is travelled by the particle along the curve as it moves from $(1, \frac{2}{3}, 1)$ to $(3, 18, 9)$.

b. What is the speed of the particle at $(1, \frac{2}{3}, 1)$?

10. Suppose T is the region between the spheres $x^2 + y^2 + z^2 = 1$ and $x^2 + y^2 + z^2 = 5$. Evaluate

$$\iiint_T \frac{e^{-(x^2+y^2+z^2)}}{\sqrt{x^2+y^2+z^2}}\,dV.$$

11. Find the centroid of the region R shown. Assume $\rho(x, y) = 1$.

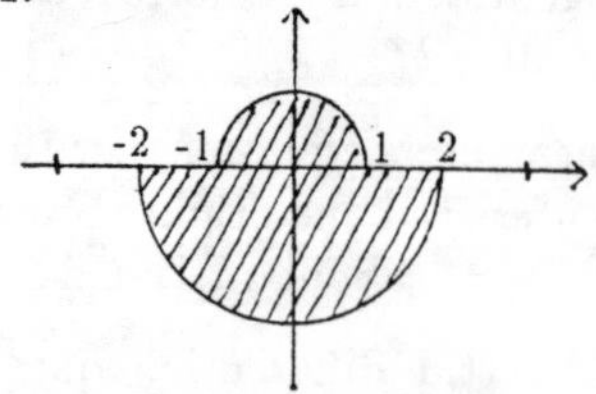

12. Show that the surface area of the region S on the cone $z = a\sqrt{x^2 + y^2}$ which lies about the region R, satisfies the following relationship:

$$\text{Area of } S = \sqrt{a^2 + 1}\ (\text{Area of } R).$$

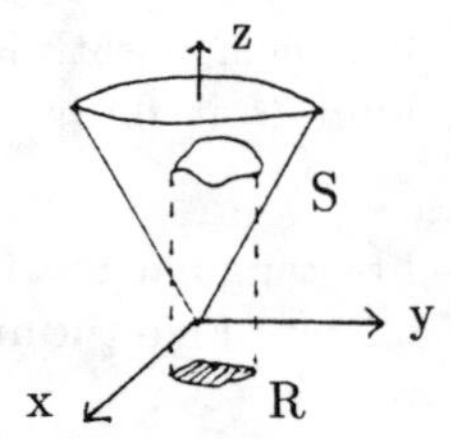

13. Consider the surface $\sqrt{x} + \sqrt{y} + \sqrt{z} = 1$.

a. Find the plane tangent to the surface at (a, b, c) (where $a > 0$, $b > 0$, $c > 0$).

b. Show that the sum of the x-intercept, y-intercept, and z-intercept of this plane is 1.

INSTITUTION: *An eastern liberal arts college for women with 2600 students that in 1987 awarded 25 bachelor's degrees in mathematics.*

EXAM: *A two and one-half hour exam (calculators allowed) from one section of third term calculus for freshmen and sophomore science and mathematics majors. Of the 70 students originally enrolled in the course, approximately 90% passed, including 67% who received grades of A or B. 30 students were in the section that took this particular exam.*

1. Suppose the temperature at the point (x, y, z) is given by the function $T(x, y, z) = y^2e^{2x+3z}$. What is $T_z(0, 2, 1)$? What is the significance of this number in terms of how the temperature changes?

2. Let $f(x, y) = x/(x + y)$ and suppose a particle is at the point $(1, 0)$. What direction should the particle travel in order that the function f decrease as rapidly as possible?

3. What is the equation of the plane tangent to the surface $z = f(x, y)$ at the point $(x_0, y_0, f(x_0, y_0))$?

4. Let $f(x, y) = |1 + x^2 - y|$. Does $f_x(0, 1)$ exist? Does $f_y(0, 1)$ exist? Be sure to give reasons for your answers.

5. Suppose $f(x, y) = e^{y(x^2-1)}$. What does the contour line for $f = 1$ look like?

6. Find all the critical points of $xy^4 + \cos x$ but do *not* classify them.

7. Use the method of Lagrange multipliers to find the point (or points) (x, y) on the circle $x^2+y^2 = 4$ where the value of the function $x^3 + y^3$ is greater than or equal to its value at any other points on the given circle.

8. Evaluate $\int\int_R xy\,dA$ over the region enclosed by the curves $y = \frac{1}{2}x$ and $y = \sqrt{x}$.

9. Below are some values of a differentiable function $f(x, y)$. The numbers are placed directly over the corresponding points so, for example, $f(2.5, 2) = 1.899$, $f(2.5, 2.5) = 1$, $f(4, 1) = 4$, $f(4, 2) = 2.203$, etc. Assuming that nothing unexpected happens between the given values, answer the following questions.

   a. If $\vec{u}$ is a unit vector in the northwest direction, which of the following numbers best approximates $f_{\vec{u}}(3, 2)$?

   (A) $-\sqrt{2}$ (B) -1 (C) 0 (D) 1 (E) $\sqrt{2}$

   b. Which of the following numbers best approximates $f_x(3, 2)$?

   (A) 0 (B) 0.1 (C) 0.2 (D) 0.5 (E) 1

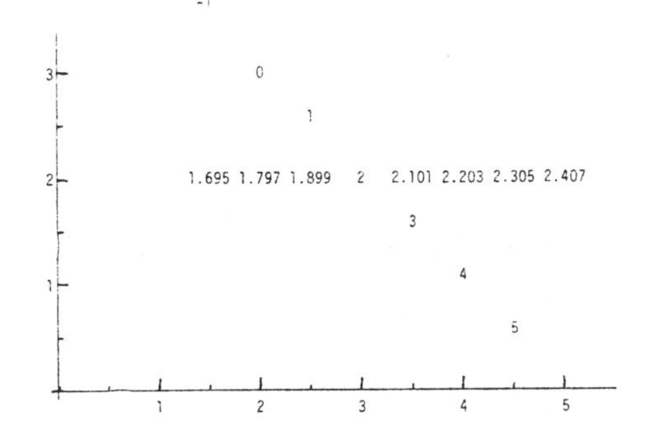

INSTITUTION: *A midwestern public university with 8500 students that in 1987 awarded 30 bachelor's and 5 master's degrees in mathematics.*

EXAM: *A two-hour final exam (calculators allowed) for one section of the third term of a calculus course for engineering, physical science, and mathematics students. Of the 106 students who enrolled in the course, 71% passed, including 25% who received grades of A or B.*

1. Find a rectangular equation of the curve $F(t) = \sqrt{t+4}\,\vec{\imath} + 2t\vec{\jmath}$ and sketch the curve in the xy-plane.

2. Find the length of the curve whose parametric equations are $x = 3t$, $y = 2t^{3/2}$ where $0 \le t \le 4$.

3. Find the equation for the plane through the points $P(1, 2, 3)$ and $Q(2, 4, 2)$ that is parallel to $\vec{a} = \langle -3, -1, -2 \rangle$.

4. For $4x^2 + y^2 - z^2 = 16$,

   a. Find and name the traces in the coordinate planes and in other planes as needed;

   b. Sketch the surface;

   c. Name the surface.

5. Find parametric equations for the tangent line to the curve $x = e^t$, $y = te^t$, $z = t^2 + 4$ at the point $P(1, 0, 4)$.

6. Find the critical points and then the extrema of $f(x, y) = x^2y - 6y^2 - 3x^2$.

7. Let $f(x, y, z) = x^2 + 3yz + 4xy$

   a. Find the directional derivative of $f(x, y, z)$ at the point $P(1, 0, -5)$ in the direction of $\vec{a} = \langle 2, -3, 1 \rangle$.

   b. Find a vector in the direction in which f increases most rapidly at P and find the rate of change of f in that direction.

8. Reverse the order of integration and evaluate the resulting integral: $\int_0^2 \int_{y^2}^4 y \cos(x^2)\,dx\,dy$.

9. Find the volume of the solid in the first octant bounded by the sphere $\rho = 2$, the coordinate planes and the cones $\phi = \pi/6$ and $\phi = \pi/3$. (Draw appropriate pictures.)

10. Use the change of variables $x = 2u$, $y = 3v$ to express the integral $\int\int_R (x^2/4 + y^2/9)\,dx\,dy$, where $R = \{(x, y) \mid (x^2/4 + y^2/9) \le 1\}$, as a double integral over a region S in the uv-plane. *(Do not evaluate.)*

READINGS

PARTICIPANT RESPONSES

Question What can be done to improve calculus instruction?

Answers

- I don't know yet.
- Don't call it "calculus."
- Hire enough instructors so that personal contact is possible.
- Make students write mathematics well.
- Ask students to justify one answer rather than "find" six thousand.
- No more answers in the back of the book
- Get the private sector involved. Educate the public.
- Involve mathematicians in teacher education — a really difficult task!
- How can we guarantee that the first innovative text won't rapidly go out of print?
- Take the show on the road.

A Special Calculus Survey: Preliminary Report

Richard D. Anderson and Donald O. Loftsgaarden

LOUISIANA STATE UNIVERSITY
UNIVERSITY OF MONTANA

A preliminary report from a special survey of calculus, conducted during October 1987 by the MAA-CBMS Survey Committee:

The MAA Survey Committee has, over the past month and with support from NSF (under Grant SRS-8511733), conducted a survey among four-year colleges and universities to get much more detailed statistical information on calculus than that available elsewhere. The (CBMS) Surveys conducted every five years traditionally are concerned with the fall term only and lump all calculus courses in just two categories with no differentiation by level.

The same survey questionnaire was sent to a stratified random sample of about one-sixth of the almost 1500 mathematics departments. The preliminary results reported below are projections to the total indicated four-year college and university populations. A much more detailed report including breakdowns by class of institution is to be submitted shortly to *Focus* and the *Notices* of the AMS.

This survey sought information on *mainstream* (M) calculus, i.e., freshman-sophomore calculus courses or sequences designed as a basis for eventual student access to upper-division mathematics courses, and *non-mainstream* (N-M) calculus, i.e., calculus courses or sequences (such as most business calculus) not intended specifically for student access to upper-division mathematics courses. The calculus courses considered were Calculus I, II, III, and IV with Calculus I the beginning semester or quarter course. Unless otherwise specified, all data here refer to phenomena in the two semesters or three quarters of the academic year AY 1986-87.

At the time of the preparation of this preliminary report, the full data were still being analyzed. It is believed that the percentage figures cited here are quite accurate, but some of the total enrollment figures may be off by a few percent (probably low). The over-all response rate from the sample was a little over 50%, with a higher response rate from the public university category and a somewhat lower response rate from the private college category.

Enrollments and Success Rates

For all of AY 1986-87, there were a total of slightly more than 300,000 enrollments in mainstream Calculus I and just under 260,000 in non-mainstream Calculus I. Thus 54% of all Calculus I enrollments were in mainstream calculus. There were also 16,000 to 17,000 enrollments in each of Calculus I (M) and Calculus I (N-M) in the summer of 1987. Total semester or quarter enrollments in all of Calculus I to Calculus IV in AY 1986-87 were 975,000.

In AY 1986-87, there were a little more than 140,000 students who completed (with a D or better) the final course of the first year of mainstream calculus (Calculus II for semester schools and Calculus III for quarter schools) and there were about 113,000 students who enrolled in the next term of calculus beyond the first year of mainstream calculus. About 85% of all enrollments in non-mainstream calculus were in Calculus I whereas less than 50% of mainstream calculus enrollments were in Calculus I.

Slightly over 20% of all students in Calculus I (M or N-M) were enrolled in lecture-recitation section formats. Almost three-fourths of all enrollees in Calculus I passed the course with a D or better (with 11-12% of all enrollees getting a D). Enrollees in lecture-recitation sections had a 3% better passing rate than enrollees in single-instructor sections.

More than four-fifths of all enrollments in Calculus I were in institutions using the semester calendar.

Section Sizes

The average of sections in calculus taught on the semester basis in AY 1986-87 are shown below:

	Mainstream			Non-Mainstream	
Section	I	II	III	I	II
Single Instr.	29	26	26	38	35
Lecture	126	108	95	153	115

In mainstream calculus the typical lecture section was split into about four recitation sections and in non-mainstream calculus the figure was about five.

Who Teaches Calculus?

The table below shows the percentage distribution by type of faculty for all *single-instructor* sections in AY 1986-87 for institutions on the semester schedule:

Faculty Type:	Mainstream I	II	III	Non-Mainstream I	II
FT Professor	70%	73%	82%	47%	45%
FT Instructor	9%	14%	10%	13%	12%
Part-time	6%	4%	3%	13%	20%
Teach. Assist.	15%	9%	5%	25%	23%

For classes taught in the lecture-recitation section format, almost 100% of the recitation sections were handled by TA's. Almost 100% of the lecture sections in mainstream calculus were taught by professorial faculty, as contrasted to 80% in non-mainstream calculus.

From this data as well as section size data, it is clear that, as expected, departments generally give priority to mainstream calculus.

Characteristics of Calculus Classes

A little more than 60% of all single-instructor mainstream semester calculus sections met four hours per week whereas about 50% of all mainstream quarter calculus sections met five hours per week.

About 3% of all calculus students have some computer use required in homework assignments.

About 55% of all calculus students on semester schedules rarely or never had their homework picked up and graded. About 30% of calculus students on semester schedules were not given short quizzes. If they were, almost all the quizzes were instructor designed. About one-eighth of such calculus students took group or departmental-designed hour exams and about 50% took group or departmental finals.

Two Proposals for Calculus

Leonard Gillman

UNIVERSITY OF TEXAS, AUSTIN

A reprint of "From the President's Desk" in the September 1987 issue of FOCUS, *the Newsletter of the Mathematical Association of America:*

In 1946, in the town of Sullivan, Indiana, a man was accused of murdering his estranged wife in what was an open and shut case. The county put its greenest attorney in charge, and no lawyer seemed willing to undertake the defense. Finally, Norval Kirkhan Harris (later Judge Harris), a well-known local attorney, agreed to take it on. He decided on the line that the whole thing was an *unfortunate accident,* and he played up the phrase unsparingly. ("You say you were at the grocery the morning of this *unfortunate accident.* Where were you the afternoon of this *unfortunate accident?*") At the final summation, the prosecutor got up and began, "I intend to show that this *unfortunate accident* ..." (The entire courtroom from the judge down burst into a guffaw, and the defendant got away with a measly 2-1/2 years for manslaughter.)

I am reminded of the incident every time I encounter the phrase *the crisis in calculus.* There is no crisis in calculus. Students come into the course unprepared—yes. Textbooks are too big—of course. Emphasis should be modified to reflect the world of computers—certainly. Crisis—no.

Any crisis that may exist is in education, or in society. Up to World War II, only a minor portion of college-age youth went on to college, and of those, only a small fraction took calculus. Students typically began with a full semester of analytic geometry or a preceding semester of trigonometry. Calculus was big stuff; at Columbia University, where I studied, the final requirement for the mathematics major was a comprehensive exam in calculus and analytic geometry. In those days, a class of 40 students was considered huge. There was little concern with "motivating" students; when a professor told you something was important, you learned it.

The Problem

Today, we bellow at 100 poorly prepared students at a time and "cover" in two terms what used to use up three. Students are reluctant to ask questions in front of so many people, and no sensible exchange of ideas is possible anyhow. Faced with student evaluations, we play to the gallery, giving easy quizzes and grades that students did not earn. (Why should I split a gut trying to buck the system? If they didn't study, that's their problem. I'll go back to my research or my garden or whatever.)

I am unpersuaded by results purporting to show that learning is independent of class size. To me, the experiments prove once again that the standard tests are insufficient. We test only a small, easily quantifiable part of what we hope the student is learning. We leave out subject matter that cannot be easily adapted to the test, as well as searching questions that require thoughtful responses and equally thoughtful, time-consuming grading. And we perforce omit a host of delicate intangibles, such as the little ways the instructor responded to a question or attacked a problem, which can make a lasting impression on students and shape their attitudes. Certainly I can never forget a discussion in George Adam Pfeiffer's class at Columbia when one of the students—a very bright one, by the way—was wrestling with epsilon and delta. Finally, in desperation, he blurted out, "But suppose I choose epsilon *large?*" "Ah," said Pfeiffer, "but you *don't* choose epsilon. *I* choose epsilon."

American students spend hours watching television. That most of what they look at is without merit is the minor crime. The major one is the fact of passive looking, encouraging them to sit back and let things come to them. "Good" television programs are still television programs: Sesame Street and Square One are still passive entertainment. There is no opportunity for viewers to hold up the show while they sit back and reflect, to mull over ideas and express them in their own words—as they can when reading a book. The constant, mindless blare trains people not to listen. Mathematics requires intense concentration; television encourages nonconcentration. I sometimes wonder how many of my students are capable of concentrating on one idea, uninterrupted, for ten full minutes.

Today there are vast numbers of families where both parents work, or in which there is only one parent (who perforce works). Parents therefore have less time to spend with their kids, stimulating their curiosity, answering their questions, reading to them, relaxing with them, inculcating a love of books. (If they can. Allan Bloom, in *The Closing of the American Mind,* asserts that though families may eat, play, and travel together, they do not think together.) As a result, students don't know *anything*. They don't know who Grant and Lee were (reported by E.D. Hirsch in *Cultural Literacy*). A history colleague tells of a student's question: "I keep forgetting. Which side was Hitler on?" I remember a college algebra class being tripped by a problem because they don't know a revolution of the moon takes about a month. (Can they never have remarked on the similarity of the words?) In 1980 I gave a counting problem that depended on knowing the number of days in that year, but the class stumbled because they didn't know it was a leap year; the day of the test was February 29th.

Neither can we count on mathematical prerequisites, on the elementary facts. What we really hope for is some true mathematical understanding; but you can't understand ideas without knowing the facts they rest on. It is always exciting to me to announce to a calculus class that we are about to enter a new realm of mathematical power—in computing areas, we will jump from parallelograms and trapezoids or other polygons to *curved* boundaries. Many students fail to share my excitement. The big jump in power I am so excited about is to them a confused blur. They have no clear picture of what they have been able to do thus far. They are not confident about computing the area within a parallelogram. As for trapezoids, they are not ever sure what they are.

Our students live in a world of morale-sapping hypocrisy. (When was the last time you braked on the yellow light?) They see America being run by crooks. (Where is Richard Nixon now that his country needs him?) The glamorizing of senseless violence by the movies and TV pays off in real life: during 1986, the city of Detroit averaged one child murder per day. What our young people see about them are not incentives to scholarship and learning but causes for despair. If all our values are a mess, what's the point of clean living?

Two Proposals

Mathematicians cannot single-handedly solve the problems of society, but we can do better than leave out related rates. Updating the curriculum is a worthy goal but addresses only one of the variables under our control. I suppose it is an improvement to go from outmoded methods of teaching ill-prepared students a ragged curriculum to outmoded methods of teaching ill-prepared students a spruced-up curriculum, but we can set our sights higher. I have two proposals, both simple, although to put them into effect may require some unglamorous hard work. But I think they would go a long way toward setting our classes on a more realistic footing.

Proposal 1: Let computers handle the drill. In learning a subject, there are two things you have to do—absorb the ideas, and acquire skill in the routines. The appropriate setting for learning ideas is some thoughtful give and take with a teacher. For skill with the routines, you have to have a lot of just plain drill. Today we don't need humans to oversee routine drill. That task should be taken over by computers. This requires some truly first-rate programs; but such things are possible.

Advantages of computer drill are well known, but I will mention some anyhow. Students do their practicing at times convenient to them. They work at their own pace. They not only get feedback but instant feedback. (In contrast, homework papers are often graded without comment by a teaching assistant and returned several days after being handed in.) Students work in privacy, with no one scolding or laughing at them or chiding them for being slow. A well-designed program, with thoughtful conditional branching, will offer guidance while at the same time allowing students to pick the topics they need practice on. The instructor is freed to devote full time to the exchange of ideas. Finally, classes can meet less often, and large classes can be divided into smaller ones.

Proposal 2: Enforce the prerequisites. Not only do we award grades that were not earned, but we do a disservice at the beginning when we admit students who are not qualified for the course. These students usually do poorly and end up soured on math. Instructors feel obliged to review background material in class, cutting into time for the regular syllabus, degrading the character of the course, and shortchanging the better-prepared students.

I propose we all be brave and enforce the prerequisites. This is consistent with the MAA-NCTM resolution of last fall on calculus in high school. Just remember to check your plan with your engineering and business colleagues, pointing out that they too will gain from the new standards—otherwise, they may put in their own mathematics courses.

The resolution just referred to lists algebra, trigonometry, analytic geometry, complex numbers, and elementary functions, studied in depth, as prerequisites for the high school calculus course. For the college precalculus course, I would say be sure to include a thorough treatment of the conic sections (with the byproduct of freeing up the calculus course from that heretofore obligatory chapter). It's a difficult course to handle, because the students know some of the material and become easily bored; but that's a poor reason for putting them directly into calculus, where the material is assumed to be known. Do an honest job, make the course exciting—there is plenty of exciting material—and entrust it to your conscientious teachers. And don't inflate the grades.

Calculus in Secondary Schools

Text of a letter endorsed by the governing boards of the Mathematical Association of America and the National Council of Teachers of Mathematics concerning calculus in the secondary schools:

Memo

TO: Secondary School Mathematics Teachers

FROM: The Mathematical Association of America
The National Council of Teachers of Math.

DATE: September, 1986

RE: Calculus in the Secondary School

Dear Colleague:

A single variable calculus course is now well established in the 12th grade at many secondary schools, and the number of students enrolling is increasing substantially each year. In this letter, we would like to discuss two problems that have emerged.

The first problem concerns the relationship between the calculus course offered in high school and the succeeding calculus courses in college. *The Mathematical Association of America (MAA) and the National Council of Teachers of Mathematics (NCTM) recommend that the calculus course offered in the 12th grade should be treated as a college-level course.* The expectation should be that a substantial majority of the students taking the course will master the material and will not then repeat the subject upon entrance to college. Too many students now view their 12th grade calculus course as an introduction to calculus with the expectation of repeating the material in college. This causes an undesirable attitude on the part of the student both in secondary school and in college. In secondary school all too often a student may feel "I don't have to study this subject too seriously, because I have already seen most of the ideas." Such students typically have considerable difficulty later on as they proceed further into the subject matter.

MAA and NCTM recommend that all students taking calculus in secondary school who are performing satisfactorily in the course should expect to place out of the comparable college calculus course. Therefore, to verify appropriate placement upon entrance to college, students should either take one of the Advanced Placement (AP) Calculus Examinations of the College Board, or take a locally-administered college placement examination in calculus. Satisfactory performance on an AP examination carries with it college credit at most universities.

The second problem concerns preparation for the calculus course. *MAA and NCTM recommend that students who enroll in a calculus course in secondary school should have demonstrated mastery of algebra, geometry, trigonometry, and coordinate geometry.* This means that students should have at least four full years of mathematical preparation beginning with the first course in algebra. The advanced topics in algebra, trigonometry, analytic geometry, complex numbers, and elementary functions studied in depth during the fourth year of preparation are critically important for students' latter courses in mathematics.

It is important to note that at present many well-prepared students take calculus in the 12th grade, place out of the comparable course in college, and do well in succeeding college courses. Currently, the two most common methods for preparing students for a college-level calculus course in the 12th grade are to begin the first algebra course in the 8th grade or to require students to take second year algebra and geometry concurrently. Students beginning with algebra in the 9th grade, who take only one mathematics course each year in secondary school, should not expect to take calculus in the 12th grade. Instead, they should use the 12th grade to prepare themselves fully for calculus as freshmen in college.

We offer these recommendations in an attempt to strengthen the calculus program in secondary schools. They are not meant to discourage the teaching of college-level calculus in the 12th grade to strongly prepared students.

LYNN ARTHUR STEEN
President
Mathematical Association
of America

JOHN A. DOSSEY
President
National Council of
Teachers of Mathematics

Calculators in Standardized Testing of Mathematics

Recommendations from a September 1986 Symposium on Calculators in the Standardized Testing of Mathematics sponsored by the The College Board and the Mathematical Association of America:

In the ten years that inexpensive hand-held calculators have been available, a great deal of consideration has been given to their proper role in mathematics instruction and testing. In September 1986, the College Board and the Mathematical Association of American arranged a Symposium on Calculators in the Standardized Testing of Mathematics to focus on a specialized but essential aspect of this debate.

The participants found in this arena of rapid change a variety of old and new issues. Many of the old issues again yield the same educational conclusions, ones that cause everyone to be frustrated with the delays that surround implementation of calculator usage.

At the same time, rapid changes in technology and price are presenting fresh issues. In particular, falling prices have essentially resolved the old equity issue of student access to calculators and have introduced instead a more important issue of student access to adequate preparation for using these devices appropriately and well.

Likewise, the earlier complications arising from different calculator capabilities that generated test administration policies which gave *upper* limits to the type of calculators permitted have been superseded. The new issue is that test makers now must specify the minimum level of sophistication necessary for a calculator to be appropriate for use in taking a given examination. Excess sophistication in a machine can be self-defeating; surplus machine capabilities may detract far more than they contribute to a student's performance.

Recommendations

1. The symposium endorses the recommendations made by the National Council of Teachers of Mathemat-

ics, the Conference Board of the Mathematical Sciences, the Mathematical Sciences Education Board, and the National Science Board that calculators be used throughout mathematics instruction and testing.

2. The symposium calls for studies to identify content areas of mathematics that have gained importance as a result of new technologies. How achievement and ability in these areas are measured should be studied, and new testing techniques should be considered.
3. The symposium points to a need for research and development on:
   a. item types and formats;
   b. characteristics of new item types;
   c. student responses to items that allow the use of a calculator; and
   d. instructional materials that require the use of a calculator.
4. The symposium believes that a mathematics achievement test should be curriculum based and that no questions should be used that measure *only* calculator skills or techniques.
5. The symposium recognizes that choosing whether or not to use a calculator when addressing a particular test question is itself an important skill. Consequently, not all questions on a calculator-based mathematics achievement test should require the use of a calculator.
6. The symposium gives strong support to the development of examinations in mathematics that require the use of a calculator for some questions. In particular, we support The College Board in its study of mathematics achievement examinations that require the use of calculators, and we commend the Mathematical Association of America for its intention to develop a new series of "calculator-based" placement examinations in mathematics.
7. The symposium recommends that nationally developed mathematics achievement tests requiring the use of a calculator should provide descriptive materials and sample questions that clearly indicate the level of calculator skills needed. However, there should be no attempt to define an upper limit to the level of sophistication that calculators used on such tests should have. Any calculator capable of performing the operations and functions required to solve the problems on a particular examination should be allowed.
8. The symposium notes that different standardized tests in mathematics are used to serve different educational purposes. Therefore, some tests need to be revised as soon as possible to allow for the use of calculators, whereas others may not need to change very soon. In every case, however, the integrity of the test must be maintained in order for its relevance to the mathematics taught in schools and colleges to be sustained.
9. Because of the importance of the SAT in the college admission process, as well as the nature of its mathematical content, the symposium carefully examined the use of calculators on that test. We recommend that calculators *not* be used on the SAT at this time. However, this issue should be reconsidered periodically in light of the status of school mathematics preparation.

Participants:

College Board Staff Members:

JAMES HERBERT, Executive Director, Office of Academic Affairs

GRETCHEN W. RIGOL, Executive Director, Office of Access Services

HARLAN P. HANSON, Director, Advanced Placement Program

ROBERT ORRILL, Associate Director, Office of Academic Affairs

Educational Testing Staff Members:

CHANCEY O. JONES, Associate Area Director

JAMES BRASWELL, Senior Examiner, Test Development

BEVERLY R. WHITTINGTON, Senior Examiner, Test Development

The College Board's Mathematical Sciences Advisory Committee:

JEREMY KILPATRICK, MSAC Chair: University of Georgia

J.T. SUTCLIFFE, St. Mark's School, Texas

CAROL E. GREENES, Boston University

THOMAS W. TUCKER, Colgate University

EDWARD SIEGFRIED, Milton Academy, Massachusetts

R.O. WELLS, JR., Rice University

The MAA Committee on Placement Examinations:

JOHN W. KENELLY, COPE Chair: Clemson University

JOAN R. HUNDHAUSEN, Colorado School of Mines

BILLY E. RHOADES, Indiana University

LINDA H. BOYD, DeKalb College

JOHN G. HARVEY, University of Wisconsin

JACK M. ROBERTSON, Washington State University

JUDITH CEDERBERG, St. Olaf College

MARY MCCAMMON, Pennsylvania State University

Invited Speakers:

JOAN P. LEITZEL, Ohio State University
BERT K. WAITS, Ohio State University
JAMES W. WILSON, University of Georgia

Invited Participants:

GEOFFREY AKST, Chair: AMATYC Education Committee; CUNY, Manhattan Community College

DONALD KREIDER, Treasurer: Mathematical Association of America; Dartmouth College

ALFRED B. WILLCOX, Executive Director, Mathematical Association of America

MICHAEL J. CHROBAK, Texas Instruments Corporation

GERALD MARLEY, UC/CSU Project; California State University, Fullerton

STEVE WILLOUGHBY, NCTM: New York University

SHIRLEY HILL, Chairman: Mathematical Sciences Education Board of the National Research Council

TAMMY RICHARDS, Texas Instruments Corporation

NSF Workshop on Undergraduate Mathematics

Text of a report prepared in June 1986 as a sequel to the National Science Board Task Committee Report on Undergraduate Science, Mathematics, and Engineering Education:

The mathematical sciences are both an enabling force and a critical filter for careers in science and engineering. Without quality education in mathematics we cannot build strong programs in science and engineering. NSF policy for science and engineering education—both precollegiate and collegiate—must be built on this central fact: mathematics is not just one of the sciences, but is *the* foundation for science and engineering.

In our view the most serious problem facing undergraduate mathematics is the quality of teaching and learning for the three million students in all fields who study undergraduate mathematics each term. We are concerned about the professional vitality of the faculty, the "currency" of the curriculum, and the shortage of mathematically-trained students, especially those preparing for careers in mathematics teaching or research. These problems are all interrelated and must be addressed collectively and simultaneously.

The explosion of new applications of mathematics and the impact of computers require major change in undergraduate mathematics. Moreover, recent trends towards unification of basic theory must be integrated into the curriculum. These trends reinforce the urgency of an NSF initiative in undergraduate mathematics based on the themes of leadership and leverage elaborated in the report of the NSB Task Committee on Undergraduate Science and Engineering Education.

We propose programs in four categories: faculty, student support, curriculum development, and global projects. Within each category we list programs in order of priority, but overall there should be at least one program supported from each category.

Faculty

A. SUMMER FACULTY SEMINARS. A nation-wide, centrally coordinated, continuing series of attractive seminars linking new mathematics with curricular reform. Seminars should cover a wide spectrum of issues and levels, from two-year college concerns to exposition of recent research. Some seminars should be taught by industrial mathematicians to facilitate the transfer of mathematical models into the curriculum. To be effective there must be enough seminars of sufficient duration to provide a significant renewal opportunity to at least 50% of the Nation's two- and four-year college and university faculties at least once every five years. (That suggests positions for 10%—about 3,000 people—per summer in a steady state.) Budget Estimate: $3-$4 million per year.

B. FACULTY-INDUSTRY LINKAGES. A special program of limited duration in which NSF would help initiate nation-wide models for faculty to become acquainted with how mathematics is used in industry and government. These models could include faculty summer internships and arrangements for faculty-student teams to work on real problems. The goal of NSF should be to establish a nation-wide system of industry programs to be sustained indefinitely by support from industry. Budget Estimate: Start-up funds of $1 million over three years.

C. INDIVIDUAL TEACHING GRANTS. Like research grants, teaching grants would provide support to individuals with innovative plans to renew specific mathematics courses or create new ones. At least 100 grants (say two per year per state) is a minimum number to have significant effect.

D. FACULTY FELLOWSHIPS. To supplement sabbaticals, perhaps matched by a teaching fellowship from the host departments. Proposed level: At least 100 per year.

Students

A. NSF UNDERGRADUATE FELLOWSHIPS. A program of stipends for talented undergraduates, administered by departments, to provide active mathematical experiences to undergraduates outside class, e.g., working with precollege students, summer opportunities for research apprenticeships, assistants in high school summer institutes, individual study, and special seminars. We are especially concerned about the continued low number of women and the almost total absence of minority students among those who pursue careers in the mathematical sciences. Special efforts to encourage talented women and minority students should be included under this program. In steady state, this program should provide support for approximately the top 25% of undergraduate mathematics majors (about 4,000 students). Budget Estimate: $4-$5 million per year.

B. STUDENT INTERNSHIPS. To provide undergraduates with opportunities in industry or government to employ mathematics in a realistic setting. Initial Recommendation: 10-12 pilot programs.

Curriculum

A. CALCULUS RENEWAL. A multi-year special undertaking, perhaps involving several consortia of institutions, to transform both texts and teaching practice in calculus—the major entry point (and impediment) to college mathematics, science, and engineering. This is of immediate priority. Budget Estimate: $500,000 per consortium per year.

B. MODEL PROGRAMS. A continuing series of projects to identify and develop examples of courses, mathematics majors, instructional environments, and curricular experiments that are right now successfully stimulating interest in undergraduate mathematics, backed up by dissemination activities (perhaps linked with the Summer Seminars) to stimulate others to develop their own programs. Budget Estimate: $2 million per year.

Global Needs

A. ASSESSMENT OF COLLEGIATE MATHEMATICS. A major study of standards, human resources, career patterns, curriculum patterns, and related issues must be undertaken to enable all institutions responsible for undergraduate mathematics to understand clearly the nature and magnitude of the problems facing collegiate mathematics. This study must deal both with introductory courses that represent the primary undergraduate mathematics experience for most students, and with the mathematics major that forms the base for a wide variety of post-baccalaureate careers. Because of preparations already underway for such an assessment, it is very important that this project be supported from the FY 1987 budget. Budget Estimate: $1 million over two years.

B. Consider the establishment of INSTITUTES FOR UNDERGRADUATE EDUCATION IN THE MATHEMATICAL SCIENCES to develop professional and research expertise in issues related to undergraduate mathematics education.

Current Programs

Three current NSF programs are also very important to the renewal of collegiate mathematics—research support for mathematics, instructional scientific equipment, and graduate fellowships. For different reasons, all three programs are now making inadequate impact on undergraduate mathematics:

- As documented in the David Report, the NSF research budget for mathematics is out-of-balance with respect to support for other disciplines, thus insuring that many capable young investigators get cut off early from the frontiers of the discipline. Availability of research support for a larger number of mathematicians is crucial to regenerating the professional vitality of the faculty which is so important to education.

- For various reasons, mathematics departments do not seek resources for such things as computers as vigorously as they should. Consequently, they do not apply for their fair share of support from programs such as CSIP. Both NSF and the mathematics profession must work to insure that significant resources for computing are available and utilized by collegiate mathematics departments.
- Because of the decline in interest among Americans for graduate study in mathematics, too few mathematics graduate students receive support from NSF Graduate Fellowships. The community must work to reverse this trend, and NSF must be prepared to support increased numbers of graduate students in mathematics.

Other Needs

These program recommendations do not address all the problems facing collegiate mathematics. The burden of remedial mathematics and the lack of public understanding that mathematics is pervasive and important significantly impede the ability of colleges and universities to maintain high-quality undergraduate programs in mathematics—the former by draining resources, the latter by limiting resources. These issues are able to be addressed by other NSF programs, so we recommend that new undergraduate resources not be used in these areas.

Making Programs Effective

Finally, we make a few recommendations for how NSF can most effectively develop undergraduate mathematics programs in a way that resonates with the existing professional structures of the mathematical sciences:

1. INTERNAL STRUCTURE. There is a need that funding be specifically directed to support and improve undergraduate mathematics education. Such funding should not divert funds from mathematical research nor should such funds be diverted to mathematical research or to other educational problems. **We propose that a well-defined unit staffed by mathematical scientists knowledgeable about educational problems be established to deal with problems in undergraduate mathematics education.** The unit should have a specific budget of sufficient size to reflect the very significant role undergraduate mathematics plays in science and engineering education.
2. Generally, but not exclusively, programs supported should have implications national in scope. Projects should be undertaken by broadly based consortia, networks, professional societies, and other national organizations. Programs calling for proposals should be focused in intent but not be so specific as to rule out innovative and imaginative approaches. Proposals addressing significant local problems should receive significant local funding or contributions in kind. Proposals addressing broad national concerns should not necessarily be expected to attract significant funding other than from NSF.
3. Professional societies and other national organizations need to assume a significant responsibility in defining and obtaining consensus as to national concerns, informing their members of NSF programs, disseminating results, and identifying necessary human resources. In light of the low response rate of mathematicians to NSF educational proposals, we recommend that a consortium of professional societies provide the community with proposal consultants. It may be necessary in the short run for NSF to fund such an activity.
4. The results of course and curriculum development need to be widely disseminated. There should be both short-term and long-term evaluative follow ups of the effectiveness of various programs supported under this effort.

Participants:

RICHARD D. ANDERSON, Louisiana State University
KENNETH COOKE, Pomona College
JOSEPH CROSSWHITE, Ohio State University
RONALD G. DOUGLAS, SUNY at Stony Brook
HARVEY KEYNES, University of Minnesota
ANNELI LAX, Courant Institute, New York University
DONALD J. LEWIS, University of Michigan
BERNARD MADISON, University of Arkansas
INGRAM OLKIN, Stanford University
HENRY O. POLLAK, Bell Communications Research
THOMAS W. TUCKER, Colgate University
LYNN ARTHUR STEEN (Chair), St. Olaf College

Transition from High School to College Calculus

Donald B. Small

COLBY COLLEGE

Report of a CUPM subcommittee concerning the transition from high school to college calculus, reprinted from THE AMERICAN MATHEMATICAL MONTHLY, *October, 1987:*

There is a widespread and growing dissatisfaction with the performance in college calculus courses of many students who had studied calculus in high school. In response to this concern, in the fall of 1983, the Committee on the Undergraduate Program in Mathematics (CUPM) formed a Panel on Calculus Articulation to undertake a three-year study of questions concerning the transition of students from high school calculus to college calculus and submit a report to CUPM detailing the problems encountered and proposals for their solution.

The seriousness of the issues involved in the Panel's study is underscored by the number of students involved and their academic ability. During the ten-year period 1973 to 1982, the number of students in high school calculus courses grew at a rate exceeding 10% annually. Of the 234,000 students who passed a high school calculus course in 1982, 148,600 received a grade of B- or higher [2]. Assuming a continuation of the 10% growth rate and a similar grade distribution there were approximately 200,000 high school students in the spring of 1985 who received a grade of B- or higher in a calculus course. Thus possibly a third or more of the 500,000 college students who began their college calculus program (in Calculus I, Calculus II, or Calculus III) in the fall of 1985 had already received a grade of B- or higher in a high school calculus course.

The students studying calculus in high school constitute a large majority of the more mathematically capable high school students. (In 1982, 55% of high school students attended schools where calculus was taught [2].) Students who score a 4 or 5 on an Advanced Placement (AP) Calculus examination normally do well in maintaining their accelerated mathematics program during the transition from high school to college. However, this is a very small percentage of the students who take calculus in high school. For example, in 1982, of the 32,000 students who took an Advanced Placement calculus examination, just over 12,000 received scores of 4 or 5, which represents only 6% of all high school students who took calculus that year. The primary concern of the Panel was with the transition difficulties associated with the remaining almost 94% of the high school calculus students.

Problem Areas

Past studies and the Panel's surveys of high school teachers, college teachers, and state supervisors suggest that the major problems associated with the transition from high school calculus to college calculus are:

1. High school teacher qualifications and expectations.
2. Student qualifications and expectations.
3. The effect of repeating a course in college after having experienced success in a similar high school course.
4. College placement.
5. Lack of communication between high schools and colleges.

(Copies of the Panel's Report including the surveys and summaries of the responses can be obtained from the Washington Office of the MAA.)

These problems were addressed by first considering accelerated programs in general, high school calculus (successful, unsuccessful), and the responsibilities of the colleges.

Accelerated Programs

Accelerated mathematics programs, usually beginning with algebra in eighth grade, are now well established and accepted in most school systems. The success of these programs in attracting the mathematically capable students was documented in the 1981-82 testing that was done for the "Second International Mathematics Study." The Summary Report [9] states with reference to a comparison between twelfth grade precalculus students and twelfth grade calculus students in the United States:

> We note furthermore that in every content area (sets and relations, number systems, algebra, geometry, elementary functions/calculus, probability and statistics, finite mathematics), the end-of-the-year average achievement of the precalculus classes was less (and in many cases considerably less) than the beginning-of-the-year achievement of the calculus students.

The report continues:

> It is important to observe that the great majority of U.S. senior high school students in fourth and fifth year mathematics classes (that is, those in precalculus

> classes) had an average performance level that was at or below that of the lower 25% of the countries. The end-of-year performance of the students in the calculus classes was at or near the international means for the various content areas, with the exception of geometry. Here U.S. performance was below the international average.

Thus those students in accelerated programs culminating in a calculus course perform near the international mean level while their classmates in (non-accelerated) programs culminating in a precalculus course perform in the bottom 25% in this international survey. The poor performance in geometry by both the precalculus and calculus students correlates well with the statistic that 38% of the students were never taught the material contained in the geometry section of the test [9, p. 59]. The test data underscores the concern expressed by many college teachers that more emphasis needs to be placed on geometry throughout the high school curriculum. This data does not, however, indicate that accelerated programs emphasize geometry less than non-accelerated programs.

The success of the accelerated programs in completing the "normal" four year high school mathematics program by the end of the eleventh grade presents schools with both an opportunity and a challenge for a "fifth" year program. There are two acceptable options:

1. Offer college-level mathematics courses that would continue the students' accelerated program and thus provide exemption from one or two semesters of college mathematics.
2. Offer high school mathematics courses that would broaden and strengthen a student's background and understanding of precollege mathematics.

Not offering a fifth year course or offering a watered-down college level course with no expectation of students earning advanced placement are not considered to be acceptable options.

A great deal of prestige is associated with offering calculus as a fifth year course. Communities often view the offering of calculus in their high school as an indication of a quality educational program. Parents, school board officials, counselors, and school administrators often demonstrate a competitive pride in their school's offering of calculus. This prestige factor can easily manifest itself in strong political pressure for a school to offer calculus without sufficient regard to the qualifications of teachers or students.

It is important that this political pressure be resisted and that the choice of a fifth year program be made by the mathematics faculty of the local school and be made on the basis of the interest and qualifications of the mathematics faculty and the quality and number of accelerated students. School officials should be encouraged to develop public awareness programs to extend the prestige and support that exists for the calculus to acceleration programs in general. This would help diffuse the political pressure as well as broaden school support within the community.

Schools that elect the first option of offering a college level course should follow a standard college course syllabus (e.g., the Advanced Placement syllabus for calculus). They should use placement test scores along with the college records of their graduates as primary measures of the validity of their course.

For schools that elect the second option, a variety of courses is possible. The following course descriptions represent four possibilities.

ANALYTICAL GEOMETRY. This course could go well beyond the material normally included in second year algebra and precalculus. It could include Cartesian and vector geometry in two- and three-dimensions with topics such as translation and rotation of axes, characteristics of general quadratic relations, curve sketching, polar coordinates, and lines, planes, and surfaces in three-dimensional space. Such a course would provide specific preparation for calculus and linear algebra, as well as give considerable additional practice in trigonometry and algebraic manipulations.

PROBABILITY AND STATISTICS. This course could be taught at a variety of levels, to be accessible to most students, or to challenge the strongest ones. It could cover counting methods and some topics in discrete probability such as expected values, conditional probability, and binomial distributions. The statistics portion of the course could emphasize exploratory data analysis including random sampling and sampling distributions, experimental design, measurement theory, measures of central tendency and spread, measures of association, confidence intervals, and significance testing. Such an introduction to probability and statistics would be valuable to all students, and for those who do not plan to study mathematics, engineering, or the physical sciences, probably more valuable than a calculus course.

DISCRETE MATHEMATICS. This type of course could include introductions to a number of topics that are either ignored or treated lightly within a standard high school curriculum, but which would be stimulating and widely useful for the college-bound high school student. Suggested topics include permutations, combinations, and other counting techniques: mathematical induction; difference equations; some discrete probability; elementary number theory and modular arithmetic; vector and matrix algebra, perhaps with an introduction to linear or dynamic programming; and graph theory.

MATRIX ALGEBRA. This course could include basic arithmetic operations on matrices, techniques for finding matrix inverses, and solving systems of linear equations and their equivalent matrix equations using Gaussian elimination. In addition, some introduction to linear programming and dynamic programming could be included. This course could also emphasize three-dimensional geometry.

High School Calculus

There are many valid reasons why a fifth year program should include a calculus course. Four major reasons: (1) calculus is generally recognized as the starting point of a college mathematics program, (2) there exists a (nationally accepted) syllabus, (3) the Advanced Placement program offers a nation wide mechanism for obtaining advanced placement, and (4) there is a large prestige factor associated with offering calculus in high school. Calculus, however, should not be offered unless there is a strong indication that the course will be successful.

Successful Calculus Courses

The primary characteristics of a successful high school calculus course are:

1. A qualified and motivated instructor with a mathematics degree that included at least one semester of a junior-senior real analysis course involving a rigorous treatment of limits, continuity, etc.
2. Administrative support, including provision of additional preparation time for the instructor (e.g., as recommended by the North Central Accreditation Association).
3. A full year program based on the Advanced Placement syllabus.
4. A college text should be used (not a watered-down high school version).
5. Advanced placement for students (rather than mere preparation for repeating calculus in college) is a major goal.
6. Course evaluation based primarily on college placement and the performance of its graduates in the next higher level calculus course.
7. Restriction of course enrollment to only qualified and interested students.
8. The existence of an alternative fifth year course that students may select who are not qualified for or interested in continuing in an accelerated program.

The bottom line of what makes a high school calculus course successful is no surprise to anyone. A qualified teacher with high but realistic expectations, using somewhat standard course objectives, and students who are willing and able to learn result in a successful transition at any level of our educational process. Problems appear when any of the above ingredients are missing.

Unsuccessful Calculus Courses

Two types of high school calculus courses have an undesirable impact on students who later take calculus in college.

One type is a one semester or partial year course that presents the highlights of calculus, including an intuitive look at the main concepts and a few applications, and makes no pretense about being a complete course in the subject. The motivation for offering a course of this kind is the misguided idea that it prepares students for a *real* course in college.

However, such a preview covers only the glory and thus takes the excitement of calculus away from the college course without adequately preparing students for the hard work and occasional drudgery needed to understand concepts and master technical skills. Professor Sherbert has commented: "It is like showing a ten minute highlights film of a baseball game, including the final score, and then forcing the viewer to watch the entire game from the beginning—with a quiz after each inning."

The second type of course is a year-long, semi-serious, but watered-down treatment of calculus that does not deal in depth with the concepts, covers no proofs or rigorous derivations, and mostly stresses mechanics. The lack of both high standards and emphasis on understanding dangerously misleads students into thinking they know more than they really do.

In this case, not only is the excitement taken away, but an unfounded feeling of subject mastery is fostered that can lead to serious problems in college calculus courses. Students can receive respectable grades in a course of this type, yet have only a slight chance of passing an Advanced Placement Examination or a college-administered proficiency examination. Those who place into second term calculus in college will find themselves in heavy competition with better prepared classmates. Those who elect (or are selected) to repeat first term calculus believe they know more than they do, and the motivation and willingness to learn the subject are lacking.

College Programs

Several studies ([1], [3], [5], [6], [7]) have been conducted on the performance in later courses by students who have received advanced placement (and possibly college credit) by virtue of their scores on Advanced Placement Calculus examinations. The studies show that, overall, students earning a score of 4 or 5 on either the AB or BC Advanced Placement Calculus examination do as well or better in subsequent calculus courses than the students who have taken all their calculus in college. It is therefore strongly recommended that colleges recognize the validity of the Advanced Placement Calculus program by the granting of one semester advanced placement with credit in calculus for students with a 4 or 5 score on the AB examination, and two semesters of advanced placement with credit in calculus for students with a 4 or 5 score on the BC examination.

The studies reviewed by the Panel do not indicate any clear conclusions concerning performance in subsequent calculus courses by students who have scored a 3 on an Advanced Placement Calculus examination. The treatment of these students is a very important transition problem since approximately one-third of all students who take an Advanced Placement Calculus examination are in this group and many of them are quite mathematically capable.

It is therefore recommended that these students be treated on a special basis in a manner that is appropriate for the institution involved. For example, several colleges offer a student who has earned a 3 on an Advanced Placement Calculus examination the opportunity to upgrade this score to an "equivalent 4" by doing sufficiently well on a Department of Mathematics placement examination. Another option is to give such students one semester of advanced placement with credit for Calculus I upon successful completion of Calculus II. A third option is to give one semester of advanced placement with credit for Calculus I and provide a special section of Calculus II for such students.

Other important transition problems are associated with students who have studied calculus in high school, but have not attained advanced placement either through the Advanced Placement Calculus program or effective college procedures. These students pose an important and difficult challenge to college mathematics departments, namely: How should these students be dealt with so that they can benefit from their accelerated high school program and not succumb to the negative and (academically) destructive attitude problems that often result when a student repeats a course in which success has already been experienced? There are three major factors to consider with respect to these students.

1. The lack of uniformity of high school calculus courses. The wide diversity in the backgrounds of the students necessitates that a large review component be included in their first college calculus course to guarantee the necessary foundation for future courses.
2. The mistaken belief of most of these students that they really know the calculus when, in fact, they do not. Thus they fail to study enough at the beginning of the course. When they realize their mistake (if they do), it is often too late. These students often become discouraged and resentful as a result of their poor performance in college calculus, and believe that it is the college course that must be at fault.
3. The "Pecking Order" syndrome. The better the student, the more upsetting are the understandable feelings of uncertainty about his or her position relative to the others in the class. Although this is a common problem for all college freshmen, it is compounded when the student appears to be repeating a course in which success had been achieved the preceding year. This promotes feelings of anxiety and produces an accompanying set of excuses if the student does not do at least as well as in the previous year.

   The uncertainty of one's position relative to the rest of the class often manifests itself in the student not asking questions or discussing in (or out of) class for fear of appearing *dumb*. This is in marked contrast to the highly confident high school senior whose questions and discussions were major components in his or her learning process.

The unpleasant fact is that the majority of students who have taken calculus in high school and have not clearly earned advanced placement do not *fit* in either the standard Calculus I or Calculus II course. The students do not have the level of mastery of Calculus I topics to be successful if placed in Calculus II and are often doomed by attitude problems if placed in Calculus I. In modern parlance, this is the *rock and hard place.*

An additional factor to consider is the negative effect that a group of students who are repeating most of the content of Calculus I has on the rest of the class as well as on the level of the instructor's presentations.

What is needed are courses designed especially for students who have taken calculus in high school and have not clearly earned advanced placement. These courses need to be designed so that they:

1. Acknowledge and build on the high school experiences of the students;
2. Provide necessary review opportunities to ensure an

acceptable level of understanding of Calculus I topics;

3. Are *clearly different* from high school calculus courses (in order that students do not feel that they are essentially just repeating their high school course);
4. Result in an equivalent of one semester advanced placement.

Altering the traditional lecture format or rearranging and supplementing content seem to be two promising approaches to developing courses that will satisfy the above criteria. For example, Colby College has successfully developed a two semester calculus course that fulfills the four conditions. The course integrates multivariable with single variable calculus, and thereby covers the traditional three semester program in two semesters [10].

Of course, the introduction of a new course entails an accompanying modification of college placement programs. However, providing new or alternative courses should have the effect of simplifying placement issues and easing transition difficulties that now exist.

Recommendations

1. School administrators should develop public awareness programs with the objective of extending the support that exists for fifth year calculus courses to accelerated programs including all of the fifth year options.
2. A fifth year program should offer a student a choice of courses (not just calculus).
3. The choice of fifth year options should be made by the high school mathematics faculty on the basis of their interest and qualifications and the quality and number of the accelerated students.
4. If a fifth year course is intended as a college level course, then it should be treated as a college level course (text, syllabus, rigor).
5. A fifth year college level course should be taught with the expectation that successful graduates (B- or better) would not repeat the course in college.
6. A fifth year program should provide an alternative option for the student who is not qualified to continue in an accelerated program.
7. A mathematics degree that includes at least one semester of a junior-senior real analysis course involving a rigorous treatment of limit, continuity, etc., is strongly recommended for anyone teaching calculus.
8. A high school calculus course should be a full year course based on the Advanced Placement syllabus.
9. The instructor of a high school calculus course should be provided with additional preparation time for this course.
10. High school calculus students should take either the AB or BC Advanced Placement calculus examination.
11. The evaluation of a high school calculus course should be based primarily on college placement and the performance of its graduates in the next level calculus course.
12. Only interested students who have successfully completed the standard four year college preparatory program in mathematics should be permitted to take a high school calculus course.
13. Colleges should grant credit and advanced placement out of Calculus I for students with a 4 or 5 score on the AB Advanced Placement calculus examination, and credit and advanced placement out of Calculus II for students with a 4 or 5 score on the BC Advanced Placement calculus examination. Colleges should develop procedures for providing special treatment for students who have earned a score of 3 on an Advanced Placement calculus examination.
14. Colleges should individualize as much as possible the advising and placement of students who have taken calculus in high school. Placement test scores and personal interviews should be used in determining the placement of these students.
15. Colleges should develop special courses in calculus for students who have been successful in accelerated programs, but have clearly not earned advanced placement.

Colleges have an opportunity and responsibility to develop and foster communication with high schools. In particular:

16. Colleges should establish periodic meetings where high school and college teachers can discuss expectations, requirements, and student performance.
17. Colleges should coordinate the development of enrichment programs (courses, workshops, institutes) for high school teachers in conjunction with school districts and state mathematics coordinators.

Members of the Panel:

GORDON BUSHAW, Central Kitsap High School, Silverdale, Washington

DONALD J. NUTTER, Firestone High School, Akron, Ohio

RONALD SCHNACKENBURG, Steamboat Springs High School, Colorado

BARBARA STOTT, Riverdale High School, Jefferson, Louisiana

JOHN H. HODGES, University of Colorado
DONALD R. SHERBERT, University of Illinois
DONALD B. SMALL (Chair), Colby College

References

[1] C. Cahow, N. Christensen, J. Gregg, E. Nathans, H. Strobel, G. Williams. Undergraduate Faculty Council of Arts and Sciences Committee on Curriculum; Subcommittee on Advanced Placement Report, Trinity College, Duke University, 1979.

[2] C. Dennis Carroll. "High school and beyond tabulation: Mathematics courses taken by 1980 high school sophomores who graduated in 1982." National Council of Education Statistics, April 1984 (LSB 84-4-3).

[3] P.C. Chamberlain, R.C. Pugh, J. Schellhammer. "Does advanced placement continue throughout the undergraduate years?" *College and University*, Winter 1968.

[4] "Advanced Placement Course Description, Mathematics." The College Board, 1984.

[5] E. Dickey. "A study comparing advanced placement and first-year college calculus students on a calculus achievement test." Ed.D. dissertation, University of South Carolina, 1982.

[6] D.A. Frisbie. "Comparison of course performance of AP and non-AP calculus students." Research Memorandum No. 207, University of Illinois, September 1980.

[7] D. Fry. "A comparison of the college performance in calculus-level mathematics courses between regular-progress students and advanced placement students." Ed.D. dissertation, Temple University, 1973.

[8] C. Jones, J. Kenelly, D. Kreider. "The advanced placement program in mathematics—Update 1975." *Mathematics Teacher*, 1975.

[9] *Second International Mathematics Study Summary Report for the United States.* Champaign, IL: Stipes Publishing, 1985.

[10] D. Small, J. Hosack. *Calculus of One and Several Variables: An Integrated Approach.* Colby College, 1986.

[11] D.H. Sorge, G.H. Wheatley. "Calculus in high school—At what cost?" *American Mathematical Monthly* 84 (1977) 644-647.

[12] D.M. Spresser. "Placement of the first college course." *International Journal Mathematics Education, Science, and Technology* 10 (1979) 593-600.

Calculators with a College Education?

Thomas Tucker

COLGATE UNIVERSITY

Reprint of a lead article that appeared in the January 1987 issue of FOCUS, *the Newsletter of the Mathematical Association of America:*

The title of this article should sound familiar; it is a variation on Herbert Wilf's "The Disk with the College Education," which appeared in the *Monthly* in January, 1982. In that article, Wilf sent a "distant early-warning signal" that powerful mathematical computer environments like muMATH, which were just then becoming available on microcomputers, would someday soon appear in pocket computers.

That day may have arrived. The Casio fx7000-G, introduced in early 1986, is no bigger than the usual $10 hand calculator, but plots functions on its small dot matrix screen. In January 1987, Hewlett-Packard released its HP-28C; again, a normal-sized hand calculator that not only plots curves but also does matrix operations, equation-solving, numerical integration, and, last but not least, symbolic manipulation!

Neither the Casio ($55-$90) nor the HP-28C (around $180) are cheap by calculator standards, but any student who has bought a calculus textbook without flinching can afford a Casio and in a couple of years I would expect the textbook to cause more flinching than the HP-28C. The questions these calculators raise for mathematics educators are the same Wilf asked in 1982 (after which he "beat a hasty retreat"). The answers are not any clearer today.

Here is a more detailed look at the two calculators. The Casio fx7000-G differs from other key-stroke programmable, scientific calculators by having a larger dot matrix screen that makes graphing possible and that allows the user to see clearly the expressions being evaluated. The plotter can graph many functions at once. The window (range of x- and y-coordinates) can be controlled by the user and can be easily magnified to zoom in on a particular portion of the graph. A moving pixel tracing out a given curve can be stopped at any time to specify a zooming-in point. The plotter can be used in this way to find the points of intersection of two curves.

The calculator, however, has no built-in routine to solve an equation numerically, nor does it have a Simpson's rule key for numerical integration like some other scientific calculators. Since every key has three functions, the keyboard is cluttered with symbols and abbreviations.

The HP-28C does a lot more than the Casio, and it is perhaps unfair to compare them at all. It is truly the first in a new generation of calculators. Although thin enough to fit easily in a breast pocket, it folds open to reveal two keyboards. On the right is a four-line screen and below it an array of 37 buttons that looks vaguely like a standard calculator keyboard. On the left is a 35-button alphabetic keyboard—there is only a single "shift key" so each key has only two functions.

But where are the sin, cos, and exp buttons? And why does the top row of six buttons on the right keyboard have no labels at all? Because the HP-28C is menu-driven, and the top row contains all-purpose function keys! If you want sin, press the TRIG button, which activates a trigonometry menu at the bottom of the screen, and then press the function button directly below SIN (press NEXT to see six more functions on the TRIG menu).

There are other menus for logs and exponentials, equation-solving, user-defined functions, statistics, plotting, matrix operations and editing, binary arithmetic, complex arithmetic, string operations, stack operations, symbolic manipulation, program control (DO UNTIL, etc.), special real arithmetic (modulo, random number generator, etc.), printing (yes, one can buy a thermal printer with infrared remote control), and, of course, a catalog of all operations.

Like other Hewlett-Packard calculators, the HP-28C is a stack machine with operators and operands entered in reverse Polish. Stack entries can be commands, real or complex numbers, lists, strings, matrices, or algebraic expressions such as '2 * 3 + 5' or 'X * SIN(X)'. Expressions can be evaluated by pressing the EVAL key, which can be used like the "=" key on algebraic calculators by those averse to reverse Polish.

If 'X * SIN(X)' is at the top of the stack and you want it symbolically differentiated, put the variable of differentiation 'X' on the stack and press the derivative button. To compute its degree 5 Taylor polynomial centered at 0, enter 5 and press TAYLR on the algebraic menu. To compute its definite integral from 0 to 1 enter the list 'X', 0, 1 and a tolerance .0001 and press the integral button.

To plot 'X * SIN(X)', press DRAW on the plotting menu: Window parameters are controlled by the user as on the Casio fx7000-G; x- and y-coordinates of any point on the screen can be found by moving cross-hairs to the desired point and pressing INS.

To find a root of the equation 'X = TAN(X)', put it on the stack followed by an initial guess, or an interval, or an interval and a guess; then press ROOT on the equation-solving menu.

To invert the 2 by 2 matrix with rows [1, 2] and [3, 4], enter [[1, 2] [3, 4]] on the stack and press $1/x$. To find its determinant, press DET on the array menu. To solve a system of equations, put the left-side coefficient matrix on the stack, followed by the right-side vector, and then press $\div$. To multiply two matrices on the stack, press $\times$.

The numerical routines are high quality. There are 12 digits of accuracy displayed and 16 digits internal. For example, with display set at three digits to the right of the decimal point, '2∧39 EVAL' yields 549755813888; multiply the result by 2 and 1.100 E 12 appears; divide that by 2 and 549755813888 reappears. The routines are also fast. A short program written to multiply a matrix times itself 100 times ran in about 2 minutes for a 6 by 6 matrix!

There are some problems. Memory is limited compared to a microcomputer: 8 by 8 matrix multiplication is about all the calculator can handle and a request for the degree 5 Taylor polynomial for $e^{e^{(x-1)}}$ overflows the symbolic differentiator (try computing the fifth derivative without regrouping to see why—Wilf's micro took 4 minutes to get the degree 9 polynomial in his 1982 article). Although the matrix algebra routines are accurate and fast, the HP-28C has never met a square matrix it couldn't invert (presumably, matrix entries such as 1 E 500 should tell the user something is wrong).

How hard are these calculators to use? Although just clearing the Casio fx7000-G memory can be a challenge without the manual, students used to a scientific calculator should feel comfortable after an hour or two. Hewlett-Packard designs more for engineers than for precalculus students, and the HP-28C is no exception. It takes ten hours to become proficient enough to begin to realize the potential of this calculator, and one could spend weeks exploring the nooks and crannies of the machine. Luckily, documentation isn't too bad. Basically, if a student can learn PASCAL, he or she can learn to use the HP-28C.

Who will use the HP-28C? It is not powerful enough to help a professional mathematician do symbolic manipulation the way MACSYMA helped Neil Sloane. (See his January 1986 *Notices* article "My Friend MACSYMA.") But plenty of calculus students would find the HP-28C handy when they are asked on their next test to differentiate $(1+x^2)^{\sin x}$, and so would I if I had to graph that function. The calculator can't solve a twenty-variable linear program by the simplex algorithm, but it could be programmed to do eight or nine variables, and I can already see using the HP-28C in my linear algebra class to find eigenvectors and eigenvalues for arbitrary 5 by 5 matrices using the power method.

One might cringe at a student integrating $1/(1+x^2)$ from 0 to 1 by pressing a button and getting .785 (actually, it takes about 20 button-pushes to get the expression on the stack and another 10 to enter the integration parameters and perform the integration, although it all takes less than a minute). But what about the arc length of $y = x^3$ from $x = 0$ to $x = 1$? Something lost, something gained.

Mathematicians are traditionally wary of technology. Perhaps their qualms are justified. To think of the area under the curve $y = 1/(1+x^2)$ from 0 to 1 as .785 and not $\pi/4$ is to miss all the beauty of a mysterious relationship between circles and triangles and areas and rates of change. Mathematicians are notoriously slow to come to grips with technology.

At the Sloan Conference on Calculus at Tulane last January, a syllabus was proposed that recommended the use of programmable calculators with a Simpson's rule button, although such calculators have existed for five years. The participants at that conference had no idea that a Casio fx7000-G or an HP-28C was looming on the horizon. How long will it take to recognize pedagogically the existence of these calculators, which many students will already have? Must it be "something gained, something lost?" Can't it just be "something gained?" Good questions to ask, but like Wilf in 1982, it is probably time to beat a hasty retreat.

Who Still Does Math with Paper and Pencil?

Lynn Arthur Steen

ST. OLAF COLLEGE

Reprint of a "Point of View" column that appeared in THE CHRONICLE OF HIGHER EDUCATION *on October 14, 1987:*

Mathematics is now so widely used in so many different fields that it has become the most populated—but not the most popular—undergraduate subject. Each term an army of three million students labors with primitive tools to master the art of digging and filling intellectual ditches: instead of using shovels and pick axes, they use paper and pencil to perform millions of repetitive calculations in algebra, calculus, and statistics. Mathematics, the queen of the sciences, has become the serf of the curriculum.

People who use mathematics in the workplace—accountants, engineers, and scientists, for example—rarely use paper and pencil any more, certainly not for significant or complex computations. Electronic spreadsheets, numerical-analysis packages, symbolic-algebra systems, and sophisticated computer graphics have become the power tools of mathematics in industry. Even research mathematicians use computers to help them with exploration, conjecture, and proof. In the college classroom, however, mathematics has—with few exceptions—remained in the paper-and-pencil era.

Academic inertia alone is not a sufficient explanation for this state of affairs. Other disciplines—notably chemistry, economics, and physics—have adapted their undergraduate curricula to include appropriate use of computers.

In contrast, many mathematicians believe that computers are rarely appropriate for mathematics instruction; theirs is a world of mental insight and abstract constructions, not of mechanical calculation or concrete representation. Most mathematicians, after all, choose mathematics at least in part because it depends only on the power of mind rather than on a variety of computational contrivances.

All that is going to change in the next two or three years, which in education are the equivalent of a twinkling of an eye. The latest pocket calculators with computer-like capabilities can perform at the touch of a few buttons many of the laborious calculations taught in the first two years of college mathematics. They can, among other things, graph and solve equations, perform symbolic differentiation as well as numerical and some symbolic integration, manipulate matrices, and solve simultaneous equations. Although such computations do not form the heart of the *ideal* curriculum as it exists in the eye of the mathematician, they do account for the preponderance of the *achieved* curriculum that is actually mastered by the typical undergraduate.

Computation has become significant for mathematics because of a major change not just in scale but of methods: the transition from numerical mathematics, the province of scientists, to symbolic and visual mathematics, the province of mathematicians. Large computers have been doing "real" mathematics for years, but cost and relative scarcity kept them out of the classroom. No more. Mathematics-speaking machines are about to sweep the campuses, embodied both as computer disks and as pocket calculators. Beginning this fall, college students will be able to use such devices to find the answers to most of the homework they are assigned.

Much as professors like to believe that education standards are set by the faculty, the ready availability of powerful computers will enable students to set new ground rules for college mathematics. Template exercises and mimicry mathematics—staples of today's texts—will vanish under the assault of computers that specialize in mimicry. Teachers will be forced to change their approach and their assignments. They will no longer be able to teach as they were taught in the paper-and-pencil era.

Change always involves risk as well as benefit. We have no precedents for learning in the presence of mathematics-speaking calculators. No one knows how much "patterning" with paper-and-pencil methods is essential to provide a foundation for subsequent abstractions. Preliminary research suggests that it may not be as necessary as many mathematics teachers would like to believe.

On the other hand, many students tolerate (and survive) mathematics courses only because they can get by with mastery of routine, imitative techniques. A mathematics course not built on the comfortable foundation of mindless calculation would almost surely be too difficult for the student whose sole reason for taking mathematics is that it is required.

Despite such risks, mathematics—and society—has much to gain from the increasing use of pocket computers in college classes:

- Undergraduate mathematics will become more like real mathematics, both in the industrial work place and in academic research. By using machines to expedite calculations, students can experience mathematics as it really is—as a tentative, exploratory discipline in which risks and failures yield clues to success. Computers change our perceptions of what is possible and what is valuable. Even for unsophisticated users, computers can rearrange the balance between "working" and "thinking" in mathematics.
- Weakness in algebra skills will no longer prevent students from pursuing studies that require college mathematics. Just as spelling checkers have enabled writers to express ideas without the psychological block of worrying about their spelling, so the new calculators will enable students weak in algebra or trigonometry to persevere in calculus or statistics. Computers could be the democratizer of college mathematics.
- Mathematics learning will become more active and hence more effective. By carrying most of the computational burden of mathematics homework, computers will enable students to explore a wider variety of examples, to study graphs of a quantity and variety unavailable with pencil-and-paper methods, to witness the dynamic nature of mathematical processes, and to engage realistic applications using typical—not oversimplified—data.
- Students will be able to explore mathematics on their own, without constant advice from their instructors. Although computers will not compel students to think for themselves, these machines can provide an environment in which student-generated mathematical ideas can thrive.
- Study of mathematics will build long-lasting knowledge, not just short-lived strategies for calculation. Most students take only one or two terms of college mathematics, and quickly forget what little they learned of memorized methods for calculation. Innovative instruction using a new symbiosis of machine calculation and human thinking can shift the balance of mathematical learning toward understanding, insight, and mathematical intuition.

Mathematics-capable calculators pose deep questions for the undergraduate mathematics curriculum. By shifting much of the computational burden from students to machines, they leave a vacuum of time and emphasis in the undergraduate curriculum. No one yet knows what, if anything, will replace paper-and-pencil computation, or whether advanced mathematics can be built on a computer-reliant foundation. What can be said with certainty, however, is that the era of paper-and-pencil mathematics is over.

Computing in Undergraduate Mathematics

Paul Zorn

ST. OLAF COLLEGE

An "issues paper" prepared in conjunction with a June 1987 conference organized by the Associated Colleges of the Midwest to examine the role of computing in undergraduate mathematics as part of an effort by the staff of the National Science Foundation to increase the impact of computing on undergraduate departments of mathematics. Views expressed in this paper are those of the author, and do not necessarily reflect the views of the National Science Foundation.

Modern computing raises unprecedented opportunities, needs, and issues for undergraduate mathematics. The relation between computing and mathematics is too young, and changing too quickly, to admit definitive positions. None are taken here. In mathematical language, this paper is not an authoritative monograph but a topical survey with many open questions.

We do not assert that computing serves every worthwhile purpose of undergraduate mathematics. The interesting issues are genuine questions: Where in the curriculum is computing appropriate, and why? What does computing cost—in time, money, and distraction from other purposes? If computers handle routine mathematical manipulations, what will students do instead? Will students' manipulative skills and intuition survive? Should we teach things machines do better?

So much said, it would be disingenuous to deny the viewpoint that motivates this paper, and is implicit throughout—that we can and should use modern computing more than we have done to improve mathematical learning and teaching. Although we will argue that the computing resource has scarcely been tapped, this paper is not simply a plea for computers in the classroom. Mathematical computing is educationally valuable only as it alters and serves curricular goals of undergraduate mathematics. It follows that curricular goals should guide teaching uses of computing, not the other way around.

What is Computing?

Until recently, computing in undergraduate mathematics usually meant writing or running programs (in Basic or Pascal) for floating point numerical operations. Much more is now possible: symbolic algebra, sophisticated graphics, interactive operating modes—all with little or no programming required of the user. Computing should be understood broadly, comprising hardware, software, and peripheral equipment.

Other forms of educational technology, such as videodiscs, might someday become important teaching tools. They are not addressed here. Covering everything that now exists would be difficult; anticipating what may exist is impossible. Our scope is comprehensive only in the sense that many kinds of educational technology, like computing, amount to new ways of representing and manipulating mathematical information.

Mathematics and Computing

Computing drives the modern mathematical revolution. As Gail Young puts it in [3],

> ...[W]e are participating in a revolution in mathematics as profound as the introduction of Arabic numerals into Europe, or the invention of the calculus. Those earlier revolutions had common features: hard problems became easy, and solvable not only by an intellectual elite but by a multitude of people without special mathematical talent; problems arose that had not been previously visualized, and their solutions changed the entire level of the field.

Like Arabic numerals and the calculus, computing is a sharper tool, but it is also more than that. Computers do more than help solve old problems. They lead to new problems, new approaches to old problems, and new notions of what it means to solve problems. They change fundamental balances that have defined the discipline of mathematics and how it is pursued: continuous and discrete, exact and approximate, abstract and concrete, theoretical and empirical, contemplative and experimental. Computers change what we think possible, what we think worthwhile, and even what we think beautiful.

Computing is becoming commonplace (if not ubiquitous) in mathematical research, even on classical problems. Without computers, research in many new areas would stop. Computing figured, more or less fundamentally, on the way to several recent spectacular advances, including the four color theorem and the Bieberbach conjecture. With numerical, graphical, and symbol-manipulating abilities, computers check calculations, test conjectures, process large data sets, search for structure, and represent mathematical objects in new ways. They make possible entirely new viewpoints on problems in mathematical research—viewpoints that are more empirical than deductive, more experimental than theoretical.

Computing has changed how mathematics is *used* at least as much as it has changed how mathematics is created. The changes are broad and deep, touching areas from arithmetic to statistics to differential equations. This part of the computer revolution, moreover, is happening in public. Changes at the rarefied research level may affect only a few people; changes at the "user" level reach a much broader constituency. Already, undergraduates freely—if sometimes naively—use sophisticated numerical "packages" in science and social science courses. Seeing computing all around them, students naturally expect some in mathematics courses, too.

It seems axiomatic (certainly to students!) that the profound effects of computing on research and applied mathematics should be reflected in undergraduate mathematics education. Honesty to our discipline and our own best interests as mathematics teachers both dictate so. Honesty requires at least that we keep ourselves and our students abreast of important developments in our field. Self-interest says we should do more than report what happens outside: we should avail ourselves of the enormous opportunities computers offer for teaching and learning mathematics.

Despite all this, computing has not yet changed the daily work of undergraduate mathematics very much. The standard freshman calculus course, for example, still consists largely of paper and pencil performance of mechanical algorithms—differentiation, graph-sketching, antidifferentiation, series expansion, etc.—just what machines do best. Graphical and numerical methods are usually treated as side issues. With a little computing power, they could illuminate important interplays between discrete and continuous ideas, exact and approximate techniques, geometric and analytic viewpoints.

Statistical as well as anecdotal evidence shows that mathematics lags behind the other undergraduate sciences in teaching uses of computing. In 1985 and 1986, for example, only 32 of approximately 2800 proposals to the NSF's College Science Instrumentation Program came from mathematics departments. Galling as it is to be elbowed from the trough, this paucity of mathematics proposals is only a symptom. Our real problem is not too few proposals, but the opportunities mathematicians miss to revitalize teaching and learning.

Problems of the Past

Reasons for the lag in undergraduate mathematical computing are easy to guess. The clearest difference between mathematics and the physical sciences is in the roles experiment and observations play in each. Although mathematics has an essential (if informal) experimental aspect, especially in research, mathematics is not a laboratory discipline in the formal, ritualistic sense that applies to the other sciences. The value of "instruments," whether computers or chemicals, to support the experimental side of the natural sciences is taken for granted, but there is no similar consensus about undergraduate mathematics. This may change, but for now, the idea of mathematical "instrumentation" is still a novelty. Machines to support undergraduate mathematical experimentation are just appearing, and we are just learning to use them. Unlike our colleagues in the natural sciences, we mathematicians must convince our departments and college administrators—and sometimes ourselves—not just that we need particular items of equipment, but that we need equipment at all.

Computing may reshape college mathematics slowly also because computers raise harder, more fundamental pedagogical questions in mathematics than in other disciplines. Computers can thoroughly transform activities in a chemistry laboratory, but they need not change the basic ideas studied there. By contrast, modern computers handle so much of what we mathematicians traditionally teach that we are forced to rethink not only *how* we teach, but also *what* and *why*. Ironically, computers may have contributed so little in undergraduate mathematics just because they can do so much.

The cost of computer programming, measured in time and distraction, has been another impediment to mathematical computing. Is the effort of implementing, say, a simple Riemann sum program in Basic worth the mathematical insight it offers? Similar questions might seem to apply in other sciences, but experimental data generated in natural science laboratories is well suited to routine numerical manipulations; a few programs go a long way. Mathematical computing, being less circumscribed, is harder to "package," and the programming overhead is correspondingly higher.

Prospects for the Future

Given this history, why should things change? The simplest reason might be called "manifest destiny." Like it or not, computing is already changing undergraduates' views of and experience with mathematics, and the rate of change is increasing. In freshman calculus, symbol-manipulating and graph-sketching programs on *handheld* calculators already reduce a good share of canonical exam questions to button-pushing. (See [4] and [5]; note particularly their metaphorical titles.) We mathematicians can either applaud or condemn these changes, but we can't ignore them. We will either anticipate and use computing developments, or we will have to fend them off.

Good omens for computing in undergraduate mathematics can also be seen in hardware and software improvements. Mathematical computing is becoming more powerful, cheaper, and easier to use. For example, "computer algebra systems" (Macsyma, Maple, Reduce, SMP, and others) are starting to appear on student-type machines. These systems do much more than algebra; they are actually powerful ("awesome," in studentspeak) and flexible mathematics packages that perform a host of routine operations—symbolic algebra, formal differentiation and integration, series expansions, graph-sketching, numerical computations, matrix manipulations, and much more. Because no programming is needed (one-line commands handle most operations), all this power costs virtually nothing in distraction. For good or ill, computing is changing the mix of "working" and "thinking" that determine what it is to know and do mathematics.

The possibilities and problems computing raises for teaching would be important even if undergraduate mathematics were thriving. On the contrary, too few students study mathematics; of those who do, too few learn it deeply or well. Freshman calculus is a squeaky wheel, but general complaints are heard up and down the curriculum: students can't figure, can't estimate, can't read, can't write, can't solve problems, can't handle theory and so on. Lacking computers didn't cause all these problems, and having computers won't solve them all. Nevertheless, the climate for change is favorable (see, e.g., [2]).

Benefits and Opportunities

Computing can benefit undergraduate mathematics teaching in many ways. Understanding the context is important: Our goal is not more computers, but better mathematical learning.

1. *To make undergraduate mathematics more like "real" mathematics.* Mathematics as it is really used has many parts: formal symbol manipulation, numerical calculation, conjectures and experiments, "pure" ideas, modeling, and applications. Undergraduate students, especially in beginning courses, see mainly the first two at the expense of the others. Another mixture of ingredients might give students a better sense of context, and help them calculate more knowledgeably and effectively. By handling routine operations, computing can free time and attention for other things.

2. *To illustrate mathematical ideas.* Analytic concepts such as the derivative have numerical, geometric, and dynamical (i.e., time-varying) as well as analytic meaning. Pursuing graphical, numerical, and dynamical viewpoints is tedious or impossible by hand techniques. Doing so is easy and helpful with computing, especially if algebraic, graphical, and numerical manipulation are all available. Given these, a student might compute difference quotients algebraically, tabulate numerical values as a parameter varies, and observe the geometric behavior of the associated secant lines at various scales.

3. *To help students work examples.* Mathematicians know the value of concrete examples for understanding theorems and their consequences. Students need examples, too, but the points examples make are often obscured by computational difficulties. With computers, students can work more and better examples. In matrix algebra and statistics, large-scale problems become feasible. In calculus, subtle points can be clarified. The fundamental theorem, for example, is often misunderstood because students have insufficient experience with the "left side"—the integral defined by Riemann sums. With a machine to crunch the numbers, the "left side" makes numerical and geometric sense. When the area-under-the-graph function can be tabulated, graphed, and geometrically differentiated, for several integrands, then the idea of the theorem is hard to miss.

4. *To study, not just perform, algorithms.* Algorithmic methods—for matrix operations, polynomial factor-

ization, finding gcd's, and other operations—are now a way of mathematical life. Students continually perform algorithms, but seldom study them in their own right. Rudimentary algorithm analysis (e.g., the O-notation) could be an important and timely application of elementary calculus. Recursion, iteration, and list processing, viewed as general mathematical techniques, also deserve more attention than they get. Treating these topics efficiently means implementing them, or seeing them implemented, on computers.

5. *To support more varied, realistic, and illuminating applications.* Limited by hand computations, many applications of college mathematics are contrived and trite. There are too many farm animals, rivers, and exotic fencing schemes. Computing allows both larger-scale versions of traditional applications (e.g., larger matrices, larger data samples) and new applications altogether (e.g., ones requiring numerical methods.) More flexible, less circumscribed applications not only *do* more, they also *show* more of the mathematics underlying them. Physical applications, for example, should be part of the historical, mathematical, and intuitive fiber of elementary calculus. Hand techniques restrict the scope of feasible physical problems to those few that can be solved in closed form, using elementary functions. Other conceptually simple applications, like many from economics, lead to high-order polynomial equations, and so are also taboo. Only simple numerical methods (root finding, numerical integration, etc.) are needed to make such applications tractable.

6. *To exploit and improve geometric intuition.* Graphs of all kinds give invaluable insight into mathematical phenomena. With computer graphics, attention can shift from the mechanics of obtaining graphs—a substantial topic in elementary calculus—to how graphs represent analytic information. Sophisticated graphics (surfaces in three dimensions, families of curves, fractals) practically require computing. They ease difficult learning transitions: from one variable to several, from function to family of functions, from real domain to complex domain, from pointwise to uniform convergence, from step to step in iterative constructions.

7. *To encourage mathematical experiments.* The polished theorem-proof-remark style of mathematical writing hides the fact that mathematics is created actively, by trial, error, and discovery. Students can learn the same way, if the labor of experimenting is not too great. With computing, students can discover that square matrices "usually" have full rank, that differentiable functions look straight at small scale, and that there is pattern to the coefficients of a binomial expansion.

   Mathematical experimentation is good both as a teaching tool and as an active, engaged attitude toward mathematics. We mathematicians often try to inculcate this attitude in students, begging them to "try something." Interestingly, the opposite problem—an excessively experimental attitude, or "hacking"—plagues computer science. Will computers breed *mathematical* hackers? Would that be a bad thing?

8. *To facilitate statistical analysis and enrich probabilistic intuition.* Data analysis in mathematical statistics is highly computational. Machines allow larger samples, and thus greater reliability and verisimilitude. Students see more *analysis,* and more of its power, with less distraction. Computing in undergraduate statistics is already becoming routine. As *methods* of data analysis becomes easier, *choosing* methods and interpreting their results becomes harder. Informed statistical analysis requires sound probabilistic intuition. Probabilistic viewpoints are also essential, of course, in classical analysis and in modern physical applications. By simulating random phenomena, computers illustrate probabilistic viewpoints concretely. Monte Carlo integration methods, for example, combine ideas from elementary calculus and probability, and show relations between them.

9. *To teach approximation.* The idea of approximation is important throughout mathematics and its applications. When students use mathematics—in other courses and in careers—they will certainly use numerical methods. Yet students' learning experience, especially in beginning courses, is almost entirely based on exact, algebraic methods: explicit functions, closed-form solutions, elementary antiderivatives, and the like. Numerical and graphical illustrations of approximation ideas are computationally expensive, but essential for understanding. Machine computing makes them possible.

   To use approximation effectively, students need some idea of error analysis. Error estimate formulas are especially intractable for hand computation because they usually involve higher derivatives, upper bounds, and other mysterious ingredients. High-level computing (root-finding, symbolic differentiation, numerical methods) helps students understand error estimation without foundering in distracting calculations.

10. *To prepare students to compute effectively—but skeptically—in careers.* Applied mathematics students who pursue technical professions (engineering, actuarial science, business, and industry) will use mathematical computing in many forms. Some kinds of software (e.g., differential equation solvers, statistical analysis packages) are standard; students should see some of them in advance. Even more important than working knowledge of particular programs is a sound *mathematical* understanding of how those—or any—programs address the problems they purport to solve. Arbitrary choices—estimates, simplifications, stopping rules—are always implicit in applications software. Duly skeptical users must understand these choices and how they affect computed results.

11. *To show the mathematical significance of the computer revolution.* The relation between mathematics and computer science offers excellent object lessons in the interconnectedness of knowledge. As a matter of general education, and for practical reasons, students should learn something about the mathematical foundations of computing, and about the mathematical problems computing raises.

12. *To make higher-level mathematics accessible to students.* Undergraduates in the natural sciences have always participated in serious research. Computing offers similar opportunities in mathematics as it strengthens the concrete, empirical side of mathematical research. With powerful graphics, students might investigate the fine structure of Mandelbrot sets, observe evolution of dynamical systems, and explore geodesics on complex surfaces.

Resource Requirements

No one doubts that educational uses of computing require hardware and software. It is less well understood that resource requirements only begin there. Chemistry departments require more than chemicals and equipment to support their laboratory courses. In the same way, more than hardware and software is needed if the benefits of having hardware and software are to be realized.

College mathematics teachers who use computing face common problems. Some problems are local (e.g., securing institutional support) and some are national. Many stem from the fact that computing is relatively new to mathematics. Mathematicians are just beginning the resource management tasks our scientific colleagues have worked at for decades: convincing administrators, purchasing and maintaining equipment, modifying time-hardened courses, developing curricular rationales, and articulating what we are doing. Undergraduate science departments write proposals, carry out supported projects, and administer grants as a matter of habit. In mathematics departments, these habits are less ingrained, and "machinery" to support them—administrative help, program and deadline information, accounting procedures—is usually primitive.

1. *Technical support.* Natural science departments maintain a complete apparatus of support services for their laboratory courses: equipment maintenance, classroom demonstration equipment, dedicated space, and paid student assistants. As mathematics departments develop and use their own versions of "instrumentation," the same support needs arise.

2. *Institutional support.* Most colleges have computers, but not necessarily the right ones, or in the right places, for mathematical use. Less tangible forms of institutional support are just as important: teaching loads that credit faculty time for developing and staffing mathematical laboratories, tenure and promotion procedures that reward such work, and administrative support for matching money for grant proposals.

3. *Time.* Realizing the mathematical benefits of computing, whatever they are, costs time—ours and our students'. Not all benefits will prove worth having, but for those we judge worthwhile, time should be provided, and accounted for honestly. Instructional computing, like other new things, often begins with a trail-blazing department zealot, for whom the work is its own reward. Eventually, ownership and responsibility should be shared. Unless time is made available, computing will remain the zealot's private province.

4. *Courseware.* It seems historically inevitable that computing will change mathematics courses and course materials. If we mathematicians are to manage the process, we need hardware, software, and "courseware"—instructional material (manuals, exercises, tests, discursive material, and full textbooks) that thoughtfully integrates, rather than sim-

ply appends, the computing viewpoint into substantial mathematics courses. The necessary hardware and software exist, or will soon. The laggard, inevitably, is courseware. Robust courseware is expensive and time-consuming to develop, and it must find a precarious balance between being too specific to be portable and being too general to be useful.

5. *Technical information.* Hardware and software change rapidly. In order to choose equipment wisely and use it effectively, mathematicians need technical specifications for hardware and software, critical reviews of educational software, reliable price data, and (hardest of all) a sense of the future. Because hardware usually outruns software, the naive strategy (choose software, then hardware to run it) guarantees obsolete hardware. Should our professional societies marshal the expertise we need? How?

6. *Shared experience.* Who is doing what? Where? With what equipment? What worked? How was topic X treated? Undergraduate mathematical computing, like any quickly developing field, depends on communication if the wheel is not repeatedly to be reinvented. Several models exist: The Sloan Foundation supports several college projects and a newsletter on teaching uses of computer algebra systems, and occasional conferences on the subject. The Maple group at the University of Waterloo publishes a users' group newsletter and organizes electronic communications for an interest group in teaching uses of computer algebra. The *College Mathematics Journal* carries a regular column (see[1]) on instructional computing. The MAA and its Sections sponsor minicourses nationally and short courses regionally. More such efforts are needed.

Open Questions

Technical, financial, institutional, and logistical considerations notwithstanding, the most difficult and interesting questions computing raises are mathematical and pedagogical. Although few answers are hazarded here, most users face some of these questions, at least implicitly. As a discipline, we face them all regularly.

1. *Will computers reduce students' ability to calculate by hand? If so, is that a bad thing?* When computers do routine manipulations in mathematics courses, students must do something else. How will students who have never mastered routine calculation, or those who enjoy it, fare in such courses? Will seeing the results but not the process of calculations help them understand, or further mystify them? If hand computation builds algebraic intuition—"symbol sense"—will machine computation destroy it?

2. *How should analytic and numerical viewpoints be balanced?* To estimate $\int_{-1}^{1} \frac{1}{1+x^2}\,dx$ numerically as 1.57 is routine. To see analytically that the answer is $\frac{\pi}{2}$ is memorable. Both facts are worth knowing. Will students learn them both in the calculus course of the future? To paraphrase Richard Askey, exact solutions are precious *because* they are rare. Will students learn this? Will we remember it?

3. *How does computing change what students should know?* Traditional courses are full of methods and viewpoints that arose to compensate for the limitations of hand computation. Do new computational tools render these topics obsolete? More generally, should we teach mathematical techniques machines can do better? Some things, surely, but which, and why? Partial fraction decomposition? The square-root algorithm? For topics we keep, will we forbid computers? What will replace topics we discard?

4. *Will the mechanics of computing obscure the mathematics?* We teach mathematics, not computing, and mathematics syllabi are already full. How will we use computing to teach mathematics without distracting technical excursions? How will we gauge whether computing effort is commensurate with the mathematical insight it gives? Can computing *save* teaching time? Anecdotal evidence suggests that calculus students can use high-level programs without undue difficulty or distraction. Can pre-calculus students do the same?

5. *How will computing affect advanced courses?* Mathematical computing frequently occurs in lower-level courses, like calculus, which have other educational goals than preparation for advanced courses. Will alumni of such courses be better or worse prepared for advanced mathematics? Should advanced courses change along with introductory courses? Can computing improve advanced courses in their own right?

6. *Computing and remediation.* Remedial course emphasize mechanical operations. Will relegating rou-

tine operations to machines reduce the need for remediation? Or could deeper, more idea-oriented courses require *more* remediation for weaker students? In either case, how can computing help students in remedial courses? Can such students use advanced computing, or do they lack some necessary sophistication?

7. *CAI vs. tool-driven computing.* In some applications, computers act as intelligent tutors, leading students through carefully prescribed tasks. In others, computers are flexible tools: students decide where and how to use them. Where is each model useful?

8. *Equity and access to computing.* Sophisticated computing on powerful computers is still financially expensive and, for most people, not always on hand. Calculators, for the price of four books, already handle graphical, numerical, and algebraic (including matrix) calculations. Sophisticated calculators might both radically "democratize" high-level computing, and make it as natural as hand-held arithmetic computation. Will they? Should they?

9. *Will computing help students learn mathematical ideas more deeply, more easily, and more quickly?* Conjecturally, yes, but the conclusion is not foregone. For undergraduate mathematics, this is the bottom line.

References

[1] R. S. Cunningham, David A. Smith. "A Mathematics Software Database Update." *The College Mathematics Journal,* 18:3 (May, 1987) 242-247.

[2] Ronald G. Douglas, Ed. *Toward a Lean and Lively Calculus: Report of the Conference/Workshop To Develop Curriculum and Teaching Materials for Calculus at the College Level.* Mathematical Association of America, 1986.

[3] Richard E. Ewing, Kenneth I. Gross, Clyde F. Martin, Eds. *The Merging of Disciplines: New Directions in Pure, Applied, and Computational Mathematics.* Springer Verlag, 1987.

[4] Thomas Tucker. "Calculators with a College Education?" *FOCUS,* 7:1 (January-February, 1987) 1.

[5] Herbert E. Wilf. "The Disk with a College Education." *The American Mathematical Monthly,* 89:1 (January, 1982) 4-8.

Planning Committee:

LESLIE V. FOSTER, San Jose State University
SAMUEL GOLDBERG, Alfred P. Sloan Foundation
ELIZABETH R. HAYFORD, Associated Colleges of the Midwest
RICHARD KAKIGI, California State University, Hayward
ZAVEN KARIAN, Denison University
JOHN KENELLY, Clemson University
WARREN PAGE, Ohio State University
DAVID SMITH, Duke University
LYNN ARTHUR STEEN (Chair), St. Olaf College
MELVIN C. TEWS, College of the Holy Cross
THOMAS TUCKER, Colgate University
PAUL ZORN, St. Olaf College

National Science Foundation Participants:

KENNETH GROSS, Division of Mathematical Sciences
DUNCAN E. MCBRIDE, Office of College Science Instrumentation
BASSAM Z. SHAKHASHIRI, Assistant Director for Science and Engineering Education
ARNOLD A. STRASSENBURG, Director, Division of Teacher Preparation and Enhancement
JOHN THORPE, Deputy Director, Division of Materials Development, Research, and Informal Science Education
ROBERT WATSON, Head, Office of College Science Instrumentation

PARTICIPANTS

GENERAL COMMENTS FROM PARTICIPANTS

- Students do not think of calculus as dull. Difficult yes, boring no.
- Less emphasis on praising the organizers
- The speakers don't know any more about this than we do, sometimes considerably less.
- Publishers have promised much and delivered little.
- We must be patient
- What about a "problem bank"?
- Lots of coffee and donuts are so helpful in getting people to meet and talk.

//

These responses were selected by the editor (LAS) from forms collected from the participants. Within the editor's categories the scribe (PLR) selected the order.

Calculus for a New Century: Participant List

Washington, D.C., October 28-29, 1987

ADAMS, WILLIAM. National Science Foundation, Washington DC 20550. [Program Director, 1800 G Street NW] 202-357-3695.

AGER, TRYG. Stanford University, Stanford CA 94305. [Sr. Research Associate, IMSSS-Ventura Hall] 415-723-4117.

ALBERS, DONALD. Menlo College, Menlo CA 94025. [Chairman, Dept of Math & CS 1000 El Camino Real] 415-323-6141.

ALBERTS, GEORGE. National Security Agency, Fort Meade MD 20755. [Executive Manager, PDOD] 301-688-7164.

ALEXANDER, ELAINE. California State University, Sacramento CA 95819. [Professor, Dept of Math & Stat 6000 J Street] 916-278-6534.

ALLEN, HARRY. Ohio State University, Columbus OH 43210. [Professor, Dept of Math 231 West 18th Avenue] 614-292-4975.

ALLEN, PAUL. University of Alabama, Tuscaloosa AL 35487. [Dept of Math P.O. Box 1416] 205-348-1966.

ANDERSON, R.D. Louisiana State University, Baton Rouge LA 70809. [Professor Emeritus, 2954 Fritchie Dr.]

ANDERSON, EDWARD. Thomas Jefferson High School, Alexandria VA 22312. [Chair Math Dept, 6560 Braddock Road] 703-941-7954.

ANDERSON, NANCY. University of Maryland, College Park MD 20742. [Professor, Dept of Psychology] 301-454-6389.

ANDREWS, GEORGE. Oberlin College, Oberlin OH 44074. [Professor, Dept of Mathematics] 216-775-8382.

ANDRE, PETER. United States Naval Academy, Annapolis MD 21402. [Professor, Math Dept] 301-267-3603.

ANTON, HOWARD. Drexel University, Cherry Hill NJ 08003. [304 Fries Lane] 609-667-9211.

ARBIC, BERNARD. Lake Superior State College, Sault Ste. Marie MI 49783. [Professor, Dept of Mathematics] 906-635-2633.

ARCHER, RONALD. Unviersity of Massachusetts, Amherst MA 01003. [Professor, Dept of Chemistry] 413-545-1521.

ARMSTRONG, JAMES. Educational Testing Service, Princeton NJ 08541. [Senior Examiner] 609-734-1469.

ARMSTRONG, KENNETH. University of Winnipeg, Winnipeg Man. Canada R3B2E9. [Chair, Dept of Math and Statistics] 204-786-9367.

ARNOLD, DAVID. Phillips Exeter Academy, Exeter NH 03833. [Instructor] 603-772-4311.

ARZT, SHOLOM. The Cooper Union, New York NY 10003. [Professor, Cooper Square] 212-254-6300.

ASHBY, FLORENCE. Montgomery College, Rockville MD 20850. [Professor, 109 South Van Buren St.] 301-279-5202.

AVIOLI, JOHN. Christopher Newport College, Newport News VA 23606. [50 Shoe Lane] 804-599-7065.

BAILEY, CRAIG. United States Naval Academy, Annapolis MD 21402. [Associate Professor, PMath Dept] 301-267-3892.

BAKKE, VERNON. University of Arkansas, Fayetteville AK 72701. [Professor, Dept of Mathematical Sciences] 501-575-3351.

BARBUSH, DANIEL. Duquesne University, Pittsburgh PA 15221. [Assistant Professor, 512 East End Avenue] 412-371-1221.

BARRETT, LIDA. Mississippi State University, Mississippi State MS 39762. [Dean College of Arts & Sciences, P.O. Drawer AS] 601-325-2644.

BARTH, CRAIG. Brooks-Cole Publishing Co., Pacific Grove CA 93950. [Editor, 511 Forest Lodge Rd] 408-373-0728.

BARTKOVICH, KEVIN. N.C. School of Science & Mathematics, Durham NC 27705. [Broad St. & West Club Blvd.] 919-286-3366.

BASS, HELEN. Southern Connecticut State Univ, North Haven CT 06473. [Professor, 59 Garfield Avenue] 203-397-4220.

BATRA, ROMESH. University of Missouri, Rolla MO [Professor, Dept of Engineering Mechanics] 314-341-4589.

BAUMGARTNER, EDWIN. LeMoyne College, Syracuse NY 13214. [Associate Professor, Dept of Mathematics] 315-445-4372.

BAVELAS, KATHLEEN. Manchester Community College, Wethersfield CT 06109. [Instructor, 46 Westway] 203-647-6198.

BAXTER, NANCY. Dickinson College, Carlisle PA 17013. [Assistant Professor] 717-245-1667.

BEAN, PHILLIP. Mercer University, Macon GA 31207. [Professor, 1406 Coleman Avenue] 912-744-2820.

BECKMANN, CHARLENE. Western Michigan University, Muskegon MI 49441. [Student, 1716 Ruddiman] 616-755-4300.

BEDELL, CARL. Philadelphia College of Textiles & Sci, Philadelphia PA 19144. [Associate Professor, Henry Ave. & Schoolhouse Lane] 215-951-2877.

BEHR, MERLYN. National Science Foundation, Washington DC 20550. [Program Director, 1800 G Street NW Room 638] 202-357-7048.

BENHARBIT, ABDELALI. Penn State University, York PA 17403. [Assistant Professor, 264 Brookwood Dr South] 717-771-4323.

BENSON, JOHN. Evanston Township High School,

Evanston IL 60202. [Teacher, 715 South Blvd.] 312-328-8019.

BENTON, EILEEN. Xavier University of Louisiana, New Orleans LA 70125. [Professor, 7325 Palmetto] 504-486-7411.

BERAN, DAVID. University of Wisconsin, Superior WI 54880. [Associate Professor] 715-394-8485.

BESSERMAN, ALBERT. Kishwaukee College, DeKalb IL 60115. [Instructor, 924 Sharon Dr] 815-756-5921.

BEST, KATHY. Mount Olive College, Mount Olive NC 28365. [Professor, Dept of Mathematics] 919-658-2502.

BLUMENTHAL, MARJORY. National Research Council-CSTB, Washington DC 20418. [Study Director, 2101 Constitution Ave NW] 202-334-2605.

BOBO, RAY. Georgetown University, Washington DC 20057. [Associate Professor, Dept of Mathematics] 202-625-4381.

BODINE, CHARLES. St. George's School, Newport RI 02840. [Teacher] 401-849-4654.

BORDELON, MICHAEL. College of the Mainland, Texas City TX 77591. [8001 Palmer Hwy] 409-938-1211.

BORK, ALFRED. University of California, Irvine Ca 92717. [Professor, ICS Department] 714-856-6945.

BOSSERT, WILLIAM. Harvard University, Cambridge MA 20318. [Professor, Aiken Computational Laboratory] 617-495-3989.

BOUTILIER, PHYLLIS. Michigan Technological University, Houghton MI 49931. [Director] 906-487-2068.

BOYCE, WILLIAM. Rensselear Polytechnic Institute, Troy NY 12180. [Professor, Dept of Mathematical Sciences] 518-276-6898.

BRABENEC, ROBERT. Wheaton College, Wheaton IL 60187. [1006 North Washington] 312-260-3869.

BRADBURN, JOHN. AMATYC MAA, Elgin IL 60123. [1850 Joseph Ct] 312-741-4730.

BRADEN, LAWRENCE. Iolani School, Honolulu HI 96826. [563 Kamoku St] 808-947-2259.

BRADY, STEPHEN. Wichita State University, Wichita KS 67208. [Associate Professor, Dept of Math 1845 North Fairmount] 316-689-3160.

BRAGG, ARTHUR. Delaware State College, Dover DE 19901. [Chair, Dept of Mathematics] 302-736-5161.

BRITO, DAGOBERT. Rice University, Houston TX 77251. [Professor, Department of Economics] 701-527-4875.

BROADWIN, JUDITH. Jericho High School, Jericho NY 11753. [Teacher, 6 Yates Lane] 516-681-4100.

BROWNE, JOSEPH. Onondaga Community College, Syracuse NY 13215. [Associate Professor, Dept of Mathematics] 315-469-7741.

BROWN, CLAUDETTE. National Research Council-MSEB, Washington DC 20418. [2101 Constitution Avenue NW]

BROWN, DONALD. St. Albans School, Washington DC 20016. [Chairman Dept Math, Mount St. Alban] 202-537-6576.

BROWN, JACK. University of Arkansas, Fayetteville AR 72701. [Student, Box 701 Yocum Hall] 501-575-5569.

BROWN, ROBERT. University of Kansas, Lawrence KS 66045. [Director Undergrad Studies, Dept of Mathematics] 913-864-3651.

BROWN, MORTON. University of Michigan, Ann Arbor MI 48109. [Professor, Dept of Mathematics] 313-764-0367.

BRUBAKER, MARVIN. Messiah College, Grantham PA 17027. [Professor] 717-766-2511.

BRUNELL, GLORIA. Western Connecticut State University, New Milford CT 06776. [Professor, 8 Howe Road] 203-355-1569.

BRUNSTING, JOHN. Hinsdale Central High School, Hinsdale IL 60521. [Dept. Chair, 55th and Grant Streets] 312-887-1340.

BUCCINO, ALPHONSE. University of Georgia, Athens GA 30602. [Dean, College of Education] 404-542-3030.

BUCK, CREIGHTON. University of Wisconsin, Madison WI 53705. [Professor, 3601 Sunset Drive] 608-233-2592.

BUEKER, R.C. Western Kentucky University, Bowling Green KY 42101. [Head, Dept of Mathematics] 502-745-3651.

BURDICK, BRUCE. Bates College, Lewiston ME 04240. [Assistant Professor, P.O. Box 8232] 207-786-6143.

BYHAM, FREDERICK. State Univ College at Fredonia, Fredonia NY 14063. [Associate Professor, Dept of Math and Computer Science] 716-673-3193.

BYNUM, ELWARD. National Institutes of Health, Bethesda MD 20892. [Director MARC Program, Westwood Bld. Room 9A18] 301-496-7941.

CALAMIA, LOIS. Brookdale Community College, East Brunswick NJ 08816. [Assistant Professor, 12 South Drive] 201-842-1900.

CALLOWAY, JEAN. Kalamazoo College, Kalamazoo MI 49007. [Professor, 1200 Academy Street] 616-383-8447.

CAMERON, DWAYNE. Old Rochester Regional School District, Mattapoisett MA 02739. [Coordinator, 135 Marion Rd.] 617-758-3745.

CAMERON, DAVID. United States Military Academy, West Point NY 10996. [Head, Dept of Mathematics] 914-938-2100.

CANNELL, PAULA. Anne Arundel Community College, Annapolis MD 21401. [1185 River Bay Rd] 301-260-4584.

CANNON, RAYMOND. Baylor University, Waco TX 76798. [Professor, Mathematics Dept] 817-755-3561.

CAPPUCCI, ROGER. Scarsdale High School, Scarsdale NY 10583. [Teacher] 914-723-5500.

CARLSON, CAL. Brainerd Community College, Brainerd NY 56401. [Teacher, College Drive] 218-829-5469.

CARLSON, DONALD. University of Illinois-Urbana, Urbana IL 61801. [Professor, 104 South Wright Street] 217-333-3846.

CARNES, JERRY. Westminster Schools, Atlanta GA 30327. [Chair Math Dept, 1424 West Paces Ferry Road] 404-355-8673.

CARNEY, ROSE. Illinois Benedictine College, Lisle IL 60532. [Professor, Dept of Math 5700 College Road] 312-960-1500.

CARR, JAMES. Iona College, New Rochelle NY 10801. [Assistant Professor, PDept of Mathematics] 914-633-2416.

CASCIO, GRACE. Northeast Louisiana University, Monroe LA 71209. [Assistant Professor, Dept of Mathematics] 318-342-4150.

CASE, BETTYE-ANNE. Florida State University, Tallahassee FL 32306. [Professor, Department of Mathematics] 904-644-2525.

CASTELLINO, FRANCIS. University of Notre Dame, Notre Dame IN 46556. [Dean, College of Science]

CAVINESS, B.F. National Science Foundation, Washington DC 20550. [Program Director, 1800 G Street NW Room 304] 202-357-9747.

CHABOT, MAURICE. University of Southern Maine, Portland ME 04103. [Chair, Dept of Math 235 Science Bldg] 207-780-4247.

CHANDLER, EDGAR. Paradise Valley Community College, Phoenix AZ 85032. [Instructor, 18401 North 32nd Street] 602-493-2803.

CHAR, BRUCE. University of Tennessee, Knoxville TN 37996. [Associate Professor, Dept of Computer Sci] 615-974-4399.

CHILD, DOUGLAS. Rollins College, Winter Park FL 32789. [Professor, Dept Math Sciences 100 Holt Ave] 305-646-2667.

CHINN, WILLIAM. 539 29th Avenue, San Francisco CA 94121. [Emeritus, City College of San Francisco] 415-752-1637.

CHIPMAN, CURTIS. Oakland University, Rochester MI 48309. [Associate Professor] 313-370-3440.

CHRISTIAN, FLOYD. Austin Peay State University, Clarksville TN 37044. [Associate Professor, Dept of Mathematics] 615-648-7821.

CHROBAK, MICHAEL. Texas Instruments Inc., Dallas TX 75265. [Product Manager, P.O. Box 655303] 214-997-2010.

CLARK, ROBERT. Macmillan Publishing Co., New York NY 10022. [Editor, 866 3rd Avenue 6th floor] 212-702-6773.

CLARK, CHARLES. University of Tennessee, Knoxville TN 37996. [Professor, Dept of Mathematics] 615-974-4280.

CLEAVER, CHARLES. The Citadel, Charleston SC 29409. [Head, Dept of Math & Computer Science] 803-222-8069.

CLIFTON, RODNEY. Brown University, Providence RI 02912. [Professor, Division of Engineering] 401-863-2855.

COALWELL, RICHARD. Lane Community College, Eugene OR 97401. [356 Paradise Court] 503-747-4501.

COCHRAN, ALLAN. University of Arkansas, Fayettevile AR 72701. [Professor, Dept of Mathematical Sciences SE] 501-575-3351.

COHEN, MICHAEL. Bay Shore High School, Merrick NY 11566. [Teacher, 2036 Dow Avenue] 516-379-5356.

COHEN, SIMON. New Jersey Institute of Technology, Newark NJ 07102. [Dept of Mathematics] 201-596-3491.

COHEN, JOEL. University of Denver, Denver CO 80208. [Associate Professor, Dept Math & Computer Science] 303-871-3292.

COHEN, MICHAEL. University of Maryland, College Park MD 20742. [Associate Professor, PSchool of Public Affairs Morrill Hall] 301-454-7613.

COLLINS, SHEILA. Newman School, New Orleans LA 70115. [Chair, Math Dept 1903 Jefferson Ave.] 503-899-5641.

COLLUM, DEBORAH. Oklahoma Baptist University, Shawnee OK 74801. [Assistant Professor, 500 West University] 405-275-2850.

COMPTON, DALE. National Academy of Engineering, Washington DC 20418. [Senior Fellow, 2101 Constitution Avenue NW] 202-334-3639.

CONNELL, CHRIS. Associated Press.

CONNORS, EDWARD. University of Massachusetts, Amherst MA 01003. [Professor, Dept of Mathematics & Statistics] 413-545-0982.

COONCE, HARRY. Mankato State University, Mankato MN 56001. [Professor, Box 41] 507-389-1473.

COPES, LARRY. Augsburg College, Minneapolis MN 55075. [Chair, 731 21st Avenue South] 612-330-1064.

CORZATT, CLIFTON. St. Olaf College, Northfield MN 55057. [Associate Professor] 507-633-3415.

COVENEY, PETER. Harper & Row Publishers Inc., New York NY 10022. [Editor, 10 East 53rd Street] 212-207-7304.

COX, LAWRENCE. National Research Council-BMS, Washington DC 20418. [Staff Director, 2101 Constitution Avenue NW] 202-334-2421.

COZZENS, MARGARET. Northeastern University, Boston MA 02115. [Associate Professor, Dept of Math 360 Huntington Ave] 617-437-5640.

CRAWLEY, PATRICIA. Nova High School, Sunrise FL 33322. [Dept Head, 2021 N.W. 77th Avenue] 305-742-6452.

CROOM, FREDERICK. University of the South, Sewanee TN 37375. [Professor, University Station] 615-598-1248.

CROWELL, RICHARD. Dartmouth College, Hanover NH 03755. [Professor, 16 Rayton Road] 603-646-2421.

CROWELL, SHARON. O'Brien & Associates, Alexandria VA 22314. [Associate, Carriage House 708 Pendleton St] 703-548-7587.

CUMMINGS, NORMA. Arapahoe High School, Littleton CO 80122. [Teacher, 2201 East Dry Creek Rd] 303-794-2641.

CURTIS, PHILIP. UCLA, Los Angeles CA 90024. [Professor, Dept of Mathematics] 213-206-6901.

CURTIS, EDWARD. University of Washington, Seattle WA 98195. [Professor, Padelford Hall] 206-543-1945.

CUTLER-ROSS, SHARON. Dekalb College, Clarkston GA 30021. [Associate Professor, 555 North Indian Creek] 404-299-4163.

CUTLER, ARNOLD. Moundsview High School, New Brighton MN 55112. [Teacher, 1875 17th Street NW] 612-633-4031.

DANCE, ROSALIE. District of Columbia Public Schools, Washington DC 20032. [Teacher, Ballou High School 3401 4th St. SE] 202-767-7071.

DANFORTH, KATRINE. Corning Community College, Corning NY 14830. [Associate Professor, P.O. Box 252] 607-962-4034.

DANFORTH, ERNEST. Corning Community College, Corning NY 14830. [Associate Professor, P.O. Box 252] 607-962-9243.

DARAI, ABDOLLAH. Western Illinois University, Macomb IL 61455. [Assistant Professor, Dept of Math Morgan Hall] 309-298-1370.

DAVIDON, WILLIAM. Haverford College, Haverford PA 19041. [Professor, Mathematics Dept] 215-649-0102.

DAVIES, RICHARD. OTG U.S. Congress, Washington DC 20510. [Analyst, Office of Technology Assessment] 202-228-6929.

DAVISON, JACQUE. Anderson College, Anderson SC 29621. [Instructor, 316 Boulevard] 803-231-2165.

DAVIS, NANCY. Brunswick Technical College, Shallotte NC 28459. [Instructor, Rt 2 Box 143-2A] 919-754-6900.

DAVIS, RONALD. Northern Virginia Community College, Alexandria VA 22205. [Professor, 3001 North Beauregard St] 703-845-6341.

DAVIS, FREDERIC. United States Naval Academy, Annapolis MD 21402. [Professor, Mathematics Dept] 301-267-2795.

DAWSON, JOHN. Penn State York, York PA 17403. [Professor, 1031 Edgecomb Avenue] 717-771-4323.

DE-COMARMOND, JEAN-MARC. French Scientific Mission, Washington DC 20007. [Scientific Attache, 4101 Reservoir Road NW] 202-944-6230.

DEETER, CHARLES. Texas Christian University, Fort Worth TX 76129. [Professor, Dept of Math Box 32903] 817-921-7335.

DEKEN, JOSEPH. National Science Foundation, Washington DC 20550. [Program Director, 1800 G Street NW Room 310] 202-357-9569.

DELIYANNIS, PLATON. Illinois Institute of Technology, Chicago IL 60616. [Associate Professor, Dept of Mathematics] 312-567-3170.

DELLENS, MICHAEL. Austin Community College, Austin TX 78768. [Instructor, P.O. Box 2285] 512-495-7256.

DEMANA, FRANKLIN. Ohio State University, Columbus OH 43210. [Professor, Dept of Math 231 West 18th Avenue] 614-292-0462.

DEMETROPOULUS, ANDREW. Montclair State College, Upper Montclair NJ 07043. [Chair, Dept of Math & Computer Science] 201-893-5146.

DENLINGER, CHARLES. Millersville University, Millersville PA 17551. [Professor, Dept of Math & Computer Science] 717-872-4476.

DEVITT, JOHN. University of Saskatchewan, Saskatoon Sask. Canada S7N0W0. [Asociate Professor, Dept of Mathematics College Drive] 306-966-6114.

DIFRANCO, ROLAND. University of the Pacific, Stockton CA 95211. [Professor, Mathematics Dept] 209-946-3026.

DICK, THOMAS. Oregon State University, Corvallis OR 97331. [Assistant Professor, Mathematics Dept] 503-754-4686.

DIENER-WEST, MARIE. Johns Hopkins University, Baltimore MD 21205. [550 North Broadway 9th floor] 301-955-8943.

DION, GLORIA. Penn State Ogontz, Abington PA 19001. [1600 Woodland Avenue] 215-752-9595.

DIX, LINDA. National Research Council-OSEP, Washington DC 21408. [Project Officer, 2101 Constitution Avenue NW] 202-334-2709.

DJANG, FRED. Choate Rosemary Hall, Wallingford CT 06492. [Chairman Math, P.O. Box 788] 203-269-7722.

DODGE, WALTER. New Trier High School, Winnetka IL 60093. [Teacher, 385 Winnetka Ave] 312-446-7000.

DONALDSON, JOHN. Amer. Society for Engineering Education, Washington DC 20036. [Deputy Executive Director, 11 Dupont Circle Suite 200] 202-293-7080.

DONALDSON, GLORIA. Andalusia High School, Andalusia AL 36420. [Chair Math Science Division, P.O. Box 151] 205-222-7569.

DORNER, GEORGE. Harper College, Palatine IL 60067. [Dean, Algonquin/Ruselle Roads] 312-397-3000.

DORNER, BRYAN. Pacific Lutheran University, Tacoma WA 98447. [Associate Professor, Dept of Mathematics] 206-535-8737.

DOSSEY, JOHN. Illinois State University, Eureka IL 61530. [Professor, RR #1 Box 33] 309-467-2759.

DOTSETH, GREGORY. University of Northern Iowa, Cedar Falls IA 50613. [Dept of Math & Computer Science] 319-273-2397.

DOUGLAS, RONALD. SUNY - Stony Brook, Stony Brook NY 11794. [Dean, Physical Sciences & Math] 516-632-6993.

DREW, JOHN. College of William and Mary, Williamsburg VA 23185. [Associate Professor, Math Dept] 804-253-4481.

DUTTON, BRENDA. Spring Hill College, Mobile AL 36608. [4000 Dauphin St.] 205-460-2212.

DWYER, WILLIAM. University of Notre Dame, Notre Dame IN 46556. [Chair, Dept of Mathematics]

DYER, DAVID. Prince George's College, Largo MD 20772. [Associate Professor, Math Dept 301 Largo Road] 301-322-0461.

DYKES, JOAN. Edison Community College, Ft. Myers FL 33907. [Instructor, 8099 College Parkway SW] 813-489-9255.

DYMACEK, WAYNE. Washington and Lee University, Lexington VA 24450. [Associate Professor] 703-463-8805.

EARLES, GAIL. St. Cloud State University, St. Cloud MN 56301. [Chair, Dept of Math & Statistics] 612-255-3001.

EARLES, ROBERT. St. Cloud State University, St. Cloud MN 56301. [Professor, Dept of Math & Statistics] 612-255-2186.

EBERT, GARY. University of Delaware, Newark DE 19716. [Professor, Math Sci Dept] 302-451-1870.

EDISON, LARRY. Pacific Lutheran University, Tacoma WA 98447. [Professor] 206-535-8702.

EDLUND, MILTON. Virginia Polytech & State Univ, Blacksburg VA 24061. [Professor, Dept of Mechanical Engineering] 703-951-1957.

EDWARDS, CONSTANCE. IPFW, Ft. Wayne IN 46805. [Associate Professor, Math Dept] 219-481-6229.

EDWARDS, BRUCE. University of Florida, Gainesville FL 32611. [Associate Chair, Dept of Math 201 Walker Hall] 904-392-0281.

EGERER, GERALD. Sonoma State University, Rodnert Park CA 94928. [Professor, Dept of Economics] 707-664-2626.

EHRET, ROSE-ELEANOR. Holy Name College, Oakland CA 94619. [Professor, 3500 Mountain Blvd] 415-436-0111.

EIDSWICK, JACK. University of Nebraska, Lincoln NE 68588. [Professor, Dept of Mathematics and Statistics] 402-472-3731.

EMANUEL, JACK. University of Missouri-Rolla, Rolla MO 65401. [Professor, Dept of Civil Engineering] 314-341-4472.

ERDMAN, CARL. Texas A & M University, College Station TX 77843. [Associate Dean, 301 Wisenbaker Engr Res Center] 409-845-5220.

ESLINGER, ROBERT. Hendrix College, Conway AR 72032. [Associate Professor, Dept of Mathematics] 501-450-1254.

ESTY, EDWARD. Childrens Television Workshop, Chevy Chase MD 20815. [4104 Leland Street] 301-656-7274.

ETTERBEEK, WALLACE. Calif State University, Sacramento CA 95819. [Professor, Math Dept 6000 J Street] 916-278-6361.

FAIR, WYMAN. University of North Carolina, Asheville NC 28804. [Professor, Mathematics Dept 1 University Heights] 704-251-6556.

FAN, SEN. University of Minnesota, Morris MN 56267. [Associate Professor, Math Discipline] 612-589-2211.

FARMER, THOMAS. Miami University, Oxford OH 45056. [Associate Professor, Dept of Math & Stat] 513-529-5822.

FASANELLI, FLORENCE. National Science Foundation, Washington DC 20550. [Associate Program Director, 1800 G Street NW Room 635] 202-357-7074.

FERRINI-MUNDY, JOAN. University of New Hampshire, Durham NH 03824. [Associate Professor, Dept of Mathematics] 603-862-2320.

FERRITOR, DANIEL. University of Arkansas, Fayetteville AK 72701. [Chancellor, 425 Administration Building] 501-575-4148.

FIFE, JAMES. University of Richmond, Richmond VA 23173. [Professor, Dept of Math/Computer Sci] 804-289-8083.

FINDLEY-KNIER, HILDA. Univ of the District of Columbia, Washington DC 20008. [Associate Professor, 2704 Woodley Place NW] 202-282-7465.

FINK, JAMES. Butler University, Indianapolis IN 46208. [Head, Dept of Mathematical Sciences] 317-283-9722.

FINK, JOHN. Kalamazoo College, Kalamazoo MI 49007. [Associate Professor] 616-383-8447.

FISHER, NEWMAN. San Francisco State University, San Francisco CA 94132. [Chairman, Mathematics Dept 1600 Holloway Ave] 415-338-2251.

FLANDERS, HARLEY. University of Michigan, Ann Arbor MI 48109. [Professor, Dept of Mathematics] 313-761-4666.

FLEMING, RICHARD. Central Michigan University, Mt. Pleasant MI 48859. [Professor, Dept of Mathematics] 517-774-3596.

FLINN, TIMOTHY. Tarleton State University, Stephenville TX 76402. [Associate Professor, Dept of Mathematics Box T-519] 817-968-9168.

FLOWERS, PEARL. Montgomery County Public Schools, Rockville MD 20850. [Teacher Specialist, 850 Hungerford Dr CESC 251] 301-279-3161.

FLOWERS, JOE. Northeast Missouri State Univ, Kirksville MO 63501. [Professor, Div of Math Violette Hall 287] 816-785-4284.

FOLIO, CATHERINE. Brookdale Community College, Lincroft NJ 07738. [Math Dept Newman Springs Road] 201-842-1900.

FRAGA, ROBERT. Ripon College, Ripon WI 54971. [Box 248] 414-748-8129.

FRANCIS, WILLIAM. Michigan Technological University, Houghton MI 49931. [Associate Professor, Dept of Mathematical Sciences] 906-487-2146.

FRANKLIN, KATHERINE. Los Angeles Pierce College, Northridge CA 91324. [Associate Professor, 8827 Jumilla Ave] 818-700-9732.

FRAY, BOB. Furman University, Greenville SC 29613. [Professor, Mathematics Dept] 803-294-2105.

FRIEDBERG, STEPHEN. Illinois State University, Normal IL 61761. [Professor, Dept of Math 119 Doud Drive] 309-438-8781.

FRIEL, WILLIAM. University of Dayton, Dayton OH 45469. [Assistant Professor, Mathematics Dept] 513-229-2099.

FRYXELL, JAMES. College of Lake County, Grayslake IL 60030. [Professor, 19351 W. Washington Street] 312-223-6601.

FULTON, JOHN. Clemson University, Clemson SC 29634. [Professor, Dept of Mathematics] 803-656-3436.

GALOVICH, STEVE. Carleton College, Northfield MN 55057. [Professor, Dept of Math & Computer Science] 507-663-4362.

GALWAY, ALISON. O'Brien & Associates, Alexandria VA 22314. [Associate, Carriage House 708 Pendleton St] 703-548-7587.

GASS, FREDERICK. Miami University, Oxford OH 45056. [Associate Professor, Dept of Math and Statistics] 513-529-3422.

GEGGIS, DAVID. PWS-Kent Publishing Co., Boston MA 02116. [Managing Editor, 20 Park Plaza] 617-542-3377.

GENKINS, ELAINE. Collegiate School, New York NY

10024. [Head Math Dept, 241 West 77th St] 212-873-0677.

GETHNER, ROBERT. Franklin and Marshall College, Lancaster PA 17604. [Professor] 717-291-4051.

GEUTHER, KAREN. University of New Hampshire, Durham NH 03824. [Assistant Professor, Dept of Mathematics] 603-862-2320.

GIAMBRONE, AL. Sinclair College, Dayton OH 45402. [Professor, Math Dept] 513-226-2585.

GILBERT, JOHN. Mississippi State University, Mississipi State MS 39762. [Associate Professor, Dept of Math P.O. Drawer MA] 601-325-3414.

GILBERT, WILLIAM. University of Waterloo, Waterloo Ont. Canada N2L3G1. [Professor, Pure Math Dept] 519-888-4097.

GILFEATHER, FRANK. University of Nebraska, Lincoln NE 68588. [Professor, Department of Mathematics] 402-472-3731.

GILMER, GLORIA. Math-Tech Connexion Inc., Milwaukee WI 53205. [President, 2001 West Vliet Street] 414-933-2322.

GLEASON, ANDREW. Harvard University, Cambridge MA 02138. [Professor, Dept of Math 110 Larchwood Drive] 617-495-4316.

GLEICK, JIM. New York Times.

GLENNON, CHARLES. Christ Church Episcopal School, Greenville SC 29603. [Instructor, P.O. Box 10128] 803-299-1522.

GLUCHOFF, ALAN. Villanova University, Villanova PA 19085. [Assistant Professor, Dept of Math Sciences] 215-645-7350.

GOBLIRSCH, RICHARD. College of St. Thomas, St. Paul MN 55105. [Professor, 2115 Summit Avenue] 612-647-5281.

GODSHALL, WARREN. Susquehanna Twp High Shool, Harrisburg PA 17109. [Teacher, 414A Amherst Drive] 717-657-5117.

GOLDBERG, SAMUEL. Alfred P. Sloan Foundation, New York NY 10111. [630 Fifth Avenue Suite 2550]

GOLDBERG, MORTON. Broome Community College, Binghamton NY 13902. [Professor, P.O. Box 1017] 607-771-5165.

GOLDBERG, DOROTHY. Kean College of New Jersey, Union NJ 07083. [Chairperson, Dept of Math Morris Ave.] 201-527-2105.

GOLDBERG, DONALD. Occidental College, Los Angeles CA 90041. [Assistant Professor, Dept of Math 1600 Campus Road] 213-259-2524.

GOLDSCHMIDT, DAVID. University of California, Berkeley CA 94720. [Professor, 970 Evans Hall] 415-642-0422.

GOLDSTEIN, JEROME. Tulane University, New Orleans LA 70118. [Dept of Mathematics] 504-865-5727.

GOODSON, CAROLE. University of Houston, Houston TX 77004. [Associate Dean, 4800 Calhoun] 713-749-1341.

GORDON, SHELDON. Suffolk Community College, East Northport NY 11731. [Professor, 61 Cedar Road] 516-451-4270.

GRAF, CATHY. Thomas Jefferson High School, Burke VA 22015. [Teacher, 6101 Windward Drive] 703-354-9300.

GRANDAHL, JUDITH. Western Connecticut State University, Danbury CT 06810. [Associate Professor, 181 White Street] 203-797-4221.

GRANLUND, VERA. University of Virginia, Charlottesville VA 22901. [Lecturer, Thornton Hall] 804-924-1032.

GRANTHAM, STEPHEN. Boise State University, Boise ID 83725. [Assistant Professor, Dept of Mathematics] 208-385-3369.

GRAVER, JACK. Syracuse University, Syracuse NY 13244. [Professor, Dept of Mathematics] 315-472-5306.

GRAVES, ELTON. Rose-Hulman Institute of Technology, Terre Haute IN 47803. [Associate Professor, Box 123] 812-877-1511.

GREEN, EDWARD. Virginia Tech, Blacksburg VA 24061. [Professor, Dept of Mathematics] 703-961-6536.

GROSSMAN, MICHAEL. University of Lowell, Lowell MA 01851. [Associate Professor, 185 Florence Road] 617-459-6423.

GROSSMAN, STANLEY. University of Montana, Missoula MT 59801. [Professor, 333 Daly Avenue] 406-549-3819.

GUILLOU, LOUIS. Saint Mary's College, Winona MN 55987. [Dept of Mathematics and Statistics] 507-457-1487.

GULATI, BODH. Southern Connecticut State Univ, Cheshire CT 06410. [Professor, 954 Ott Drive] 203-397-4486.

GULICK, DENNY. University of Maryland, College Park MD 20742. [Professor, Dept of Mathematics] 301-454-3303.

GUPTA, MURLI. George Washington University, Washington DC 20052. [Professor, Dept of Mathematics] 202-994-4857.

GUSEMAN, L.F. Texas A & M University, College Station TX 77843. [Professor, Dept of Mathematics] 409-845-3261.

GUSTAFSON, KARL. University of Colorado, Boulder CO 80309. [Professor, Campus Box 426-Mathematics] 303-492-7664.

GUYKER, JAMES. SUNY College at Buffalo, Buffalo NY 14222. [Professor, Math Dept 1300 Elmwood Ave.] 716-837-8915.

HABER, JOHN. Harper & Row, New York NY 10022. [Editor, 10 East 53rd Street] 212-207-7243.

HAINES, CHARLES. Rochester Institute of Technology, Rochester NY 14623. [Associate Dean, College of Engineering] 716-475-2029.

HALLETT, BRUCE. Jones & Bartlett Publishers, Brookline MA 02146. [Editor, 208 Fuller St] 617-731-4653.

HALL, LEON. University of Missouri, Rolla MO 65401. [Associate Professor, Dept of Math & Statistics] 314-341-4641.

HAMBLET, CHARLES. Phillips Exeter Academy, Exeter NH 03833. [Instructor] 603-772-4311.

HAMMING, RICHARD. Naval Postgraduate School, Monterey CA 93943. [Code 52Hg] 408-646-2655.

HAMPTON, CHARLES. The College of Wooster, Wooster OH 44691. [Chair, Mathematical Sciences Dept] 216-263-2486.

HANCOCK, DON. Pepperdine University, Malibu CA 90265. [Associate Professor, Math Dept Natural Science Division] 213-456-4241.

HANSON, ROBERT. James Madison University, Harrisonburg VA 22807. [Coordinator, Dept of Math & Computer Science] 703-568-6220.

HARTIG, DONALD. Calif. Polytechnic State University, San Luis Obispo CA 93407. [Professor, Math Dept] 805-756-2263.

HART, THERESE. National Research Council-MS 2000, Washington DC 20418. [2101 Constitution Ave. NW] 202-334-3740.

HARVEY, JOHN. University of Wisconsin-Madison, Madison WI 53706. [Professor, Dept Math 480 Lincoln Drive] 608-262-3746.

HAUSNER, MELVIN. CIMS/NYU, New York NY 10012. [Professor, 251 Mercer Street] 212-998-3190.

HAYNSWORTH, HUGH. College of Charleston, Charleston SC 29424. [Associate Professor, Dept of Mathematics] 803-792-5735.

HEAL, ROBERT. Utah State Unversity, Logan UT 84322. [Associate Dept Head, Mathematics Dept] 801-750-2810.

HECKENBACH, ALAN. Iowa State University, Ames IA 50011. [Associate Professor, Dept of Mathematics] 515-294-8164.

HECKLER, JANE. MAA - JPBM, Washington DC

HECKMAN, EDWIN. Central New England College, Westboro MA 01581. [Professor, 4 Lyman Street] 617-366-5527.

HEID, KATHLEEN. Pennsylvania State University, University Park PA 16802. [Assistant Professor, 171 Chambers Building] 814-865-2430.

HELLERSTEIN, SIMON. University of Wisconsin, Madison WI 53706. [Professor, Dept of Math Van Vleck Hall] 608-263-3302.

HENDERSON, JIM. Colorado College, Colorado Springs CO 80903. [Assistant Professor, PDept of Mathematics] 303-473-2233.

HENSEL, GUSTAV. Catholic University of America, Washington DC 20064. [Assistant Dean, Dept of Mathematics] 202-635-5222.

HERR, ALBERT. Drexel University, Philadelphia PA 19104. [Associate Professor, Dept Math/Comp Sci 32nd & Chestnut Sts] 215-895-2672.

HILDING, STEPHEN. Gustavus Adolphus College, St. Peter MN 56082. [Professor, Dept of Mathematics] 507-931-7464.

HILLTON, THOMAS. Educational Testing Service, Princeton NJ

HILL, THOMAS. Lafayette College, Easton PA 18042. [Dept of Mathematics] 215-250-5282.

HILL, SHIRLEY. University of Missouri, Kansas City MO 64110. [Professor, Dept of Mathematics] 816-276-2742.

HIMMELBERG, CHARLES. University of Kansas, Lawrence KS 66045. [Chair, Dept of Mathematics] 913-864-3651.

HINKLE, BARBARA. Seton Hill College, Greensburg PA 15601. [Chair, Dept of Math & Computer Science] 412-834-2200.

HODGSON, BERNARD. Universite Laval, Quebec Canada G1K7P4. [Professor, Dept De Maths Stat. & Actuariar] 418-656-2975.

HOFFER, ALAN. National Science Foundation, Washington DC 20550. [1800 G Street NW]

HOFFMAN, DALE. Bellevue Community College, Bellevue WA 98005. [Professor, 12121 S.E. 27th Street] 206-747-8515.

HOFFMAN, KENNETH. Massachusetts Institute of Technology, Washington DC 20036. [Professor, 1529 18th Street NW] 202-334-3295.

HOFFMAN, ALLAN. National Academy of Sciences, Washington DC 20418. [Executive Director COSEPUP, 2101 Constitution Avenue NW]

HOLMAY, KATHLEEN. JPBM Public Information, Washington DC

HORN, P.J. Northern Arizona University, Flagstaff AZ 86011. [Assistant Professor, Box 5717] 602-523-6880.

HORN, HENRY. Princeton University, Princeton NJ 18544. [Professor, Dept of Biology] 609-452-3000.

HOWAT, KEVIN. Wadsworth Publishing Co., Belmont CA 94002. [Publisher, 10 Davis Drive] 415-595-2350.

HSU, YU-KAO. University of Maine, Bangor ME 04401. [Professor, Room 111 Bangor Hall] 207-581-6138.

HUANG, JANICE. Milligan College, Milligan College TN 37682. [Associate Professor, Dept of Mathematics]

HUDSON, ANNE. Armstrong State College, Savannah GA 31413. [Professor, Dept of Math & CS 11935 Abercorn St] 912-927-5317.

HUGHES-HALLETT, DEBORAH. Harvard University, Cambridge MA 02138. [Senior Preceptor, Dept of Mathematics] 617-495-5358.

HUGHES, RHONDA. Bryn Mawr College, Bryn Mawr PA 19010. [Associate Professor, Dept of Mathematics] 215-645-5351.

HUGHES, NORMAN. Valparaiso University, Valparaiso IN 46383. [Associate Professor] 219-464-5195.

HUNDHAUSEN, JOAN. Colorado School of Mines, Golden CO 80401. [Dept of Mathematics] 303-273-3867.

HUNSAKER, WORTHEN. Southern Illinois University, Carbondale IL 62901. [Professor, Dept of Mathematics] 618-453-5302.

HUNTER, JOYCE. Webb School of Knoxville, Knoxville TN 37923. [Teacher, 9800 Webb School Drive] 615-693-0011.

HURLEY, SUSAN. Siena College, Loudonville NY 12209. [Science Division] 518-783-2459.

HURLEY, JAMES. University of Connecticut, Storrs CT

06268. [Professor, 196 Auditorium Road Rm 111] 203-486-4143.

HVIDSTEN, MICHAEL. Gustavus Adolphus College, St. Peter MN 56082. [Assistant Professor, Dept of Mathematics] 507-931-7480.

INTRILIGATOR, MICHAEL. UCLA, Los Angeles CA 90024. [Professor, Dept of Economics] 213-824-0604.

JACKSON, ALLYN. American Mathematical Society, Providence RI 02940. [Staff Writer, P.O. Box 6248] 401-272-9500.

JACKSON, MICHAEL. Earlham College, Richmond IN 47374. [Assistant Professor, PP] 317-983-1620.

JACOB, HENRY. University of Massachusetts, Amherst MA 01003. [Professor, Dept Math/Stat Lederle Research Tower] 413-545-0510.

JANSON, BARBARA. Janson Publications Inc, Providence RI 02903. [President, 222 Richmond Street Suite 105] 401-272-0009.

JAYNE, JOHN. Univ. of Tennessee at Chattanooga, Chattanooga TN 37413. [Professor, Math Dept] 615-755-4545.

JENKINS, FRANK. John Carroll University, University Heights OH 44118. [Assistant Professor, Mathematics Department] 216-397-4682.

JENKINS, JOE. SUNY University at Albany, Albany NY 12222. [Chair Dept of Math, 1400 Washington Ave.] 518-422-4602.

JENSEN, WALTER. Central New England College, Dudley MA 01570. [Head, 10 Shepherd Avenue] 617-943-3053.

JEN, HORATIO. W.C.C. College, Youngwood PA 15601. [Professor] 412-925-4184.

JOHNSON, DAVID. Lehigh University, Bethlehem PA 18015. [Associate Professor, Mathematics Dept #14] 215-758-3730.

JOHNSON, JERRY. Oklahoma State University, Stillwater OK 74078. [Professor, Dept of Mathematics] 405-624-5793.

JOHNSON, LEE. Virginia Polytech & State University, Blacksburg VA 24061. [Professor, Dept of Mathematics 460 McBryde Hall] 703-961-6536.

JOHNSON, ROBERT. Washington and Lee University, Lexington VA 24450. [Professor, Dept of Mathematics] 703-463-8801.

JONES, LINDA. National Research Council-MS 2000, Washington DC 20418. [2101 Constitution Ave NW] 202-334-3740.

JONES, ELEANOR. Norfolk State University, Norfolk VA 23504. [Professor, Dept of Mathematics]

JONES, WILLIAM. Univ of the District of Columbia, Washington DC 20011. [Assistant Professor, Dept of Math4200 Connecticut Avenue NW] 202-282-3171.

JUNGHANS, HELMER. Montgomery College, Gaithersburg MD 20878. [Professor, 220 Gold Kettle Drive] 301-926-4403.

KAHN, ANN. Mathematical Science Education Board, Washington DC 20006. [Consultant, 818 Connecticut Ave NW Suite 325] 202-334-3294.

KALLAHER, MICHAEL. Washington State University, Pullman WA 99164. [Professor, Mathematics Dept] 509-335-4918.

KANIA, MAUREEN. Earl Swokowski. LTD, West Allis WI 53227. [Executive Assistant, 12124 West Ohio Avenue] 414-546-3860.

KAPUT, JAMES. Educational Tech Center-Harvard Univ., North Dartmouth MA 02747. [473 Chase Road] 617-993-0501.

KARAL, FRANK. NYU - Courant Institute, New York NY 10012. [Professor, 251 Mercer Street] 212-998-3162.

KARIAN, ZAVEN. Dension University, Granville OH 43023. [Dept of Math Sciences] 614-587-6563.

KASPER, RAPHAEL. National Research Council-CPSMR, Washington DC 20418. [Executive Director, 2101 Constitution Avenue NW]

KATZEN, MARTIN. New Jersey Institute of Technology, West Paterson NJ 07424. [Associate Professor, 11 Washington Drive]

KATZ, VICTOR. Univ of the District of Columbia, Washington DC 20011. [Professor, Dept of Math4200 Connecticut Avenue NW] 202-282-7465.

KAY, DAVID. University of North Carolina, Asheville NC 28804. [Professor, Mathematics Dept 1 University Heights] 704-251-6556.

KEEVE, MICHAEL. Norfolk State University, Norfolk VA 23504. [Instructor, Math & Computer Sci Dept] 804-623-8820.

KEHOE-MOYNIHAN, MARY. Cape Cod Community College, West Barnstable MA 02668. [Professor] 617-362-2131.

KENELLY, JOHN. Clemson University, Clemson SC 29631. [Alumni Professor, 327 Woodland Way] 803-656-5217.

KEYNES, HARVEY. University of Minnesota, Minneapolis MN 55455. [Professor, School of Math 127 Vincent Hall] 612-625-2861.

KIMES, THOMAS. Austin College, Sherman TX 75090. [Chairman, Dept of Mathematics] 214-892-9101.

KING, ELLEN. Anderson College, Anderson SC 29621. [Instructor, 316 Boulevard] 803-231-2162.

KING, ROBERT. Westmar College, LeMars IA 51031. [Assistant Professor, 115 7th Street SE] 712-546-6117.

KIRKMAN, ELLEN. Wake Forest University, Winston-Salem NC 27109. [Associate Professor, Box 7311 Reynolds Station] 919-761-5351.

KLATT, GARY. Unversity of Wisconsin, Whitewater WI 53190. [Professor, Math Dept 800 West Main] 414-472-5162.

KNIGHT, GENEVIEVE. Coppin State College, Columbia MD 21045. [Professor, 2500 W. North Ave., Baltimore, 21216] 301-333-7853.

KOKOSKA, STEPHEN. Colgate University, Hamilton NY 13346. [Assistant Professor, Dept. of Mathematics] 315-824-1000.

KOLMAN, BERNARD. Drexel University, Philadelphia PA 19104. [Professor] 215-895-2683.

KOPERA, ROSE. National Research Council-BMS, Washington DC 20418. [2101 Constitution Ave NW] 202-334-2421.

KRAMAN, JULIE. National Research Council-MSEB, Washington DC 20006. [818 Connecticut Ave NW Suite 325] 202-334-3294.

KRAUS, GERALD. Gannon University, Erie PA 16541. [Chair, Math Dept University Square] 814-871-7595.

KREIDER, DON. Dartmouth College, Sharon VT 05065. [Vice Chairman, Math & CS Dept RR #1 Box 487]

KUHN, ROBERT. Harvard University, Cambridge MA 02138. [Lecturer, Dept of Mathematics] 617-495-1610.

KULM, GERALD. Amer Assoc for Advancement of Science, Washington DC 20005. [Associate Program Director, 1333 H Street NW] 202-326-6647.

KULNARONG, GRACE. National Research Council-MSEB, Washington DC 20006. [818 Connecticut Ave NW Suite 325] 202-334-3294.

KUNZE, RAY. University of Georgia, Athens GA 30602. [Chair, Dept of Mathematics] 404-542-2583.

LATORRE, DONALD. Clemson University, Clemson SC 29631. [Professor, Dept of Mathematical Sciences] 803-656-3437.

LACEY, H.E. Texas A & M University, College Station TX 77843. [Head, Dept of Mathematics]

LAMBERT, MARCEL. Universite du Quebec a Trois-Rivieres, Trois-Rivieres Que Canada G9A5H7. [Dept Head, Department de Math-Info] 819-376-5126.

LANE, BENNIE. Eastern Kentucky University, Richmond KY 40475. [Professor, Wallace 402] 606-622-5942.

LANG, JAMES. Valencia Community College, Orlando FL 32811. [Professor, 1800 South Kirkman Road] 305-299-5000.

LAUBACHER, MICHAEL. Holland Hall School, Tulsa OK 74137. [Teacher, 5666 East 81st Street] 918-481-1111.

LAUFER, HENRY. SUNY at Stony Brook, Long Island NY 11794. [Professor, Dept of Mathematics] 516-632-8247.

LAX, PETER. NYU-Courant Institute of Math Science, New York NY 10012. [251 Mercer Street] 212-460-7442.

LECKRONG, GERALD. Brighton Area Schools, Brighton MI 48116. [Teacher, 7878 Brighton Road] 313-229-1400.

LEE, KEN. Missouri Western State College, St. Joseph MO 64507. [Professor] 816-271-4284.

LEINBACH, CARL. Gettysburg College, Gettysburg PA 17325. [Chair, Computer Science P.O. Box 506] 717-337-6735.

LEITHOLD, LOUIS. Pepperdine University, Pacific Palisades CA 90272. [Professor, 336 Bellino Drive] 213-454-2500.

LEITZEL, JAMES. Ohio State University, Columbus OH 43210. [Associate Professor, Dept of Math 231 West 18th Avenue] 614-292-8847.

LEVINE, MAITA. University of Cincinnati, Cincinnati OH 45221. [Professor, Dept of Mathematical Sciences] 513-475-6430.

LEVY, BENJAMIN. Lexington High School, Lexington MA 02173. [Teacher, 215 Waltham Street] 617-862-7500.

LEWIN, JONATHAN. Kennesaw College, Marietta GA 30061. [Associate Professor] 404-423-6040.

LEWIS, KATHLEEN. SUNY at Oswego, Oswego NY 13126. [Assistant Professor, Dept of Mathematics] 315-341-3030.

LEWIS, GAUNCE. Syracuse University, Syracuse NY 13126. [Mathematics Dept] 315-343-0788.

LINLEY, DAVID. Nature.

LIPKIN, LEONARD. University of North Florida, Jacksonville FL 32216. [Chairman, Dept of Math 4567 St. Johns Bluff Rd] 904-646-2653.

LISSNER, DAVID. Syracuse University, Syracuse NY 13210. [Professor, Math Dept] 315-423-2413.

LITWHILER, DANIEL. U.S. Air Force Academy, Colorado Springs CO 80840. [Head, Dept of Math Sciences] 303-472-4470.

LIUKKONEN, JOHN. Tulane University, New Orleans LA 70118. [Associate Professor, Mathematics Dept] 504-865-5729.

LOCKE, PHIL. University of Maine, Orono ME 04469. [Associate Professor, 236 Neville Hall] 207-581-3924.

LOFQUIST, GEORGE. Eckerd College, St. Petersburg FL 33712. [Math Dept P.O. 12560] 813-864-8434.

LOGAN, DAVID. University of Nebraska, Lincoln NE 68588. [Professor, Dept of Mathematics]

LOMEN, DAVID. University of Arizona, Tucson AZ 85721. [Professor, Mathematics Dept] 602-621-6892.

LOVELOCK, DAVID. University of Arizona, Tucson AZ 85721. [Dept of Mathematics] 602-621-6855.

LOWENGRUB, MORTON. Indiana University, Bloomington IN 47405. [Professor, Bryan Hall 104] 812-335-6153.

LUCAS, WILLIAM. National Science Foundation, Washington DC 20550. [1800 G Street NW Room 639] 202-357-7051.

LUCAS, JOHN. University of Wisconsin-Oshkosh, Oshkosh WI 54901. [Professor, Dept of Math Swart Hall 206] 414-424-1053.

LUKAWECKI, STANLEY. Clemson University, Clemson SC 29634. [Professor, Dept of Mathematical Sciences] 803-656-3449.

LUNDGREN, RICHARD. University of Colorado at Denver, Denver CO 80202. [Chairman, Math Dept 1100 14th Street] 303-556-8482.

LYKOS, PETER. Illinois Institute of Technology, Chicago IL 60616. [Consultant, Dept of Chemistry] 312-567-3430.

MADISON, BERNARD. National Research Council, Washington DC 20418. [Project Director MS 2000, 2101 Constitution Ave. NW] 202-334-3740.

MAGGS, WILLIAM. EOS - American Geophysical Union, Washington DC 20009.

MAGNO, DOMINIC. Harper College, Palatine IL 60067. [Associate Professor, Algonquin/Ruselle Roads] 312-397-3000.

MAHONEY, JOHN. Sidwell Friends School, Washington DC 20016. [3825 Wisconsin Avenue NW] 202-537-8180.

MALONE, J.J. Worcester Polytechnic Institute, Worcester MA 01609. [Professor, Dept of Math Sciences 100 Institute Rd] 617-793-5599.

MANASTER, ALFRED. University of California-San Diego, La Jolla CA 92093. [Professor, Dept of Mathematics C-012] 619-534-2644.

MANITIUS, ANDRE. National Science Foundation, Washington DC 20550. [Deputy Director Div. of Math Sc., 1800 G Street NW]

MARCOU, MARGARET. Montgomery County Schools, Chevy Chase MD 20815. [Teacher, 5 Farmington Ct.] 301-656-2789.

MARSHALL, JAMES. Western Carolina University, Cullowhee NC 28723. [Assistant Professor, P.O. Box 684] 704-227-7245.

MARSHMAN, BEVERLY. University of Waterloo, Waterloo Ontario Canada N2L3G1. [Assistant Professor, Dept of Applied Mathematics] 519-885-1211.

MARTINDALE, JOHN. Random House Inc., Cambridge MA 02142. [Editorial Director, P215 1st Street] 617-491-2250.

MARXEN, DONALD. Loras College, Dubuque IA 52001. [Professor] 319-588-7570.

MASTERSON, JOHN. Michigan State University, East Lansing MI 48824. [Professor, Math Dept 211 D Wells Hall] 517-353-4656.

MASTROCOLA, WILLIAM. Colgate University, Hamilton NY 13346. [Associate Professor, Dept of Mathematics] 315-824-1000.

MATHEWS, JEROLD. Iowa State University, Ames IA 50011. [Professor, Dept of Mathematics] 515-294-5865.

MATTUCK, ARTHUR. Massachusetts Institute of Technology, Cambridge MA 02139. [Dept of Mathematics Room 2-241] 617-253-4345.

MAYCOCK-PARKER, ELLEN. Wellesley College, Wellesley MA 02181. [Assistant Professor, Dept of Mathematics] 617-235-0320.

MAZUR, JOSEPH. Marlboro College, Marlboro VT 05344. [Professor] 802-257-4333.

MCARTHUR, JAMES. Bethesda-Chevy Chase High School, Bethesda MD 20814. [Teacher, 4301 East West Highway] 301-654-5264.

MCBRIDE, RONALD. Indiana University of Pennsylvania, Indiana PA 15701. [Professor, Mathematics Dept] 412-357-2605.

MCCAMMON, MARY. Penn State University, University Park PA 16802. [Math Dept 328 McAllister Bldg] 814-865-1984.

MCCARTNEY, PHILIP. Northern Kentucky University, Annapolis MD 21401. [Professor, 1886 Crownsville Road] 301-224-3139.

MCCLANAHAN, GREG. Anderson College, Anderson SC 29621. [Instructor, 316 Boulevard] 803-231-2165.

MCCOLLUM, MARY-ANN. Jefferson County Gifted Program, Birmingham AL 35226. [Teacher, 1707 Kestwick Cir.] 205-879-0531.

MCCOY, PETER. United States Naval Academy, Annapolis MD 21402. [Professor, Mathematics Dept] 301-267-2300.

MCCRAY, LAWRENCE. National Research Council-CPSMR, Washington DC 22050. [Associate Executive Director, 2101 Constitution Ave. NW] 202-334-3061.

MCDONALD, KIM. Chronicle of Higher Education.

MCDONALD, BERNARD. National Science Foundation, Washington DC 20550. [1800 G StreetNW]

MCGEE, IAN. University of Waterloo, Waterloo Ontario Canada N2L3G1. [Professor, Applied Math Dept] 519-885-1211.

MCGILL, SUZANNE. University of South Alabama, Mobile AL 36688. [Chair, Dept of Mathematics and Statistics] 205-460-6264.

MCINTOSH, HUGH. The Scientist.

MCKAY, FRED. National Research Council-MS 2000, Washington DC 20418. [2101 Constitution Avenue NW]

MCKEON, KATHLEEN. Connecticut College, New London CT 06320. [Box 1561] 203-447-1411.

MCLAUGHLIN, RENATE. University of Michigan-Flint, Flint MI 48502. [Professor, Dept of Mathematics] 313-762-3244.

MCNEIL, PHILLIP. Norfolk State University, Norfolk VA 23504. [Professor, Dept. Math 2401 Corprew Avenue] 804-623-8820.

MELLEMA, WILBUR. San Jose City College, San Jose CA 95128. [Instructor, Math Dept 2100 Moorpark] 408-298-2181.

MELMED, ARTHUR. New York University, New York NY 10003. [Research Professor, SEHNAP-23 Press Bldg Washington Square] 212-998-5228.

MESKIN, STEPHEN. Society of Actuaries, Columbia MD 21044. [Actuary, 5626 Vantage Point Road] 202-872-1870.

METT, COREEN. Radford University, Radford VA 24142. [Professor, Dept of Mathematics and Statistics] 703-831-5026.

MILCETICH, JOHN. Univ of the District of Columbia, Washington DC 20011. [Professor, Dept of Math4200 Connecticut Avenue NW] 202-282-7328.

MILLER, ALICE. Babson College, Babson Park MA 02157. [Assistant Professor] 617-239-4476.

MINES, LINDA. National Research Council-MSEB, Washington DC 20006. [818 Connecticut Ave NW Suite 325] 202-334-3294.

MISNER, CHARLES. University of Maryland, College Park MD 20742. [Professor, Physics Department] 301-454-3528.

MITCHELL, GEORGE. Indiana University of Pennsylvania, Indiana PA 15701. [Professor, 120 Concord Street] 412-357-2305.

MOCHIZUKI, HORACE. Univ California - Santa Barbara, Santa Barbara CA 93106. [Professor, Dept of Mathematics] 805-961-3462.

MOODY, MICHAEL. Washington State University, Pullman WA 99164. [Assistant Professor, Mathematics Dept] 509-335-3172.

MOORE, LAWRENCE. Duke University, Durham NC

27706. [Associate Professor, PDept of Mathematics] 919-684-2321.

MOORE, JOHN. Univ California - Santa Barbara, Santa Barbara CA 93106. [Professor, Dept of Mathematics] 805-961-3688.

MORAWETZ, CATHLEEN. NYU-Courant Institute of Math Science, New York NY 10012. [Director, 251 Mercer Street] 212-460-7100.

MORLEY, LANNY. Northeast Missouri State Univ, Kirksville MO 63501. [Head, Div of Math Violette Hall 287] 816-785-4547.

MORREL, BERNARD. IUPUI, Indianapolis IN 46223. [Associate Professor, Dept of Math 1125 East 38th Street] 317-274-6923.

MORTON, PATRICK. Wellesley College, Wellesley MA 02181. [Assistant Professor, Dept of Mathematics] 617-235-0320.

MOSKOWITZ, HERBERT. Purdue University, West Lafayette IN 47907. [Professor, Krannert Graduate School of Management] 317-494-4600.

MOSLEY, EDWARD. Arkansas College, Batesville AR 72501. [Professor] 501-793-9813.

MOVASSEGHI, DARIUS. CUNY - Medgar Evers College, Brooklyn NY 11225. [Professor, 1150 Carroll Street] 717-735-1900.

MULLER, ERIC. Brock University, St. Catherine Ontario Canada L2S3A1. [Professor] 416-688-5550.

MURPHY, CATHERINE. Purdue University Calumet, Hammond IN 46323. [Head, Dept of Mathematical Sciences] 219-989-2270.

NAIL, BILLY. Clayton State College, Morrow GA 30260. [Professor, 5900 Lee Street] 404-961-3429.

NARODITSKY, VLADIMIR. San Jose State University, San Jose CA 95192. [Associate Professor, Dept of Math & Computer Science] 408-277-2411.

NEAL, HOMER. University of Michigan, Ann Arbor MI 48109. [Chair, Dept of Physics 1049 Randall Lab] 313-754-4438.

NELSON, ROGER. Ball State University, Muncie IN 47306. [Associate Professor, Dept of Math Sciences] 317-285-8640.

NELSON, JAMES. University of Minnesota, Duluth MN 55812. [Associate Professor, 10 University Avenue] 218-726-7597.

NEWMAN, ROGERS. Southern University, Baton Rouge LA 70813. [Professor, Dept of Mathematics] 504-771-4500.

NORDAI, FREDERICK. Shippensburg University, Shippensburg PA 17257. [Associate Professor, P621 Glenn Street] 717-532-1642.

NORFLEET, SUNNY. St. Petersburg Junior College, Tarpon Springs FL 34689. [Teacher, 1309 Vermont Avenue] 813-938-7049.

NORTHCUTT, ROBERT. Southwest Texas State University, San Marcos TX 78666. [Professor, Mathematics Department] 512-245-2551.

NOVAK, CAROLYN. Syracuse University, Utica NY 13502. [Student, 213 Richardson Avenue] 315-733-4590.

NOVIKOFF, ALBERT. New York University, New York NY 10012. [Dept of Math 251 Mercer Street] 212-982-5019.

NOVINGER, PHIL. Florida State University, Tallahassee FL 32306. [Associate Professor, Dept of Mathematics] 904-644-1479.

O'BRIEN, RUTH. O'Brien & Associates, Alexandria VA 22314. [President, Carriage House 708 Pendleton St.] 703-548-7587.

O'DELL, RUTH. County College of Morris, Randolph NJ 07869. [Associate Professor, PRoute 10 and Center Grove Road] 201-361-5000.

O'DELL, CAROL. Ohio Northern University, Ada OH 45810. [Associate Professor, Dept of Math & Computer Science] 419-772-2354.

O'MEARA, TIMOTHY. University of Notre Dame, Notre Dame IN 46556. [Provost, Administration Building Room 202] 219-239-6631.

O'REILLY, MICHAEL. University of Minnesota, Morris MN 56267. [Math Discipline] 612-589-2211.

OFFUTT, ELIZABETH. Springbrook High School, Bethesda MD 20814. [Teacher, 9304 Elmhirst Dr.] 301-530-6238.

ORTIZ, CARMEN. Inter American Univ of Puerto Rico, Humacao PR 00661. [Lecturer, P.O. Box 204] 809-758-8000.

OSER, HANS. SIAM News.

OSTEBEE, ARNOLD. St. Olaf College, Northfield MN 55057. [Associate Professor] 507-663-3420.

OST, LAURA. Orlando Sentinel, Orlando FL

PAGE, WARREN. NYC Technical College SUNY, Brooklyn NY 10705. [Professor, 30 Amberson Ave. Younkers NY] 914-965-3893.

PALLAI, DAVID. Addison-Wesley Publishing Co., Reading MA 01867. [Senior Editor, Route 128] 617-944-3700.

PALMER, CHESTER. Auburn University-Montgomery, Montgomery AL 36193. [Professor, Dept of Mathematics] 205-271-9317.

PAOLETTI, LESLIE. Choate Rosemary Hall, Wallingford CT 06492. [Teacher, P.O. Box 788] 203-269-7722.

PARTER, SEYMOUR. University of Wisconsin, Madison WI 53706. [Dept of Mathematics Van Vleck Hall] 608-263-4217.

PASSOW, ELI. Temple University, Bala Cynwyd PA 19004. [Professor, 30 North Highland Avenue] 215-664-6854.

PATTON, CHARLES. Hewlett-Packard Co., Corvallis OR 97330. [Software Engineer, 1000 N.E. Circle Blvd. MS 34-L9] 503-757-2000.

PAUGH, NANCY. Woodbridge Township School District, Woodbridge NJ 07095. [Supervisor, P.O. Box 428 School Street] 201-750-3200.

PEACOCK, MARILYN. Tidewater Community College, Portsmouth VA 23703. [Assistant Professor, State Rte. 135] 804-484-2121.

PENNEY, DAVID. University of Georgia, Bogart GA 30622. [Associate Professor, 235 West Huntington Road] 404-542-2610.

PENN, HOWARD. United States Naval Academy, Annapolis MD 21402. [Professor, Mathematics Dept] 301-267-3892.

PETERSEN, KARL. University of North Carolina, Chapel Hill NC 27514. [Professor, Dept of Mathematics] 919-962-2380.

PETERSON, DORN. James Madison University, Harrisonburg VA 22807. [Physics Dept] 703-568-6487.

PETERSON, BRUCE. Middlebury College, Middlebury VT 05753. [Professor] 802-388-3711.

PETERSON, IVARS. Science News.

PETZINGER, KEN. College of William and Mary, Williamsburg VA 23185. [Professor] 804-253-4471.

PHUA, MEE-SEE. Univ of the District of Columbia, Washington DC 20011. [Dept of Math4200 Connecticut Avenue NW] 202-282-7465.

PICCOLINO, ANTHONY. Dobbs Ferry Public Schools, Yonkers NY 10710. [Math Coordinator, 33 Bonnie Briar Rd] 914-793-2645.

PIRTLE, ROBERT. John Wiley & Sons, New York NY 10158. [Editor, 605 3rd Avenue 5th floor] 212-850-6348.

PLOTTS, RANDOLPH. St. Petersburg Junior College, St. Petersburg FL 33733. [Instructor, 6605 5th Avenue North] 813-341-4738.

POIANI, EILEEN. Saint Peter's College, Jersey City NJ 07306. [Professor, 2641 Kennedy Boulevard] 201-333-4400.

POLLAK, HENRY. , Summit NJ 17901. [40 Edgewood Road] 201-277-1143.

POLUIKIS, JOHN. St. John Fisher College, Rochester NY 14618. [Professor, 3497 East Avenue] 716-586-4600.

PONZO, PETER. University of Waterloo, Waterloo Ontario Canada N2L3G1. [Professor, Applied Math Dept] 519-885-1211.

PORTER, JACK. University of Kansas, Lawrence KS 66045. [Professor, Dept of Mathematics] 913-864-4367.

POSTNER, MARIE. St. Thomas Aquinas College, Sparkill NY 10976. [Assistant Professor, PRoute 340] 914-359-9500.

POWELL, WAYNE. Oklahoma State University, Stillwater OK 74075. [Associate Professor, Dept of Mathematics] 405-624-5790.

PRESS, FRANK. National Academy of Sciences, Washington DC 20418. [President, 2101 Constitution Avenue NW] 202-334-2100.

PRICE, CHIP. Addison-Wesley Publishing Company, Reading MA 01867. [Editor-in-Chief, Route 128] 617-944-3700.

PRICE, ROBERT. Addison-Wesley Publishing Company, Reading MA 01867. [Editor-in-Chief, Route 128] 617-944-3700.

PRICHETT, GORDON. Babson College, Wellesley MA 02157. [Vice President, Babson Park] 617-239-4316.

PRIESTLEY, W.M. University of the South, Sewanee TN 37375. [Professor] 615-598-5931.

PROSL, RICHARD. College of William and Mary, Williamsburg VA 23185. [Chair, Dept of Computer Science] 804-253-4748.

PROTOMASTRO, GERARD. St. Peter's College, Bloomfield NJ 07003. [Professor, 96 Lindbergh Blvd.] 201-333-4400.

PURZITSKY, NORMAN. York University, Downsview Ontario Canada M3J1P3. [Associate Professor, Dept of Mathematics] 416-736-5250.

QUIGLEY, STEPHEN. Scott Foresman and Co., Glenview IL 60025. [Editor, 1900 East Lake Avenue] 312-729-3000.

QUINE, J.R. Florida State University, Tallahassee FL 32306. [Professor, Dept of Mathematics] 904-644-6050.

QUINN, JOSEPH. University of North Carolina, Charlotte NC 28223. [Chairman, Dept of Mathematics] 704-547-4495.

RADIN, ROBERT. Wentworth Institute of Technology, West Hartford CT 06119. [Professor, 781 Farmington Avenue] 203-233-8106.

RAGER, KEN. Metropolitan State College, Denver CO 80204. [Professor, 1006 11th Street] 303-556-3284.

RAJAH, MOHAMMED. Miracosta College, Oceanside CA 92056. [Professor, 1 Barnard Drive] 619-757-2121.

RALSTON, ANTHONY. SUNY - Buffalo, Buffalo NY 14260. [Professor, Dept of Computer Science 226 Bell Hall] 716-878-4000.

RAMANATHAN, G.V. University of Illinois-Chicago, Chicago IL 60680. [Professor, Dept of Math Statistics CS; Box 4348] 312-996-3041.

RAMSEY, THOMAS. University of Hawaii, Honolulu HI 96822. [Associate Professor, PMath Dept 2565 The Mall] 808-948-7951.

RAPHAEL, LOUISE. National Science Foundation, Washington DC 20550. [Program Director, DMS 1800 G Street NW] 202-357-7325.

RASMUSSEN, DOUG. Chemeketa Community College, Salem OR 97309. [Instructor, P.O. Box 14007] 503-399-5246.

RAY, DAVID. Bucknell Unviersity, Lewisburg PA 17837. [Professor, Dept of Mathematics] 717-524-1343.

REDISH, EDWARD. University of Maryland, College Park MD 20742. [Professor, Dept of Physics] 301-454-7383.

REED, MICHAEL. Duke University, Durham NC 27706. [Chair, Dept of Mathematics] 919-684-2321.

REED, ELLEN. Trinity School at Greenlawn, South Bend IN 46617. [107 South Greenlawn Avenue] 219-287-5590.

REICHARD, ROSALIND. Elon College, Elon College NC 27244. [Assistant Professor, Dept of Mathematics P.O. Box 2163] 919-584-2285.

RENZ, PETER. Mathematical Association of America, Washington DC 20036. [1529 18th Street NW]

RICE, PETER. University of Georgia, Athens GA 30602. [Professor, Mathematics Dept] 404-542-2593.

RIESS, RONALD. Virginia Polytech & State University, Blacksburg VA 24061. [Dept of Mathematics 460 McBryde Hall] 703-961-6536.

RISEBERG, JOYCE. Montgomery College, Rockville MD 20850. [Professor, 51 Mannakee Street] 301-279-5203.

ROBERTS, WAYNE. Macalester College, St. Paul MN 55113. [Professor, 1500 Grand Avenue] 612-696-6337.

RODGERS, PAMELA. O'Brien & Associates, Alexandria VA 22314. [Senior Associate, Carriage House 708 Pendleton St.] 703-548-7587.

RODI, STEPHEN. Austin Community College, Austin TX 78723. [Chair, Dept of Math & Phys Sci 2008 Lazybrook] 512-495-7222.

ROECKLEIN, PATRICIA. Montgomery College, Rockville MD 20850. [Associate Professor, 51 Mannakee Street] 301-279-5199.

ROGERS, LAUREL. University of Colorado, Colorado Springs CO 80933. [Assistant Professor, PDept of Math P.O. Box 7150] 303-593-3311.

ROITBERG, JOSEPH. Hunter College, New York NY 10021. [Professor, 695 Park Avenue] 212-772-5300.

ROITBERG, YAEL. New York Institute of Technology, Old Westbury NY 11568. [Associate Professor] 516-686-7535.

ROLANDO, JOSEFINA. St. Thomas University, Miami FL 33054. [Professor, 16400 N.W. 32nd Avenue] 305-625-6000.

ROLANDO, TOMAS. St. Thomas University, Miami FL 33054. [Professor, 16400 N.W. 32nd Avenue] 305-625-6000.

ROLWING, RAYMOND. University of Cincinnati, Cincinnati OH 45221. [Professor, Dept of Mathematical Sciences] 513-475-6430.

ROSENHOLTZ, IRA. University of Wyoming, Laramie WY 82071. [Dept of Mathematics] 307-766-3192.

ROSENSTEIN, GEORGE. Franklin and Marshall College, Lancaster PA 17604. [Professor, Box 3003] 717-291-4227.

ROSENSTEIN, JOSEPH. Rutgers University, New Brunswick NJ 08904. [Professor, Dept of Mathematics] 201-932-2368.

ROSENTHAL, WILLIAM. Ursinus College, Collegeville PA 19426. [Assistant Professor, Dept of Math and Computer Science] 215-489-4111.

ROSEN, LINDA. National Research Council-MSEB, Washington DC 20006. [Project Officer, 818 Connecticut Avenue NW Suite 325] 202-334-3294.

ROSS, KENNETH. University of Oregon, Eugene OR 97403. [Professor, Dept of Mathematics] 503-686-4721.

ROUSSEAU, T.H. Siena College, Loudonville NY 12180. [Head, Dept of Mathematics] 518-783-2440.

ROXIN, EMILIO. University of Rhode Island, Kingston RI 02881. [Professor] 401-792-2709.

RUBENSTEIN, PATRICIA. Montgomery College, Gaithersburg MD 20879. [Professor, 19038 Whetstone Circle] 301-948-2737.

RUSSO, PAULA. Trinity College, Hartford CT 06106. [Assistant Professor, PDept of Mathematics] 203-527-3151.

RYFF, JOHN. National Science Foundation, Washington DC 20550. [Program Director, 1800 G Street NW Room 339] 202-357-3455.

SACHDEV, SOHINDAR. Elizabeth City State University, Elizabeth City NC 27909. [Chairman, Dept of Math & Computer Sci Box 951] 919-335-3243.

SADLOWSKY, ROGER. Columbia Heights High School, New Brighton MN 55112. [Teacher, 2393 Pleasant View Dr.] 612-574-6530.

SAHU, ATMA. University of Maryland, Princess Anne MD 21853. [Assistant Professor] 301-651-2200.

SALAMON, LINDA. Washington University, St. Louis MO 63130. [Dean, College of Arts & Science] 314-889-5000.

SALZBERG, HELEN. Rhode Island College, Providence RI 02908. [Professor, Dept of Math & Computer Sci] 401-456-8038.

SAMPSON, KIRSTEN. JPBM, Washington DC

SANDEFUR, JAMES. Georgetown University, Washington DC 20057. [Professor, Mathematics Dept] 703-687-6145.

SATAGOPAN, K.P. Shaw University, Raleigh NC 27606. [Associate Professor, 4303-3 Avent Ferry Road] 919-755-4877.

SAYRAFIEZADEH, MAHMOUD. Medgar Evers College-CUNY, Brooklyn NY 11225. [Associate Professor, 1150 Carroll St.] 718-735-1897.

SCHEPPERS, JAMES. Fairview High School, Boulder CO 80303. [Chair Math Dept, 1515 Greenbriar Blvd.] 303-499-7600.

SCHICK-LENK, JUDITH. Ocean County College, Toms River NJ 08723. [College Dr] 201-255-0400.

SCHLAIS, HAL. University of Wisconsin Centers, Janesville WI 53534. [2909 Kellogg Avenue] 608-755-2811.

SCHMEELK, JOHN. Virginia Commonwealth University, Richmond VA 23284. [Associate Professor, 1015 West Main St] 804-257-1301.

SCHMIDT, HARVEY. Lewis and Clark College, Portland OR 97219. [Associate Professor, Campus Box 111] 503-293-2743.

SCHNEIDER, DAVID. University of Maryland, College Park MD 20742. [Associate Professor, Dept of Mathematics] 301-454-5002.

SCHREMMER, ALAIN. Community College of Philadelphia, Philadelphia PA 19130. [Associate Professor, 1700 Spring Garden Street] 215-751-8413.

SCHROEDER, BERNIE. Univ. of Wisconsin-Platteville, Platteville WI 53818. [Associate Professor] 608-342-1746.

SCHURRER, AUGUSTA. Univ of Northern Iowa, Cedar Falls IA 50614. [Professor, Dept of Math & Computer Science] 319-273-2432.

SCHUTZMAN, ELIAS. National Science Foundation, Washington DC 20550. [Program Director, 1800 G Street NW] 202-357-9707.

SEIDLER, ELIZABETH. Mercy High School, Baltimore MD 21239. [Teacher, 1300 East Northern Parkway] 301-433-8880.

SEIFERT, CHARLES. University of Central Arkansas, Conway AK 72032. [Chairman, Math & Computer Sci Dept Main 104] 501-450-3147.

SELDEN, JOHN. Tennessee Technological University, Cookeville TN 38505. [Assistant Professor, Math Dept Box 5054] 615-372-3441.

SELDON, ANNIE. Tennessee Technological University, Cookeville TN 38505. [Assistant Professor, PPMath Dept Box 5054] 615-372-3441.

SELIG, SEYMOUR. National Research Council-BMS, Washington DC 20418. [2101 Constitution Ave NW] 202-334-2421.

SESAY, MOHAMED. Univ of the District of Columbia, Silver Spring MD 20910. [Professor, 8750 Georgia Avenue #118A] 301-565-2623.

SESSA, KATHLEEN. D.C. Heath and Co., Lexington MA 02173. [Developmental Editor, 125 Spring Street] 617-860-1544.

SHARMA, MAN. Clark College, Atlanta GA 30314. [Professor, Dept of Math 240 James Brawley Dr] 404-577-6685.

SHARP, JACK. Floyd Junior College, Rome GA 30161. [Associate Professor, P.O. Box 1864] 404-295-6357.

SHIFLETT, RAY. Calif State Polytechnic University, Pomona CA 91768. [Dean, 3801 West Temple Avenue] 714-869-3600.

SHOOTER, WILLIAM. Gloucester County College, Sewell NJ 08080. [Coordinator] 609-465-5000.

SIEBER, JAMES. Shippensburg University, Shippensburg PA 17257. [Professor, Dept of Math & Computer Sci] 717-532-1405.

SIEGEL, MARTHA. Towson State University, Towson MD 21204. [Professor, Dept of Mathematics] 301-321-2980.

SIMPSON, DAVID. Southwest State University, Marshall MN 56258. [Professor] 507-537-6141.

SINGH, PREMJIT. Manhattan College, Riverdale NY 10471. [Assistant Professor, PDept of Math & Computer Sci] 212-920-0385.

SKIDMORE, ALEXANDRA. Rollins College, Winter Park FL 32789. [Professor] 305-646-2516.

SKITZKI, RAY. Shaker Heights High School, Shaker Heights OH 44120. [Teacher, 15911 Aldersyde Drive] 216-921-1400.

SLACK, STEPHEN. Kenyon College, Gambier OH 43022. [Associate Professor, Mathematics Department] 614-427-5267.

SLINGER, CAROL. Marian College, Indianapolis IN 46222. [Head, Dept of Math 3200 Cold Spring Road] 317-929-0281.

SLOUGHTER, DAN. Furman University, Greenville SC 29613. [Assistant Professor, Mathematics Dept] 803-294-3233.

SLOYAN, STEPHANIE. Georgian Court College, Lakewood NJ 08701. [Professor, Dept of Mathematics] 201-364-2200.

SMALL, DON. Colby College, Waterville ME 04601. [Associate Professor] 207-872-3255.

SMITH, DAVID. Duke University, Durham NC 27706. [Associate Professor, Dept of Mathematics] 919-684-2321.

SMITH, ROBERT. Millersville University, Millersville PA 17551. [Professor, Dept of Math & Computer Science] 717-872-3780.

SMITH, ROSE-MARIE. Texas Woman's University, Denton TX 76204. [Chair, Dept of Math P.O. Box 22865] 817-898-2166.

SMITH, RICK. University of Florida, Gainesville FL 32611. [Associate Professor, Dept of Mathematics] 904-392-6168.

SNODGRASS, ALICE. John Burroughs School, Webster Grove MO 63119. [Teacher, 440 East Jackson Rd] 314-993-4040.

SOLOMON, JIMMY. Mississippi State University, Missisipi State MS 39762. [Professor, Dept of Math P.O. Drawer MA] 601-325-3414.

SOLOW, ANITA. Grinnell College, Grinnell IA 50112. [Associate Professor, Dept of Mathematics] 515-269-4207.

SPANAGEL, DAVID. St. John Fisher College, Rochester NY 14618. [Instructor, Dept of Math & CS 3690 East Ave] 716-385-8190.

STAHL, NEIL. Univ Wisconsin Center-Fox Valley, Menasha WI 54952. [Associate Professor, PMidway Road] 414-832-2630.

STAKGOLD, IVAR. University of Delaware, Newark DE 19716. [Professor, Dept of Math 501 Ewing Hall] 302-451-2651.

STARR, FREDERICK. Oberlin College, Oberlin OH 44074. [President]

STEARNS, WILLIAM. University of Maine, Orono ME 04469. [Associate Professor, 228 Neville Hall] 207-581-3928.

STEEN, LYNN. St. Olaf College, Northfield MN 55057. [Professor, Dept of Mathematics] 507-663-3114.

STEGER, WILLIAM. Essex Community College, Reisterstown MD 21136. [Associate Professor, 12717 Gores Mill Rd] 301-522-1393.

STEPP, JAMES. University of Houston, Houston TX 77004. [Professor, Dept. of Mathematics] 713-749-4827.

STERN, ROBERT. Sunders College Publishing, Philadephia PA 19105. [Senior Editor, 210 West Washington Square]

STERRETT, ANDREW. Denison University, Granville OH 43023. [Professor] 614-587-6484.

STEVENSON, JAMES. Ionic Atlanta Inc, Atlanta GA 30309. [CEO, 1347 Spring Street] 404-876-5166.

STEVENS, CHRISTINE. National Science Foundation, Washington DC 20550. [Associate Program Director, 1800 G Street NW Room 635] 202-357-7074.

STODGHILL, JACK. **Dickinson College, Carlisle PA 17013. [Associate Professor] 717-245-1743.**

STONE, DAVID. Georgia Southern College, Statesboro GA 30460. [Professor, Dept of Math & CS L. Box 8093] 912-681-5390.

STONE, THOMAS. PWS-Kent Publishing Co., Boston MA 01970. [Editor, 20 Park Plaza] 617-542-3377.

STOUT, RICHARD. Gordon College, Wenham MA 01984. [Chair, Dept of Mathematics] 617-927-2300.

STRALEY, TINA. Kennesaw College, Marietta GA 30061. [Chair, Dept of Mathematics] 404-423-6104.

STRANG, GILBERT. Massachusetts Institute of Technology, Cambridge MA 02139. [Professor, Room 2-240] 617-253-4383.

STRONG, ROGER. Livonia Public Schools, Plymouth MI 48170. [39651 Mayville] 313-455-1530.

SUMMERHILL, RICHARD. Kansas State University, Manhattan KS 66506. [Associate Professor, Dept of Mathematics] 913-532-6750.

SUNLEY, JUDITH. National Science Foundation, Washington DC 20550. [Director, DMS 1800 G Street NW] 202-357-9669.

SURI, MANIL. University of Maryland, Catonsville MD 21228. [Assistant Professor, Dept of Mathematics] 301-455-2311.

SWAIN, STUART. University of Maine-Machias, Machias ME 04654. [Assistant Professor, Science Division] 207-255-3313.

SWARD, GILBERT. Montgomery College, Chevy Chase MD 20815. [Professor, 9101 Levelle Dr.] 301-657-3056.

SWARD, MARCIA. National Research Council-MSEB, Washington DC 20006. [Executive Director, 818 Connecticut Ave NW Suite 325] 202-334-3294.

SWOKOWSKI, EARL. Marquette University, Milwaukee WI 53227. [12124 West Ohio Avenue] 414-546-3860.

SZOTT, DONNA. South Campus - CCAC, West Mifflin PA 15102. [Professor, 1750 Clairton Road] 421-469-6228.

TALMAN, LOUIS. Metropolitan State College, Denver CO 80204. [Assistant Professor, Dept Math Sciences Box 38] 303-556-8438.

TELES, ELIZABETH. Montgomery College, Takoma Park MD 20912. [Associate Professor, Department of Mathematics] 301-587-4090.

TEMPLE, PATRICIA. Choate Rosemary Hall, Wallingford CT 06492. [Calculus Head, P.O. Box 788] 203-269-7722.

THESING, GARY. Lake Superior State College, Sault Ste. Marie MI 49783. [Head, Dept of Mathematics] 906-635-2633.

THOMPSON, MELVIN. Howard University, Washington DC 20059. [Director Develop. & Research Admin., Rm. 1116 2300 6th St. NW] 202-636-5077.

THOMPSON, DON. Pepperdine University, Malibu CA 90265. [Natural Science Division] 213-456-4239.

THOMPSON, THOMAS. Walla Walla College, College Place WA 99324. [Professor, Dept of Mathematics] 509-527-2181.

THONGYOO, SUTEP. Syracuse University, Syracuse NY 13210. [Student, B5 Apt #4 Slocum Heights] 315-423-2373.

THORNTON, EVELYN. Prairie View A&M University, Prairie View TX 77446. [Professor, Dept of Mathematics] 409-857-4091.

THRASH, JOE. University of Southern Mississippi, Hattiesburg MS 39406. [Associate Professor, PBox 5045 Southern Station] 601-266-4289.

TILLEY, JOHN. Mississippi State University, Mississipi State MS 39762. [Professor, Dept of Math P.O. Drawer MA] 601-325-3414.

TOLBERT, MATTHEW. U. S. Congress, Washington DC [House Committee on Sci Space & Tech]

TOLER, CHARLES. Wilmington Friends School, Wilmington DE 19803. [Teacher, 101 School Road] 302-575-1130.

TOLLE, JON. University of North Carolina, Chapel Hill NC 27514. [Professor, Curriculum in Math Phillips Hall] 919-962-0198.

TOUBASSI, ELIAS. University of Arizona, Tucson AZ 85721. [Professor, Dept of Mathematics] 602-621-2882.

TREFZGER, JIM. McHenry County College, Crystal Lake IL 60012. [Professor, Route 14 at Lucas Road] 815-455-3700.

TREISMAN, URI. University of Calif-Berkeley, Berkeley CA 94720. [PDP-230B Stephens Hall] 415-642-2115.

TRIVIERI, LAWRENCE. Mohawk Valley Community College, Utica NY 13501. [Professor, 1101 Sherman Drive] 315-792-5369.

TROYER, ROBERT. Lake Forest College, Lake Forest IL 60045. [Professor] 312-234-3100.

TUCKER, THOMAS. Colgate University, Hamilton NY 13346. [Professor, Dept of Mathematics] 315-824-1000.

TUCKER, RICHARD. Mary Baldwin College, Staunton VA 24401. [Assistant Professor] 703-887-7112.

TUFTE, FREDRIC. University of Wisconsin-Platteville, Platteville WI 53818. [Associate Professor, Dept Math 1 University Plaza] 608-342-1745.

UPSHAW, JANE. University of South Carolina-Beaufort, Beaufort SC 29928. [Chair, 800 Cartaret Street] 803-524-7112.

URION, DAVID. Winona State University, Winona MN 55987. [Professor, Dept of Mathematics and Statistics] 507-457-5379.

VANVELSIR, GARY. Anne Arundel Community College, Arnold MD 21012. [Professor, 101 College Parkway] 301-260-4565.

VAVRINEK, RONALD. Illinois Math and Science Academy, Aurora IL 60506. [1500 West Sullivan Road] 312-801-6000.

VELEZ-RODRIGUEZ, ARGELIA. Department of Education, Washington DC 20202. [Director, Room 3022ROB-3 7th & D Streets SW] 202-732-4396.

VICK, WILLIAM. Broome Community College, Binghamton NY 13902. [Professor, P.O. Box 1017] 607-771-5165.

VIKTORA, STEVEN. Kenwood Academy, Chicago IL 60615. [Chair, 5100 South Hyde Park Blvd Apt 3D] 312-947-0882.

VOBEJDA, BARBARA. Washington Post.

VONESCHEN, ELLIS. Suffolk County Community College, Selden NY 11784. [Professor, 533 College Rd] 516-451-4270.

WAGONER, RONALD. California State University, Fresno CA 93710. [Professor, 617 East Teal Circle] 209-438-5512.

WAGONER, JENETTE. California State University, Fresno CA

WAITS, BERT. Ohio State University, Columbus OH 43201. [Professor, Dept of Math 231 West 18th Avenue] 614-292-0694.

WALKER, RUSSELL. Carnegie Mellon University, Pittsburgh PA 15601. [Senior Lecturer, Dept of Mathematics] 412-268-2545.

WALSH, MARY-LU. D.C. Heath & Co., Lexington MA 02173. [Editor, 125 Spring Street] 617-860-1144.

WALSH, JOHN. Science Magazine.

WANG, AMY. Montgomery College, Vienna VA 22180. [Assistant Professor, 1834 Batten Hollow Road] 703-281-0579.

WANTLING, KENNETH. Washington College, Chestertown MD 21620. [Assistant Professor] 301-778-2800.

WARDROP, MARY. Central Michigan University, Mt. Pleasant MI 48859. [Professor, Dept of Mathematics] 517-774-3596.

WARD, JAMES. Bowdoin College, Brunswick ME 04011. [Chair, Dept of Mathematics] 207-725-3577.

WASHINGTON, HARRY. Delaware State College, Dover DE 19901. [Associate Professor, Dept of Mathematics] 302-736-3584.

WATKINS, SALLIE. American Institute of Physics, Washington DC 20009. [Sr. Education Fellow, 2000 Florida Ave NW] 202-232-6688.

WATSON, ROBERT. National Science Foundation, Washington DC 20550. [Acting Head Undergraduate Ed, 1800 G Street NW]

WATSON, MARTHA. Western Kentucky University, Bowling Green KY 42101. [Professor, Dept of Mathematics] 502-745-6224.

WATT, JEFFREY. Indiana University, Indianapolis IN 46206. [Associate Instructor, PP.O. Box 2813] 317-849-4136.

WELLAND, BOB. Northwestern University, Evanston IL 60208. [Associate Professor, Lunt Hall] 312-492-5576.

WELLS, DAN. Western Carolina University, Cullowhee NC 28723. [Associate Professor, Box 837] 704-586-5797.

WENGER, RONALD. University of Delaware, Newark DE 19716. [Director, Math Center 032 Purnell Hall] 302-451-2140.

WESTERMAN, JOAN. O'Brien & Associates, Alexandria VA 22314. [Senior Associate, Carriage House 708 Pendleton Street] 703-548-7587.

WHITAKER, PATRICIA. Elon College, Elon College NC 27244. [Assistant Professor, Dept of Mathematics P.O. Box 2163] 919-584-2285.

WHITE, ROBERT. National Academy of Engineering, Washington DC 20418. [President, 2101 Constitution Avenue NW] 202-334-3200.

WICK, MARSHALL. Univ of Wisconsin-Eau Claire, Eau Claire WI 54702. [Chair, Dept of Mathematics] 715-836-2768.

WILLARD, EARL. Marietta College, Marietta OH 45750. [Dept of Mathematics] 614-374-4811.

WILLCOX, ALFRED. Mathematical Association of America, Washington DC 20036. [1529 18th Street NW]

WILSON, JACK. University of Maryland, College Park MD 20742. [Professor, Dept of Physics] 301-345-4200.

WINGO, WALTER. Design News, 22207. [Washington Editor, 4655 N 24th St. Arlington VA] 703-524-3816.

WOLFSON, PAUL. West Chester University, West Chester PA 19383. [Associate Professor, Dept of Mathematical Sciences] 215-436-2452.

WOODS, JOHN. Oklahoma Baptist University, Shawnee OK 74801. [Professor] 405-275-2850.

WRIGHT, DONALD. University of Cincinnati, Cincinnati OH 45221. [Professor, Dept of Mathematical Sciences] 513-475-3461.

YOUNG-DAVIS, PATSY. Lake Mary High School, Lake Mary FL 32746. [Chair Math Dept, 655 Longwood-Lake Mary Road] 305-323-2110.

YOUNG, GAIL. National Science Foundation, Ossining NY 10562. [Program Director, 53B Van Cortland Ave]

YOUNG, PAUL. University of Washington, Seattle WA 98915. [Chair, Computer Science Board] 206-543-1695.

YUHASZ, WAYNE. Random House Publishing Co., Cambridge MA 02142. [Senior Editor, 215 1st Street] 617-491-3008.

ZIEGLER, JANET. UPI.

ZORN, PAUL. Purdue University, West Lafayette IN 47906. [Associate Professor of Mathematics, St. Olaf College] 317-494-1915.